797,885 Books
are available to read at

Forgotten Books

www.ForgottenBooks.com

Forgotten Books' App
Available for mobile, tablet & eReader

ISBN 978-1-334-20874-4
PIBN 10630915

This book is a reproduction of an important historical work. Forgotten Books uses state-of-the-art technology to digitally reconstruct the work, preserving the original format whilst repairing imperfections present in the aged copy. In rare cases, an imperfection in the original, such as a blemish or missing page, may be replicated in our edition. We do, however, repair the vast majority of imperfections successfully; any imperfections that remain are intentionally left to preserve the state of such historical works.

Forgotten Books is a registered trademark of FB &c Ltd.
Copyright © 2017 FB &c Ltd.
FB &c Ltd, Dalton House, 60 Windsor Avenue, London, SW19 2RR.
Company number 08720141. Registered in England and Wales.

For support please visit www.forgottenbooks.com

1 MONTH OF FREE READING

at

www.ForgottenBooks.com

By purchasing this book you are eligible for one month membership to ForgottenBooks.com, giving you unlimited access to our entire collection of over 700,000 titles via our web site and mobile apps.

To claim your free month visit: www.forgottenbooks.com/free630915

* Offer is valid for 45 days from date of purchase. Terms and conditions apply.

English
Français
Deutsche
Italiano
Español
Português

www.forgottenbooks.com

Mythology Photography **Fiction**
Fishing Christianity **Art** Cooking
Essays Buddhism Freemasonry
Medicine **Biology** Music **Ancient Egypt** Evolution Carpentry Physics
Dance Geology **Mathematics** Fitness
Shakespeare **Folklore** Yoga Marketing
Confidence Immortality Biographies
Poetry **Psychology** Witchcraft
Electronics Chemistry History **Law**
Accounting **Philosophy** Anthropology
Alchemy Drama Quantum Mechanics
Atheism Sexual Health **Ancient History**
Entrepreneurship Languages Sport
Paleontology Needlework Islam
Metaphysics Investment Archaeology
Parenting Statistics Criminology
Motivational

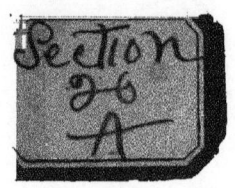

Presented to
The Library
of the
University of Toronto
by

The Engineering Institute
of Canada.

SOCIETY OF ENGINEERS

ESTABLISHED MAY 1854

Journal and TRANSACTIONS FOR 1902

AND

GENERAL INDEX, 1857 TO 1902

EDITED BY

PERRY F. NURSEY

SECRETARY

London

E. & F. N. SPON, LIMITED, 125 STRAND

New York

SPON & CHAMBERLAIN, 123 LIBERTY STREET

1903

The Society is not responsible either for the statements made or for the opinions expressed in this volume.

PAPERS AND PREMIUMS.

The Council of the Society of Engineers invites original communications from Members and Associates, as well as from gentlemen who do not belong to the Society, on subjects connected with any branch of Engineering.

For any papers that may be considered sufficiently meritorious the Council may at discretion award one or other of the following Premiums, viz.:—

1. THE PRESIDENT'S PREMIUM, given annually by the President, and consisting of a Gold Medal of the value of Five Guineas.

2. THE BESSEMER PREMIUM, provided for annually by the late Sir Henry Bessemer, F.R.S., Honorary Member, of the value of Five Guineas.

3. THE SOCIETY'S PREMIUMS, given annually by the Society, of an aggregate value not exceeding Twenty Pounds.

The number and value of the Society's Premiums are decided by the Council according to the number of meritorious papers read during the year.

By the Rules of the Society, Members of Council are disqualified from receiving Premiums for Papers.

PREMIUMS FOR 1902.

At a Meeting of the Society, held on February 2, 1903, the following Premiums were presented, viz.:—

The President's Gold Medal to:

    Thomas Andrews, F.R.S., for his paper on The Effect of Segregation on the Strength of Steel Rails.

The Bessemer Premium of Books to:

    Augustus R. Galbraith, for his paper on The Hennebique System of Ferro-concrete Construction.

A Society's Premium of Books to:

    Benjamin H. Thwaite, for his paper on British *versus* American Patent Law Practice and Engineering Invention.

A Society's Premium of Books to:

    Brierley D. Healey, for his paper on Recent Blast Furnace Practice.

CONTENTS.

	PAGE
INAUGURAL ADDRESS OF THE PRESIDENT, PERCY GRIFFITH	1
BRITISH v. AMERICAN PATENT LAW PRACTICE AND ENGINEERING INVENTION. By BENJAMIN H. THWAITE	23
AUSTRALIAN TIMBER BRIDGES AND THE WOODS USED IN THEIR CONSTRUCTION. By HERBERT E. BELLAMY	63
RECENT BLAST FURNACE PRACTICE. By BRIERLEY D. HEALEY	97
NOTES ON SOME TWENTIETH CENTURY LOCOMOTIVES. By CHARLES ROUS-MARTEN	125
VACATION VISITS	155
THE HENNEBIQUE SYSTEM OF FERRO-CONCRETE CONSTRUCTION. By AUGUSTUS R. GALBRAITH	177
THE EFFECT OF SEGREGATION ON THE STRENGTH OF STEEL RAILS. By THOMAS ANDREWS, F.R.S.	209
DEPRECIATION OF PLANT AND WORKS UNDER MUNICIPAL AND COMPANY MANAGEMENT. By CHARLES H. W. BIGGS	271
OBITUARY	301
ANNUAL GENERAL MEETING	305
ANNUAL REPORT OF COUNCIL AND STATEMENT OF ACCOUNTS, 1902	309
INDEX	317

SOCIETY OF ENGINEERS.

ESTABLISHED MAY 1854.

COUNCIL AND OFFICERS FOR 1902.

Council.

President—PERCY GRIFFITH.

Vice-Presidents
- JAMES PATTEN BARBER.
- DAVID BUTLER BUTLER.
- NICHOLAS JAMES WEST.

JOSEPH BERNAYS.	RICHARD St. GEORGE MOORE.
GEORGE AUSTIN PRYCE CUXSON.	HENRY SHERLEY-PRICE.
GEORGE ABRAHAM GOODWIN.	JOSEPH WILLIAM WILSON.
WILLIAM HENRY HOLTTUM.	MAURICE WILSON.

Members of Council, ex-officio.

Past President (1901) CHARLES MASON.
" (1900) HENRY O'CONNOR.
" (1899) JOHN CORRY FELL.
" (1898) WILLIAM WORBY BEAUMONT.
" (1897) GEORGE MAXWELL LAWFORD.
" (1896) SAMUEL HERBERT COX.

Hon. Secretary and Treasurer—GEORGE BURT.

Trustees
- SIR BENJAMIN BAKER, K.C.M.G.
- SIR DOUGLAS FOX.
- SIR WILLIAM HENRY WHITE, K.C.B., LL.D., F.R.S.

Hon. Auditor—SAMUEL WOOD, F.C.A.
Hon. Solicitors—MESSRS. BLYTH, DUTTON, HARTLEY & BLYTH.
Bankers—LLOYDS BANK (LIMITED).
Secretary—PERRY FAIRFAX NURSEY.

OFFICES:
17 VICTORIA STREET, WESTMINSTER, S.W.
Hours, 10 a.m. to 4 p.m.; *Saturdays,* 10 a.m. to 1 p.m.

PLACE OF MEETING.
THE ROYAL UNITED SERVICE INSTITUTION, WHITEHALL.

TRANSACTIONS, &c.

February 3rd, 1902.

INAUGURAL ADDRESS.

By PERCY GRIFFITH, President.

Gentlemen,—This being my first opportunity of addressing you after my election to the distinguished position of president, it is as much my duty as it is my privilege to express to you, so far as words allow, my high appreciation of the great honour you have conferred upon me. The system by which your presidents are selected is one which clearly implies that election to that high office is intended as a recognition of services rendered to the profession in general and this society in particular in the more arduous but less prominent offices of member of council and vice-president. In looking back however over my own period of probation, I am struck with the comparatively small service I have rendered previous to attaining this the highest honour the members can confer. It is only right that I should also acknowledge here the important influence exerted on my career, both in this society and professionally, by my association with and succession to our distinguished past president, the late Jabez Church, who so successfully occupied this chair twenty years ago. Under such circumstances I feel that it behoves me to accept the dignity you have conferred upon me with more than ordinary diffidence, and with the strongest possible determination to prove that your confidence has not been misplaced. I therefore assume this chair and the handsome badge, which itself seems to increase the responsibilities as it certainly does the dignity of the position, with a strong sense of indebtedness to the society and an equally firm resolve to meet the liability so far as my powers and opportunities allow.

Remembering the long and distinguished list of my predecessors, and the great and invaluable work accomplished by the society during its long career of forty-eight years, it is no light task for any new president to adequately maintain, much less to advance, its influence and usefulness. I must however point out that my success or failure rests to a large extent upon the co-operation and support of the members generally. While I am confident of receiving the moral support of every member who takes any personal interest in the affairs of the society, I should like to take this opportunity of appealing for larger attendances at our meetings and vacation visits. Especially I do appeal to those, who, through want of active interest in our proceedings, are disposed to question the advantages of membership, because by making a point of attending our various functions with greater regularity they will not only increase the usefulness and success of the society as a whole, but at the same time will themselves appreciate the advantages afforded by the society to a much greater extent than they could otherwise do.

Let me now turn from these purely personal considerations to topics of more general interest. Recalling the many valuable and interesting addresses which have been delivered by previous occupants of this chair, it will be readily appreciated that the difficulty of presenting new matter increases year by year. While however this compels me to seek your indulgence, it also provides me with many an object lesson of what a presidential address should be, and I am therefore confident that I cannot do wrong in following the general practice of dealing first with the affairs of the society, and afterwards referring briefly to those branches of engineering in which I am personally most interested.

Domestic Affairs.

As regards our own domestic concerns, the recent practice of issuing to members an annual report from the council and a statement of accounts in which the year's proceedings are briefly summed up, obviates the necessity for my detaining you with any simple record of events. It is, however, well that your attention should be drawn to some general points, which, besides being of particular importance, may perhaps serve to indicate the direction in which the influence and value of the society may be extended.

In the first place, I am sure we may congratulate ourselves that in spite of special difficulties, such as the continuance of the war in South Africa, the comparative stagnation of trade, and more particularly the growing competition of kindred societies, both in respect of members and papers, we have, after allowing for twenty resignations, three deaths and nine memberships lapsed, during the year just ended, secured a net increase of seven in our membership roll; we have also fully occupied our eight ordinary meetings with original contributions of no mean order on interesting modern features of engineering, and last, but not least, have fully maintained a financial position which compares favourably with that of any other society of a like nature. Our visits to works were also both interesting and successful, and I am confident that all those who attended them have considered the time and expense more than justified, while those who missed them have every reason to regret their abstention, whether voluntary or otherwise. On this head let me remark that the larger the support given to these visits to works, the easier is it to extend their scope and usefulness; as an illustration of this I may mention that during the past year a visit to the pumping plant in connection with the Severn Tunnel, for which special facilities were obtained by Mr. Nicholas West, our new vice-president, had to be abandoned owing to the small number of members who could promise their support. It is proposed to make another effort to arrange this visit, and in order that a more general interest may be aroused in it, a brief outline of the nature and extent of the works will be issued to members which it is hoped will secure a sufficiently large attendance to warrant the visit being undertaken.

Our annual dinner held in December last was also a worthy successor to the many successful functions of a similar character which had preceded it, and not only the numbers, but also the distinguished list of visitors present, were undeniable testimony to the dignity and importance of our society in the scientific and engineering world.

As regards our annual general meeting, a special word is necessary. For some time past the small attendances at these (our only meetings for dealing with purely domestic affairs) has been a matter of regret to the council, but it seems clear that the somewhat unattractive nature of the programme set forth in the usual notice, and the absence of any full report of the meeting in the transactions have contributed to this, and it is now under the consideration of the council to remove these objections as far as possible. Although it would be premature for me to make any more definite statement on this head, I can and do invite the members generally to pay more attention to this meeting in the future, reminding them that on these occasions the council invite suggestions and discussions upon the general operations of the society and the possibility of increasing its usefulness, and that very interesting have often occurred which have proved of great advantage to the society in general and those present in particular.

I would call attention to the scarcity of social functions in our previous programmes and to the great profit and pleasure afforded to members by opportunities of meeting in a more social way than is possible at our ordinary meetings. This question is also under the consideration of the council, and I am hopeful that in connection with our next annual general meeting, if not before, some announcement will be made on this point, which will prove interesting and acceptable, at least to those members who reside within easy reach of the metropolis.

The possibility of improving upon our ordinary programme has often been discussed by the council and many interesting and promising schemes have been carefully considered, but I think you will all admit that, with the excellent papers which have been obtained in the past, and the valuable and exhaustive discussions produced by them, there is no great urgency on this head. It is however possible that in the near future some change in the programme may be attempted, in which case I can only urge you to recognise in this the earnest endeavour of the council to extend the usefulness of the society, and ask your cordial co-operation and support.

Before leaving these domestic matters let me say a word as to the objects and position of the society in relation to the present requirements and condition of the engineering world, in many respects so different now to what they were in the year 1854 when the society was established. The chief difference arises from the large number of kindred institutions which have sprung up in recent years, each appealing to a special branch of the profession, until there is scarcely a department of engineering which is not represented by a particular society or institution. It has sometimes been urged that by this means the ground has been cut from under our feet and that we therefore have no special mission to fulfil. This however is by no means a true statement of the facts, for our worthy founders, with great prescience, defined the object of the society so broadly that in spite of the altered conditions by which we are surrounded at the present time, we can still claim to fill a not unimportant place in the engineering world, as is proved by our membership roll of about 500 and the eminent names which that roll includes. Moreover, although nearly every variety of the genus "engineer" is represented in our ranks and our general aim and purpose differ so little from those of the two older institutions, it is no small satisfaction to us to realise that the most friendly relations exist between our society and all the other kindred societies. This in itself shows that in spite of our cosmopolitan nature (and may I not say because of this?) we have a very useful work to fulfil, and it may not be out of place for me to suggest what that work is. In the first place, being unlimited in respect of the subjects available for our consideration so long as they are covered by that very broad term "engineering," we can be sure of a wide and ever varying audience for whatever matter we may select for discussion, and it is our endeavour to interest every variety of engineer from time to time and, so far as may be, to allow no branch of engineering to be neglected. Then again, while the institutions of civil and mechanical engineers provide ample scope for those whose experience enables them to deal with the largest and most important of the engineering problems and works of the times, our society affords an opening (provided nowhere else) for the discussion of minor questions which make up in numbers and difficulty for their relative lack of individual importance. In this and other respects we can claim to be an educational medium by which the younger and less prominent members of the profession can secure the airing and discussion of their theories and practice, and almost equally important, experience in preparing papers and in public speaking, accomplishments which in these days are almost essential to individual success. Having this in view, it is not impossible that, in order to develope the educational element in our proceedings, the council may some day revive the practice of substituting lectures by eminent gentlemen on various branches of engineering work for the ordinary papers and discussion. Here, as in every case of the kind, the council would be glad to ascertain the views of the members generally before making any change.

From the foregoing remarks I think it may be fairly concluded, firstly, that our society has a special and peculiar mission to fulfil; and secondly, that the council are sparing no pains to render the society's work from year to year as efficient as possible, bearing in mind the ever-changing conditions under which its operations are carried on. It is necessary that I should once more remind you that, however hard your council and officers may work, or whatever efforts may be made to maintain or extend the efficiency of the society's work, the result will be disappointing alike to the council and yourselves unless the members generally take a personal interest in the affairs of the society as a whole, and show this interest by introducing new members or associates, by preparing papers for discussion, by attending our meetings and visits in large numbers, and by doing generally all they can to assist the efforts made by those responsible for the conduct of those affairs.

Having detained you somewhat long with this reference to our internal affairs, I will now direct your attention to those branches of engineering with which I am, from my experience, most qualified to deal, viz. the supply of water and gas. I do not propose to inflict upon you any historical review of past progress, firstly, because that has been very exhaustively dealt with by others more competent than myself, and secondly, because, however fascinating the study of the past may be, it is more than ever necessary, in these days of rapid progress, to devote oneself energetically to those newer problems which arise with ever increasing frequency, rather than to follow slavishly in the steps of those who have gone before. Do not, however, conclude from this remark that I wish to depreciate the value of the work done by our predecessors. A knowledge and full appreciation of this is undoubtedly as necessary now as ever it was, but the time is past when any branch of engineering can be learned from a study of existing works alone. I therefore propose to briefly consider only those modern developments of gas and water supply which represent the lines upon which advancement is proceeding and may be expected to proceed in the near future.

Water Supply.

Before attempting to deal with mechanical details connected with the question of water supply, it is necessary that something should be said with regard to the broader aspects of this most important subject, for although this must be classed among the most ancient of the engineering arts, the rapid change of the conditions in recent years, especially as regards growth and concentration of population, renders the work of the water engineer anything but a sinecure. It must strike even the most casual observer as a matter calling for very serious consideration, that even at the present time, there are many parts of this country (which prides itself upon its engineering talent, even to claiming precedence over any other nation) without any public supply of water at all, and others where a deficiency of water is almost an annual occurrence. Considering the vast sums of money, the amount of scientific study, and the mass of literature devoted to this one subject, this fact is certainly a marvellous, and withal a most regrettable one, and as at first sight it appears to reflect anything but credit upon our profession, I feel perfectly justified in attempting to point out what, in my opinion, are some of the principal causes for the present state of affairs.

System of Control.—It is becoming every day increasingly

apparent that the present system of control, itself the outcome of a system devised many years since, which has been modified from time to time to meet new conditions, is largely responsible for many of the difficulties which face the water engineer of the present time. By this, it is left to each district, be it a large city or a small village or merely a scattered rural community, to look after itself, whether by means of the local authority or by the help of private enterprise, and although in recent years local authorities have begun to realise more fully their responsibilities in regard to such matters, it cannot be said that in every case, or even in the majority of cases, they deal adequately with the question of water supply. The initial difficulty appears to me to arise from the uncertainty and vacillation shown by Parliament in fixing the actual responsibility for obtaining a water supply in each and every case. The Public Health (Water) Act of 1878, the most drastic measure dealing with this point, merely charges the local authority with "seeing that every occupied dwelling-house in their district has within a reasonable distance an available supply of wholesome water for domestic purposes," . . . but no compulsory powers are given to render this obligation a practical one unless such a supply can be obtained "at a reasonable cost," which is defined as a maximum of 13l. per house. But the most striking omission from this and other similar Acts is the point as to by whom the water supply is to be undertaken. This is left in the first place to private enterprise, failing which there is no doubt a technical liability resting upon the local authority to get a supply by some means; but it is a well-known fact that in practice it involves a most complicated and difficult procedure to enforce the provision of a water supply on any person or body, however great the need may be. The indefinite character of existing legislation upon the relationship of private water companies to local authorities and *vice-versa*, is also a constant source of obstruction and delay to the carrying out of sorely needed improvements in this respect, provoking much needless opposition and litigation, and affording opportunities for personal jealousy and questions of political economy arising to postpone, and sometimes prevent the practical remedy of existing evils.

Areas of Supply.—Another important element in the question under review is that of areas of control. At present there is practically no universal system for regulating the limits of various water districts. Companies are, for instance, at liberty, subject only to the consent of the road authorities, to lay their mains, to undertake the supply of any area they consider suitable or likely to pay, without regard to any other question whatever, while local authorities are confined, except by an elaborate process of amalgamation requiring the consent of the Local Government Board, within the boundaries of their particular area as defined for municipal or general sanitary purposes. To go one step further and include those wealthier districts or companies who can afford to obtain parliamentary sanction for their own individual undertakings, we find some companies going direct to Parliament, others applying for a provisional order of the Board of Trade, and local authorities acting sometimes under the Public Health Act, 1875, sometimes under provisional order of the Local Government Board, and also sometimes under special Acts of Parliament. Is it strange that under such a system, inconsistencies, anomalies, injustice, and unnecessary expense and delay are the rule rather than the exception?

Physical Features of this Country.—Let me, however, present a still broader aspect of this question as regards our own country. Owing to the physical and geographical features of this island, the western portion, with its mountainous regions, impermeable surface strata and larger rainfall, is provided by nature with an ample supply of water, mostly stored at such a level that a line of pipes, and in some cases filter beds and a puddle or masonry wall, are the only works required to obtain all the water that is required, and even then vast quantities are allowed to run to waste; whereas the eastern portion, being comparatively flat, having a more or less permeable surface formation and a much smaller rainfall, has to sink wells, make deep borings, and provide expensive pumping plant in order to obtain the necessary supply, and in some districts (e.g. Essex) even these means are incapable of providing water in sufficient quantities for the rapidly increasing demands of the urban communities.

Allocation of Sources of Supply.—Another feature of the case is the not uncommon event of certain sources of supply being the object of keen competition between rival applicants and involving the various authorities in excessive expenditure to secure the allotment of their proper and legal share of the water; while on the other hand vast quantities of splendid water are often allowed to run to waste, both from the source and through defective mains and fittings, for want of proper conservation. It appears to me an indisputable contention, that the subject of water supply, more particularly the allocation of the various sources, is a matter that should not be left in the hands of individual applicants, subject only to the expensive and often inconsistent supervision of the Legislature, but should be treated as a public question and adjudicated upon by specially created Boards of Control, constituted with a view to the special requirements of the case. It is impossible within the limits of this address to carry this point further into detail, and it is perhaps unnecessary, in view of the fact that the Government have intimated their intention of promoting new legislation (the nature of which has, however, not been revealed), and both the County Councils Association and the British Association of Waterworks Engineers, have taken the matter up recently in a very practical manner.

Let it therefore suffice if I have emphasized the necessity for some public action being taken at an early date with a view to the amendment of the present condition of affairs, and let me conclude this brief survey by expressing a most earnest wish, that the day may not be far distant when every inhabitant of this ancient kingdom may be provided with an ample supply of pure water.

Municipalization.—A few words may, however, be said before leaving this very important subject, upon the more modern features of water supply that particularly affect us as engineers. To those interested in Parliamentary proceedings, the most prominent characteristic of recent practice has been the oft recurring question of municipalization. This, although not strictly an engineering matter, has, especially in the metropolis, engaged the attention of so many eminent engineers, that I can hardly pass it by unnoticed. The ethics of the subject I can happily leave to politicians and public reformers, but having had some experience with both companies and local authorities, I can safely refer to the practical bearing of the question upon the engineering features of water supply. In the first place it is within the experience of every waterworks engineer that when advising a water company one of the first points to be considered before any scheme, either for a new works or for extensions, can be carried out, is "What is the attitude of the local authority?" If this happens to be hostile the procedure, and incidentally often the scheme itself, has to be adjusted with a view to eliminating the element of opposition as far as may be; as for instance by keeping down the amount of capital applied for, or by arranging the site of the works, the area of supply and other engineering features of the scheme to suit the idiosyncrasies of those whose only basis for interference is "opposition." The capital absorbed in arguing the details of a purchase clause, or in meeting the innumerable other points of objection so frequently raised by local authorities (in many cases themselves in default as regards the provision of a water supply) is only indirectly of interest to the engineer in that it seriously increases the cost of the works and consequently the charges upon consumers. The point is, however, a serious one and calls for an immediate remedy; this can only be supplied by Parliament deciding once for all whether waterworks undertakings as a whole are to be considered as municipal or private works, or at any rate the circumstances under which the preference is to be given to the one or the other. In regard to new schemes, for instance, some limit might at least be fixed to the opposition of a local authority which is not prepared to carry out the works on its own account; and *vice versa,* some steps might be taken to prevent private companies proceeding at all so long as the local authority concerned is in a position and willing to do so. I am confident that, however much professional business may be secured owing to the present unsatisfactory state of affairs, engineers as a body would cordially welcome a change that would clear away the uncertainties, inconsistencies and expense that are now so common to waterworks practice arising from this cause.

Large Gravitation Supplies.—Turning now to matters exclusively of engineering interest, perhaps the most striking feature of modern practice is the magnitude of individual works for our large cities, and the disposition to avoid local sources of supply, and seek those which are unlimited in quantity and entirely free from risk of contamination, almost without regard to the distance between source and consumers. The cause of this is no doubt the enormous increase in the population per unit of area in our large cities, and one must admit that the want of a satisfactory solution of the sewage problem has also a considerable influence upon the question. Until we are able to render the sewage effluents from our large towns perfectly innocuous, it must always be a more or less dangerous proceeding to draw a water supply from any adjacent source open to contamination thereby. On the other hand, the practice of obtaining water from outside areas, especially when these are a considerable distance away, presents difficulties which no engineer can afford to neglect or despise. Chief among these is the question of allocation. This of course can be, and indeed has in most cases been, satisfactorily overcome by provision being made for the supply of all towns and districts within reasonable distance from either the works or the pipe line. Another difficulty which has perhaps not received sufficient attention is the risk of accident to, or interference with the line of pipes by an invading enemy, or by political or social agitators, or even by the mischief-loving lunatic. The period of anxiety through which one of our largest cities has recently passed owing to the limited capacity of the pipe line suggests another element which ought not to be overlooked, and that is, not only the heavy total outlay involved by any long distance gravitation scheme, but the large proportion of the total represented by each section of the main conduit. Thus arises the temptation to delay the construction of each additional section to the latest possible moment, and the risk that meanwhile the demands may exceed the supply and a water famine ensue, notwithstanding the existence of an unlimited supply in the storage reservoirs. I think therefore, in view of these difficulties, I may confidently suggest that in every case, local sources of supply should be developed to their utmost capacity before gigantic gravitation schemes are resorted to.

Underground Water Supplies.—The difficulties attaching to gravitation schemes are however unimportant compared to those which prevail in districts depending upon underground sources for their water supply. Here, although one is, under the existing law, practically safe from interference from rival claimants, engineering difficulties assail one in every direction, and fortunate indeed is that undertaking which can boast an unlimited supply of pure water from underground sources. It is however far beyond the scope of this address to detail the nature or to quote examples of these difficulties. They are only too familiar to all who have to deal with such cases, and unfortunately even experience of the most extensive character is unable to provide an infallible solution of each particular example as it occurs. The most eminent engineering and geological experts have sometimes found themselves at fault, and the utmost that can be done in many instances is to experiment with every available and promising source in turn until success more or less complete is achieved. Ample confirmation for this statement will be found in the past history and present condition of the water supply in the county of Essex; the latter being fully described in the Report of Dr. J. C. Thresh, M.D., County Medical Officer of Health, issued in 1900 and recently revised. My own experience has however provided abundant examples of the serious difficulties which so often face the water engineer in this part of the country. Fortunately all underground water schemes are not equally difficult in design or unsatisfactory in result, and I trust I may be pardoned for referring to two examples of deep borings in which I am personally interested, which, being unique as regards dimensions, may perhaps be submitted as representative of modern practice and experience in this connection.

Boring at Gainsborough.—The first is at Gainsborough, Lincolnshire, and was undertaken by the Urban District Council (then Local Board of Health) in the year 1894, under the advice of my predecessor, the late Jabez Church. Gainsborough is situated on the banks of the river Trent, the surface formation being red marls (of the New Red Sandstone series), having a thickness of about 700 feet. The idea of using water from the Trent was first considered, having been strongly recommended by the late Thomas Hawksley, but on the advice of several geologists, a trial boring had been sunk into the Bunter beds (underlying the red marl), and the existence of a large supply of excellent underground water satisfactorily proved. It therefore became a question of selecting one of two sources, both having points of advantage. In the end, my late partner's advice was acted upon, and the new boring commenced. This was completed after five years' anxious and difficult work, in July 1900, and now stands of the following dimensions:—12 feet from the surface lined with 36-inch internal diameter cast-iron guide pipes backed with concrete; thence to 300 feet from surface lined with 30-inch internal diameter steel tubes backed with cement; thence to 775 feet from surface lined with 18-inch internal diameter steel tubes, also backed with cement; thence to 1065 feet from surface lined with $15\frac{1}{2}$-inch tubes, partly perforated; and thence to 1515 feet from surface open boring $15\frac{1}{2}$-inch diameter. The work was carried out by Messrs. E. Timmins & Sons, Limited, of Runcorn, at a cost of 6645l.

The trial boring, which is situated on the same site, and is only 1350 feet deep, terminating 10-inch in diameter, yields 25,000 gallons per hour, and when pumping from a depth of 200 feet the water falls to 100 feet from surface. It is therefore confidently anticipated that not less than 60,000 gallons per hour will be obtained from the larger boring, in which the pumps will be fixed at a depth of 300 feet from the surface. The machinery is now in course of construction, and presents some elements of novelty, but I would prefer to leave any description of it until the working results can also be given. The work should be completed during the spring, and I shall then be only too glad to afford any member of the Society facilities for inspecting it should he wish to do so. The fact that the new boring is the largest constructed for waterworks purposes in this country, or I believe in the world, must be my excuse for referring to it on this occasion.

Boring at Lincoln.—The second boring to which I have referred is for the city of Lincoln, and was only recently commenced. It may be considered as having been undertaken more or less in consequence of the success obtained at Gainsborough, being intended to tap the same strata. Lincoln is, however, some twenty miles S.S.E. of Gainsborough, and as this is transverse of the " dip " of the strata, there is no fear of the pumping in one case affecting the water-level in the other. The boring here will have to be carried to a much greater depth (estimated at 2200 feet from surface), and some 400 feet will be through the "Lias" beds, which are more difficult to operate than the marls or the sandstone itself. The terminal diameter has therefore been reduced to 12 inches instead of 15½ inches, as at Gainsborough, and owing to the greater distance from the outcrop, arrangements are being made to pump from a depth of 400 feet from the surface, which it is anticipated will secure an ample supply of water. The contract for the boring has been let to Messrs. Chas. Chapman & Sons, Limited, at the sum of 14,605*l.*, and a term of four years has been allowed for its completion. At the time of writing 30 feet of 12 feet diameter cast-iron cylinders have just been fixed for the purpose of shutting out the water in the subsoil, and the boring proper is about to be commenced.

Other Borings.—Although I have ventured to describe these borings as representative of modern practice owing to their dimensions, I should very much have liked to refer to a number of other borings in various parts of the country which present features of interest either as regards conditions or results, but my address would attain undue length were I to attempt to do so. I must, however, in this connection call your attention to the growing necessity for obtaining supplies of water from underground sources in districts where surface sources are either insufficient, unsuitable, or totally lacking. There is also no doubt that underground water is often preferred owing to its usually excellent quality and extreme purity. Where the water is hard, as underground water often is, the process of softening can be applied at such a small cost, both in plant and working, that it need not be considered a serious disadvantage. The development of geological research and the improvement of deep-well pumping machinery have also contributed largely to the popularity of underground sources.

Borings versus Wells.—In regard to the latter, however, a word may be said on the question of the modern practice of dispensing with wells and drawing the water direct from borings. This has the double advantage of considerably reducing the cost of the sinking and of enabling the water to be drawn from a much lower level, thereby increasing the supply obtainable.

Borehole Pumps.—Special attention must, of course, be paid in such cases to the details of the pump, as its inaccessibility for repairs involves a risk of uneconomical working on the one hand or extensive outlay for repairs on the other. At the same time, I have myself proved that where a borehole pump has been carefully designed, a continuous run of twelve months and more can be secured with no appreciable sign of wear, and it is not therefore necessary to condemn the system as such on this ground. For deliveries over 10,000 gallons per hour, a double-acting "concertina" pump (that is, one barrel superimposed over the other, and the buckets working in opposite directions) is in my opinion the most suitable, as no bottom valve or "clack" is required. In the Gainsborough pumps, I have adopted the principle of the "Ashley" bucket with multiple valves, which has been very successfully applied at Brighton and elsewhere; only in this case, I have provided the necessary waterway by enlarging the barrel above and below the working part rather than by restricting the diameter of the bucket extension. I have thereby obtained a minimum waterway of 90 per cent. of the full diameter of the pump.

Air Lift Pump.—No engineer interested in pumping machinery can have failed to hear of the recent introduction into this country of the system known as the "Air Lift Pump," and as there appears to be a good deal of misconception as to its purpose and working results, I feel myself fully justified in making some reference to it. For some time past the system has been in constant use in America, mostly in connection with oil wells, but the credit of introducing it into this country is claimed by Mr. G. Y. Murray, of the British-American Well Works Company. I have myself inspected a number of installations erected by this firm, all of which appeared to be extremely successful. Other firms have now entered the field, and competition is very keen for this new class of pump. The principle involved is of the most elementary character, viz. by forcing air under pressure to a given depth below the level of the water, an upward flow is produced which affords a supply more or less proportional to the quantity of air pumped into the well or boring. Of course the very simplicity of the process has induced many people to claim for it the most extravagant and impossible results; but as further experience is obtained, these claims are gradually being discounted and a clearer insight obtained into the actual facts of the case. After very exhaustive inquiries as to the working results obtained from existing installations, I am convinced that as regards economy of coal consumption compared with the water lifted, there is no reasonable ground for comparison between the ordinary pump and compressed air. In one case I obtained a guarantee equivalent to 50 million foot-pounds in water lifted per 112 lb. of coal, but this was insufficient to warrant the adoption of the system, as an ordinary pump could be applied equally well to the case in question. Speaking generally, however, it is doubtful whether this guarantee represents the results obtained in ordinary practice, or anything approaching them. It must therefore be assumed that, as matters stand at present, there is no justification for resorting to compressed air in cases where the standard form of pump can be conveniently applied.

Application of Air Lift Pump.—On the other hand, there are no doubt circumstances (by no means uncommon) to which the system is not only applicable, but where it is an invaluable solution of an otherwise hopeless position; as, for instance, where the water is highly charged with sand or detritus, or where, owing to the fall of the natural water level, it is impossible to reach it with the ordinary form of pump; in both these cases the air lift pump will prevent the necessity for sinking a well or additional boring, the cost of which might be prohibitive and involve the abandonment of the supply altogether. Another case where the system is exceedingly useful is where the supply can only be obtained from a number of underground sources each representing but a small proportion of the total required. Here the necessity for a number of isolated pumping stations is avoided; the power plant being confined to the air-compressing station which may be situated in any convenient spot within the area of the supply; by this means not only is the cost of the steam or motive power reduced, but the land required to be purchased is practically confined to one small site for the compressing machinery instead of a number corresponding to the number of wells or borings. With some waters it is also a great advantage to have them aërated, and this is effected very thoroughly where compressed air is used for lifting. The air-lift system has been applied very successfully at the Pembury Pumping Station of the Tunbridge Wells Corporation Waterworks and at a number of private wells and borings in London and elsewhere. A large plant is also in course of erection at the Ford Pumping Station of the Birkenhead Corporation Waterworks. The test pumping at the Dartford Pumping Station of the Kent Waterworks Company was also carried out by a similar plant. Owing to the small cost of the plant, it is very suitable for testing the yield of borings, but at the present time there is a difficulty in getting air-compressors of any magnitude on hire for this purpose. However, as the system becomes more generally used, this difficulty will no doubt be overcome.

There are, of course, many other branches of the all-important subject of water supply which might be referred to with interest on account of the advances now being made in them; but I feel that on the one hand I have detained you already rather long with the few points I have attempted to deal with, and on the other hand I do not wish to appear in any way to be invading the ground covered by that kindred body, the British Association of Waterworks Engineers. I will, therefore, proceed to direct your attention to a few interesting features of modern Gas Engineering.

Gas Supply.

It is neither necessary nor fitting that I should deal at any great length with this branch of engineering, because the presidential address delivered by my worthy predecessor, Mr. H. O'Connor, as well as the transactions of the various gas engineers associations, afford information of the most exhaustive and up-to-date character upon nearly every department in which the gas engineer is interested. For this reason I do not propose to say anything with regard to the details of manufacture, all of which have been discussed at great length, and improved of recent years to an extent undreamt of twenty years back.

Competition with Electric Light.—It is, however, necessary to remind you that this rapid development of manufacturing details is due almost entirely to the rivalry of the electric light, which, being at first considered in the light of a mortal enemy, is now recognised as the greatest friend the gas industry ever had. No one would, I am sure, attempt to urge that this important industry would have made such rapid strides during recent years except for the incentive provided by the necessity for competing with such a formidable rival as electric lighting proved at first to be. Fortunately for gas engineers and gas investors, this rivalry, although still severe and of a continually extending character, has received a serious check owing to the reductions effected in the cost of making and purifying gas, the increased application of gas for heating and power purposes, the enormous extension of gas business among small consumers created by the slot-meter and "free fittings" system, and the more recent development of incandescent gas lighting for both internal and external purposes. A great deal might be said upon all these points, but I will content myself with a brief reference to the last named, this being conspicuous at the present moment for the remarkable advances made within the last few months.

Incandescent Gas Lighting.—From the remarks of Mr. O'Connor upon this subject in his presidential address delivered exactly two years ago,[*] you will obtain a very clear idea of the enormous influence exerted upon gas illumination generally by the incandescent mantle and burner; also of the

[*] Transactions, 1900, p. 17.

weak points of the system which were most conspicuous at that time. Since then, however, much has been done to remedy those defects and to develop the application of the system generally; so much so, that it is now almost the exception rather than the rule to see the ordinary old-fashioned burners in use for street lighting, and the incandescent burner is also making rapid inroads into our houses and shops. In fact, there is a very fair prospect of the near future seeing the ordinary burner entirely displaced, and lighting carried out on the incandescent principle altogether. This prospect is of more than ordinary importance to gas engineers, because it opens up the possibility of a complete change being made in the nature of the gas sent out. I mean that, if illuminating power is to be subordinated to heating power, or even eliminated altogether, the very elaborate and expensive processes of purification and enrichment now in use will be rendered unnecessary, and the cost of production reduced to a standard hitherto unknown. This state of affairs has indeed been already foreshadowed by the provisions inserted in the South Metropolitan Gas Company's Act passed last session, wherein the price charged is to be varied, within certain limits, according to the actual illuminating power of the gas supplied; and I think no great effort of the imagination is required to foreshadow the time when 14 and 16-candle power gas, at prices varying from 6s. to 2s. 3d. per 1000 cubic feet, will be a thing of the past, and the cost of artificial lighting, and incidentally the cost of motive power, will be reduced to a mere fraction of what it is at the present time. The enormous extension in the use of gas for heating and motive power which would obviously follow a change such as I have indicated is, I think, a very welcome prospect not only to the gas engineer but to the general public. The reduction of the smoke nuisance, the impetus to trade which a cheapening of motive power would necessarily involve, and the more effective heating of our living rooms, are matters of importance to everyone, and should secure increased efforts on the part of engineers towards their realisation.

This is hardly a suitable occasion for detailing the many improvements recently made in the incandescent burner and the mantle, but it is patent to all that considerable advances have been made since Mr. O'Connor delivered his address, not only in overcoming the difficulty of vibration to which he referred, but in improving and cheapening the various parts of the burner itself, and especially in rendering it independent of the chimney, a source of so much trouble with the older designs. Looking, however, to the immediate future, there is still ample scope for development in the production of a more permanent "mantle," or incandescent medium, and the further cheapening of this and the burner—improvements which must unfortunately await the expiry of existing patent and manufacturing rights, but which may be confidently anticipated before very long. It is doubtful whether any much greater efficiency (or candle-power per unit of gas consumed) can be looked for, as the present maximum of 40 candles per cubic foot per hour is of itself a sufficient advance on the 3-candle efficiency of the argand burner, and any attempt to increase this appears to involve serious and prohibitive disadvantages in the increased cost of the light and the shortened life of the mantle. Still, as I have already indicated, there seems to be a vast field for further development in the direction of supplying cheaper gas having a much smaller illuminating power than is required for the ordinary burner, and I suggest that it would be a most interesting and instructive experiment if some enterprising gas authority would secure the necessary powers and undertake the entire displacement of ordinary burners with incandescent and the supply of "low-grade" gas throughout its system at proportionately reduced charges.

Supervision of Fittings.—Arising out of this suggestion, let me call attention to a point which I consider of vital importance to the whole question of incandescent lighting, viz. supervision of mantles and burners. Many gas authorities have undertaken this on a fairly comprehensive scale with excellent results, notably the South Metropolitan Gas Company, of which I am a consumer myself. Speaking personally for a moment, I am greatly impressed with the advantages of having my incandescent burners looked after and myself relieved of any worry or inconvenience in connection with them, the charge of 4s. per annum per burner being money well spent on my part, and I should imagine a very profitable source of revenue to the company. May I urge upon all gas authorities the desirability of at once undertaking and extending this system to the utmost, as it must be obvious that the more general it becomes the more profitable it will be, owing to the reduced cost of supervision per burner. The time is past when responsibility can be said to cease on the delivery of reasonably pure gas at a statutory pressure at the consumer's meter, and no gas manager can hope to maintain or extend the success of his undertaking unless he is prepared to assist the consumers in making the best use of the gas after delivery. It is therefore but a small step to make the supervision of burners a special department of the general management, and as this can generally be combined with the lighting, cleaning and maintenance of street lamps, the extra cost of labour cannot be considered a serious objection, neither need the charge on the consumer be in any sense a prohibitive one. The recent practice of hiring out stoves, fires, and general fittings and the rapidly extending use of "slot" or "prepayment" meters, all of which are proved to be extremely profitable investments of capital, are steps in the direction I have indicated, and afford the groundwork for a further extension of the internal supervision which, in my opinion, is so desirable and in fact necessary.

High-Pressure Incandescent Lighting.—In connection with the subject of incandescent lighting there is however one feature of which the novelty is proved by the fact that it was not even mentioned by Mr. O'Connor in his address delivered only two years ago, and the importance is obvious to everyone: I refer to the high-pressure system applied to street, yard and shop lighting. At the annual meeting of the Incorporated Gas Institute held in this building last June, three very interesting and exhaustive papers on this subject were read and discussed, and I cannot do better than refer anyone anxious to study the matter to the transactions of that Institute, in which the papers and discussion are recorded in full.* It must, however, suffice for my present purpose if I call your attention briefly to the chief points of progress attained and the splendid prospect which this system has opened up for the further improvement of external lighting by gas.

Sugg's Improvements.—Although many gas engineers have interested themselves in this matter and have devoted considerable attention to the theoretical and experimental aspects of the question, there are three firms which, by their successful application of the system to actual practice, are entitled to be called pioneers in high-pressure incandescent gas lighting. The first of these is represented by our old friend and member, Mr. William Sugg, whose contribution to the transactions of the Incorporated Gas Institute last June affords a most interesting review of the history of this question from its first inception. It is unnecessary for me to occupy your time by repeating what can be found elsewhere, and I will not therefore attempt to retail this historical matter here.

I may, however, briefly summarise the various points of Mr. Sugg's work, which have contributed so largely to the successful application of high-pressure gas lighting. The chief of these is the design of a motor, driven by water pressure, for raising the pressure of the gas from that existing in the service main to whatever may be required. This is small enough to be enclosed in the base of the lamp standard, if required, is perfectly automatic and maintains a perfectly steady pressure

* Transactions of the Incorporated Gas Institute, 1901.

at the burner. It is designed more particularly for street lamps at crossings and refuges where high power is specially necessary. Another application is a motor operated by steam pressure instead of water. This also is of very small dimensions and is perfectly automatic, being more particularly suitable for factories and yards where steam power is available. A third motor is driven by belting from shafting, or by a gas engine, and is also adapted to workshop lighting. Besides these purely mechanical devices for pressure raising, Mr. Sugg has devoted special attention to the design of burners suitable to different purposes, to the automatic regulation of the gas consumption and to the most effective disposal of the light obtained at the burner. In this last particular I think it must be admitted that Mr. Sugg stands pre-eminent and almost, if not entirely, alone in the services rendered to the gas industry.

Welsbach Company's Improvements.—The second firm to which I have referred is the Welsbach Incandescent Gas Lighting Company, Limited, now unfortunately occupying a somewhat too prominent position in the financial world. In this case also I can only describe their work in the briefest possible terms. It cannot be denied that in their "Kern" burner, a great advance was made in incandescent lighting, if only in the abolition of the chimney, the most objectionable feature of the whole system. Having solved the problem of mixing the air with the gas in the most efficient manner, it became a very simple matter to restore the chimney (in this case made of metal) in order to produce the necessary draught, and to slightly modify the burner to suit the special conditions of high-pressure lighting. It is, however, as (practically) the sole manufacturers of mantles that this firm has made the most conspicuous and needful advances. The extra pressures and the more rapid gas consumption in the burner involved the necessity for greatly strengthening the structure of the mantle, and great improvements have been made in this direction. There is, however, much yet to be done before a perfect mantle is produced suitable for this modern feature of gas lighting, and it is earnestly to be hoped that the near future will see a solution of this most difficult problem.

Scott-Snell Lamp.—The third firm in question is the Scott-Snell Self-Intensifying Gas Lamp Company, Limited, who appear to have successfully overcome the necessity for any extraneous power being called into operation in order to secure the higher pressures required in this system. Their automatic pressure-raiser or "self-intensifier" was very clearly described by the patentee, Mr. Scott-Snell, in the third of the Gas Institute papers I have referred to, but as the principle is

entirely a novel one, I may be pardoned for briefly recapit
ing the leading features of it. The object of the appara
mainly to utilise the waste heat of the burner to co
either gas or air for high-pressure lighting or other si
purpose. In the first instance, and at the time Mr. Sne
paper was read (June 1901), there was no thought of compressing anything but the gas required for the burner or burners
connected with the particular apparatus, but recent experiments
have shown the much greater advantages of compressing air
only, and introducing this into the burner at the required pressure, and this is what is now done. The evolution of the motor
was not by any means a simple operation, as will be seen from
the conditions which the inventor found to be essential, viz. (1)
it must be operated by waste heat, (2) require no lubrication,
(3) have no dead centre, (4) have no fly-wheel, connecting rods,
stuffing boxes or friction-creating parts, (5) be self-controlling,
(6) inexpensive, and (7) simple in construction and repair.

The apparatus being completely enclosed in the ordinary
head of the lamp is very inconspicuous, and there are no
external fittings except one small pipe conveying the compressed
air to the burner, with a small box containing the whole of the
necessary valves. The top is formed with a circular water
reservoir or dish having a lid which keeps the supply of water
replenished by rain, etc., and which also serves to condense any
water evaporated within the reservoir before it is dissipated into
the external air. In any but very hot climates or exceptionally
dry weather in this country, this arrangement obviates the
necessity for periodical re-charging of the reservoir. The actual
pumping of air is effected by a reciprocating displacer in the
form of a relatively large hollow drum or cylinder with a very
short stroke, the weight of which is supported by a spring
acting on a vertical rod fixed in the top of the drum. This rod
operates a flexible diaphragm (which forms the equivalent of
the stuffing box) at the extreme top of the apparatus, that is,
above the water-dish but under the lid of the same. The working cylinder, which is open at the top, is slightly larger than the
displacer and is provided at the upper end with an air inlet
and the outlet for the air when compressed. The principle of
working is somewhat complex, and can only be clearly understood from diagrams or an inspection of the various parts of the
apparatus. The reciprocations average about 120 per minute,
but are more or less rapid according to the number of burners
(and therefore amount of heat applied) at the base of the
cylinder. By this means the amount of air delivered is automatically governed in proportion to the quantity required,
although the valves secure the pressure being constant at what-

ever they may be adjusted for. This brief sketch will afford some idea of the remarkable results obtained by this new form of heat engine under the very onerous conditions laid down, all of which will be found to have been successfully complied with. There can be no doubt that, with the improvements recently effected, this remarkable invention constitutes a great advance in the evolution of a perfect system of high-pressure gas lighting.

Inverted Incandescent Gas Lamp.—No review of modern progress in incandescent gas lighting would be complete without a reference to a very important improvement by which the burner and mantle can be inverted or inclined at any angle. By this apparently simple alteration gas lighting has made yet another step forward as a competitor with the electric light by rendering it possible to apply it in a decorative manner hitherto peculiar to the electric glow-lamp. The heat of this gas burner, although not yet reduced to the standard of its rival, is rendered far less objectionable through its being distributed round an annular opening at the base of the globe, instead of being concentrated at the apex of the mantle or top of a chimney. It also has the great advantage of being absolutely shadowless.

My thanks are due to several firms who have supplied me with particulars of borings carried out by them which it was impossible to refer to in the time at my disposal, also to those gas engineers who have afforded me the necessary records and illustrations of work done from which the above brief summaries have been made.

Conclusion.

My task is now completed, and I shall congratulate myself very heartily if I have succeeded in providing any matter of present-day interest either as regards our society or my particular branches of engineering work; much more shall I be gratified if I have retained your attention throughout this very imperfect survey of modern advances and future prospects in the domain of the gas and water engineer; and still further if my address is deemed worthy of association with those of my predecessors in this chair.

A cordial vote of thanks to the President for his Inaugural Address was proposed by Mr. Charles Mason, the immediate Past-President, seconded by Mr. George A. Goodwin, Past-President, and carried by acclamation.

The vote of thanks was duly acknowledged by the President, Mr. Percy Griffith.

March 3rd, 1902.

PERCY GRIFFITH, PRESIDENT, IN THE CHAIR.

BRITISH *v.* AMERICAN PATENT LAW PRACTICE AND ENGINEERING INVENTION.*

By BENJAMIN HOWARTH THWAITE.

IT is recognised by all serious historians that Britain's progress in engineering and invention constitute the greatest glory of the Victorian era. Up to the seventies all the civilised world called for British engineers to devise and construct railways and rolling stock, gas-works, docks and harbours, and sanitary works. Had any one suggested that by the end of the Victorian era the prestige of the British engineer and inventor would be eclipsed by that of those of America, he would have been laughed at. In the author's opinion, however, British industrial supremacy is lost, and what this may mean to the nation is inadequately realised; but when it is understood that the major part of England's wealth and power is distinctly traceable to the period of British engineering and inventive supremacy, the loss of such mastery is a fact of the first significance.

To-day, not only are foreign engineers carrying out work that would thirty years ago have inevitably fallen into the hands of British engineers, but many of the great projects that have sprung from the research and inventive work of one of the most illustrious Englishmen that ever lived—Michael Faraday—are being designed by foreigners and by Americans, even when such projects are to be carried out in England, and the machinery and materials of construction are to a great extent being produced in the United States. If a country like England, so peculiarly dependent on the success of her manufacturing industries for the sustenance of her people, falls off in her manufacturing efficiency and permits herself to be surpassed and superseded by a rival, then the inevitable condition of that country can be foretold. One leading American already predicts the effect, and with confidence he says, that under the stress of invincible American competition, the manufacturing

* A Society's Premium was awarded to the author for this paper.

industries of Britain will be so seriously contracted in the near future as to prevent the country from sustaining more than 15 millions of population. What this contraction and associated depreciation in values would mean can be guessed. England would descend in the scale of nations to the level of Spain or Holland—nations of merely tolerated power.

The marvellous progress of the manufacturing industries of the United States is illustrated by one item alone—that of iron—as the following statement will show. It gives the production of iron in America as compared with that of Great Britain in the years stated:

				United States. Tons.	Great Britain. Tons.
1896	.	.	.	Production 8,281,573	..
1897	.	.	.	„ 9,625,388	
1898	.	.	.	„ 12,233,579	..
1899	.	.	.	„ 13,967,727	..
1900	.	.	.	„ 13,411,531	8,908,570
1901	.	.	.	„ 16,250,727	8,108,000

The output of the United States last year exceeded that of Great Britain and Germany combined.

The causes of America's supremacy and British engineering and inventive decadence are not difficult to determine. One amongst the most conspicuous causes is, the great encouragement that the American Constitution has held out to the American inventor, and the protection always given to him. While our own Government has consistently taxed the British inventor, as if he were a nuisance to be suppressed, the American Government has realised the wisdom of applying the axiom relating to invention first declared in 1790 by George Washington, which is as follows:—"I cannot forbear intimating to you the expediency of giving effectual encouragement, as well to the introduction of new and useful inventions from abroad, as to the exertions of skill and genius in producing them at home." We shall presently realise that the American Government has, in the fullest measure, applied George Washington's wise axiom.

No nation depending for its existence upon manufacturing industry, dare stand still. The basis of industrial prosperity is the constant search for greater and more complete perfection, and for increased economy in cost of production; and in proportion as the search for perfection is accompanied with the utilisation of inventive resources, the more efficient and triumphant are the results likely to be. If we require evidence to demonstrate the truth of the axioms propounded, we have only to look across the Atlantic.

America has to thank England for that noble nucleus of her population, the Pilgrim fathers, than whom a finer band of men never associated, nor with a purer object. She has also to thank England for the basis of the manufacturing industries that American modern ingenuity and enterprise have extended to a degree that has enabled Columbia to obliterate Britannia's record of industrial supremacy, and write in the record of her own marvellous success, and especially in those industries which are the most closely allied with the practical application of engineering science. Applied ingenuity, and the confident enterprise to apply it, in the United States, perhaps more than all the other causes, except that of a fiscal character, explains why America has outstripped us in engineering and metallurgical industries.

Whether we can restore to British innate talent for invention, the virility and efficiency that are necessary to secure results akin to those of the United States, and which have in their culminating effects both dazzled and frightened Europe, is a question no one can answer. The author, however, believes that if the remedies are at once applied that will put us on a level plane with the United States—and one of the remedies is that which will be defined in this paper—then British ingenuity and resource will, at any rate, prevent England and her industries from being permanently submerged by the waves of American competition. Industrial competition is becoming daily more keen and formidable, and our dependence on our industrial strength, because of the increasing decadence of our agricultural resources, is becoming more and more complete. It is for the individual British engineer to ask himself the question whether or not the present apathy and indifference, either the product of disastrous pessimism or culpable optimism, are to be permitted to continue.

Let us analyse the leading force that has stimulated the faculty of invention of American engineers. Although the United States Government Commissioners of Patents at Washington have repeatedly declared that six or seven-eighths —or some equivalent or major proportion—of the entire manufacturing capital of the United States is directly or indirectly based on American patents, this fact, of such portentous significance, has completely escaped the attention of British statesmen. The principle of the United States patent system may be concretely stated to consist in the application in as fair a manner as it is possible of the golden rule, "ut res magis valeat quam fereat." American patents are always to be so interpreted as to uphold, and not to destroy the right of the inventor.

How different is the treatment of the inventor by the

British Government! here in England the inventor is instantly harassed by persistent taxation, and this only for what is merely a documentary registration of date, and the printed disclosure of the invention. The taxation of the inventor, too often poor, but rich in real ingenuity, eventually kills all desire to persist in improving, by his inventive genius, new or existing methods or processes of industry. It is not difficult to demonstrate in positive terms how striking is the encouragement given by the United States to inventors, and how discouraging to the inventor is the policy of Great Britain. The following financial comparison alone is adequate testimony of the negative and positive influences of the patent systems of the two countries:

STATEMENT showing the Cost of a BRITISH DATE-REGISTRATION PATENT for Fourteen Years and that of the UNITED STATES, only granted after Examination by Experts, for Seventeen Years.

BRITISH			AMERICAN (U.S.)	
When due.	Patent Fees.	£. s. d.	When due.	Patent Fees. £ s. d.
1st year	Application	1 0 0	1st year	Application . 3 2 6
Up to 4th year	Filing complete	3 0 0		
Up to 5th year	Renewal fee	5 0 0	2nd year: On issue of patent no further fee is required. The patent is good for 17 years.	4 3 4
Up to 6th year	,,	6 0 0		
Up to 7th year	,,	7 0 0		
Up to 8th year	,,	8 0 0		
Up to 9th year	,,	9 0 0		
Up to 10th year	,,	10 0 0		
Up to 11th year	,,	11 0 0		
Up to 12th year	,,	12 0 0		
Up to 13th year	,,	13 0 0		
Up to 14th year	,,	14 0 0		
	Total, 14 years	£99 0 0		Total, 17 years . £7 5 10

Six months are allowed for the American inventor in which to find the fees, whereas the British Government simply take advantage of the plea of extension to pay the renewal fees to levy an additional tax. For one month's extension of time 1*l.* is charged, for two months' extension 3*l.*, and for three months' extension 5*l.*; so that an inventor must either pay 5*l.* for three months' extension or lose his patent.

It will be realised on examining the total cost of the British patent with its duration that the British inventor is taxed 7*l.* 1*s.* 5*d.* for each year, during the 14 years the invention is kept alive; whereas the American inventor has only to pay at the rate of 8*s.* 7*d.* per year of the life of his patent. These comparative figures alone, constitute eloquent testimony to the unfair character of the taxation of the British inventor.

It might be suggested that the cheapness of the American patent, including the expert examination, must necessarily require a grant from the exchequer in addition to the fees, but as a matter of fact, the Patent Department at Washington is a profitable one. This is simply explained by the fact that the number of patents applied for is enormous, and the number of American patents actually granted is very great, as the following figures show. The average number of patents applied for in the United States in the four years 1896, 1897, 1898, and 1899, equalled 40,237. Of this number no less than 21,986 were actually sealed after search and examination. The average receipts equalled 246,212*l.* In the United Kingdom, for the same period, the average number of applications for patents equalled 28,645; out of this number 13,726 were completed on an average; the average receipts equalled 197,466*l.*

Of course it will be realised that the fee payable on application for the United States patent 3*l.* 2*s.* 6*d.* and covering the cost of search and examination, constitutes a splendid nucleus. But the American inventor knows that this fee will either secure to him a good title for a patent property or disclose to him the disagreeable fact that his invention is unfortunately not new, and thus at once enable him to cut his loss.

On page 6 is given a comparative and tabulated surplus so earned by the two departments, British and American, for a period of years, and shows at a glance how near to perfection the American patent system really is; providing as it does a test of novelty and utility, and conferring upon the inventors of the patents as granted, a title and cachet of some real value.

The total surplus of the United States Patent Department with practically twice the number of applications filed in the British Patent Office, is in the period specified less than the amount extorted from the British inventors by nearly three quarters of a million sterling.

ANNUAL SURPLUS REVENUE OF THE PATENT OFFICES OF GREAT BRITAIN AND THE UNITED STATES OF AMERICA.

Year.	United States.	Great Britain.
	£	£
1890	28,215	109,366
1891	26,315	100,339
1892	35,119	103,036
1893	20,367	79,775
1894	17,478	85,763
1895	32,150	86,341
1896	42,129	110,508
1897	50,560	111,300
1898	308	122,370
1899	22,735	102,484*
Total surplus	275,376	1,011,282

The object of the American patent system has been to give the inventor full value for the amount he expends. The American Government knows full well that the incentive to industrial enterprise represented in an American patent for a new invention of promise will amply repay the country, and this significant fact explains in a word the phenomenal expansion in the States of much of the manufacturing enterprise. The American people should be not only grateful to George Washington, who laid down the basis of a splendid patent system, but they should also not forget the framers of the American Patent Act of 1861, by which the monument of American sagacity was capped and completed—the patent life period being extended from 14 to 17 years. The American lawmakers had realised that the maturity of a great invention involved such an expenditure of time, as to leave in the 14 years of the British system an insufficient period of time for the inventor's harvest gathering, so they wisely extended the period to 17 years.

It is often the case that an invention has been perfected, say, in 9 or 10 years from the date of the application in

* Surplus for 1900 equalled 105,424*l*. The dollar is taken to equal 4*s*. 2*d*.

England. The inventor is then placed in an awkward predicament. If he is to be paid for his perseverance, expenditure and time, and the amount to cover the depreciation, the royalty for the few remaining years—if it is in such proportions as to merely return him the amount he has actually expended—will block the use of the invention, because there are many who will wait the short period remaining of the life of the patent before applying the system. The American 17 years period gives a chance to the inventor to obtain reasonable recompense.

The magnificent success of the American patent system is reflected in the buoyant confidence and pride of the average American. Wherever he goes he finds evidence of his countryman's ingenuity—in fact, the word "American" has become almost synonymous for something ingenious. Our roll-call of inventors of the nineteenth century is a long and brilliant one, but the list of inventors of American origin contains far more names than that of all Europe put together, including Great Britain.

Thanks to the stimulus of the American patent law, and to the high rate of wages paid to engineering artizans, American ingenuity has in no instance been more brilliant and more revolutionary than in effecting manual operations by automatic mechanism. The result has been almost magical. In manufacturing works of stupendous magnitude operations that once involved the use of an army of men are effected by a few men actuating switches and the like. Indeed it may almost be said that no realm of human activity has been ignored by the American inventor. In the field of agricultural machinery he has effected marvels; witness the modern American mower and harvester, which automatically, and in one operation, binds each bundle of grain, cuts the cord, and discharges the bundle; witness also the reaper which, as it gathers the straw, binds it with its own wisps. The American inventor has enormously extended the field in which engineering science can be profitably employed. Patented machinery has reduced the cost of almost every article of daily use. The examiners in the American Patent Office state that the class from which the inventions mainly come is that of men engaged in the working of machinery, who are constantly thinking out improvements.

The amount of enterprise that is directly traceable to the beneficent character of the American patent system is almost immeasurable, and is represented by the circulation of hundreds of millions sterling, whilst the number of the intelligent population profitably engaged is many millions. The effect of invention is to level up the character of labour, by reducing the area of operations that depend upon purely physical, as distinct from mental effort.

The author pointed out in 1894 that the wealth of the population of the United States per capita in different States was a direct function of the number of patents taken out in proportion to the population of the States. Thus the average value per capita of the following most inventive States, viz. New Jersey, Colorado, Rhode Island, Montana, Massachusetts, District of Columbia, and Connecticut, was $682·76, whilst the average value per capita in the following least inventive States, viz. West Virginia, Arkansas, Tennessee, Georgia, Virginia, Mississippi, and South Carolina, was only $176·73. Thus the highly inventive States per head are 400 per cent. better off than the less inventive ones.

Whilst immense factories are springing up in the United States—the basis of some new invention—our country can only show here and there signs of any extension, a considerable proportion of which is the product of American ingenuity. Before the seventies what a different picture England presented! We know that the nucleus of some of our greatest industrial establishments had for their basis the patented inventions of the founders. Numerous instances exist which prove the difference between England before the seventies and now.

The fatal defects of the British patent system are so conspicuously obvious that one is lost in astonishment on realising that Government after Government has come into power without making a serious attempt to make the necessary drastic alterations on the lines of the American system. Twenty years will soon have elapsed since the 1883 Patent Act was first put into force, and in this twenty years tens of thousands of really valuable inventions will have been lost to the country by the deadening effect of this Act. Let us examine the effect of the yearly and increasing incubus of taxation on British inventive effort. This deadening effect is graphically shown in Fig. 1, which is a diagram showing the effect of the renewal fees on British patents applied for in the year 1888. Fig. 2 is a graphic comparison of the British and American patent systems. The hatched portion shows the number of United States patents to be kept alive from date of application to the end of the 17 years, date of application 1899. The dark portions show the number of British patents kept alive from the date of application, 1889, to the end of the twelfth year.

The author has separated for analysis in the following table a group of British patents which includes all patents sealed in the years 1884 to 1888, from another group which includes the patents sealed in the years 1889 to 1893. Later patents are too young to permit of examination, but so far as they have existed, the lesson of the deadening effect of the yearly fees is proportionately the same, proving that the shortened life is not so much due to the intrinsic defects or worthlessness of the patents as simply to the burden of taxation.

EFFECT OF THE YEARLY TAXATION ON SEALED BRITISH PATENTS.

Average of the five years—1884 to 1888—shown by figures indicated in line A; and for 1889 to 1893 shown by figures in line B.

Patents Sealed.	Percentage of Original Patents Sealed.									
		5th Year.	6th Year.	7th Year.	8th Year.	9th Year.	10th Year.	11th Year.	12th Year.	13th Year.	14th Year.
A (number 9548) . .	A	29·9	22·4	18·5	15	12·4	10·6	9·1	7·9	6·7	5
B (number 11114) . .	B	31·7	23	18·16	15·12	12·4	10·4	8·6			
Equivalent to a reduction in percentage of original number . . .	A	70·1	77·6	81·5	85	87·6	89·4	90·9	92·1	93·3	95
	B	68·3	77	81·84	84·78	87·6	89·6	91·4			
Yearly percentage of falling off in annual fees paid	A	70.1	7·5	5	2·4	2·6	1·8	1·5	1·2	1·2	1·7
	B	68·3	8·7	4·84	3·64	2·72	2·0	1·8			

In Appendix A the exact position of the British patents sealed up to the end of the year 1899 and extending from 1888, as well as the patents kept alive, is given.

Appendix B shows the number of the patents that have expired in the same period through non-payment of fees.

Appendix C shows the number of applications for patents sealed, etc. under the Act of 1883.

Appendix D shows the number of patents, designs, and trade marks applied for and registered, the receipts and expenditure and the number of staff employed in the Patent Office in the United Kingdom and the United States of America in the years 1896–1899.

If we compare the British patent system prior to the Act of 1883, or, say in 1875, with that put in force in 1883, one striking fact is immediately apparent, and that is, that the comparative wealth of those who ventured to enjoy the luxury of patenting in this country before 1883 permitted them to maintain the subsequent fees in a greater measure than is done by the inventors who patented under the Act of 1883. For instance, in the year 1875 the ratio of applications to the sealed and those kept alive was as follows:—

                            1875
Applications   .   .   4561
Sealed   .   .   3112   or 68·22 per cent. of applications.

Maintained for seven years (from date of application) by payment of 50*l*., in addition to the 25*l*. paid on application   .   .   895, or 28·7 per cent. of patents sealed.

Maintained for fourteen years (from date of application) by additional payment of 100*l*.   .   .   .   295, or 9·4 per cent. of patents sealed.

Still referring to the applications for patents of 1875 we find the following results:—

No. of Patent Applications.	No. Sealed.	In Force Third Year.	In Force Seventh Year.	In Force Fourteenth Year.
4561	3112	3049	895	295
	68 per cent. of applications	98 per cent. of sealed	28·76 per cent. of sealed	9·4 per cent. of sealed
Compared with 1889	50·8 per cent.	31·6 per cent.	17·1 per cent.	..

The payments for patent and renewal in 1875 were: 25*l*. on application, 50*l*. for three years' protection, and 100*l*. for the eleven additional years.

The immediate effect of the 1883 Patent Act, was an increase in the number of applications from 5983 to no less than 17,110 or close on 200 per cent. of increase, showing clearly the influence of the reduction of the fees of the 1883 Patent Act. The patents actually sealed at the end of the fourth year from 1883 to 1887, equalled 9983 compared with number of 3,962 sealed in the year 1883, demonstrating *nem. con.* that 9983 of the applicants were quite capable of providing the amount of 4*l*.; so one may reasonably advance the argument that every British inventor who is able to provide the amount necessary to complete his British patent could find an additional 3*l*. to make up the amount that would be equivalent to that necessary to obtain an American patent, more so because he would know that he would have for this additional payment a patent title deed of value for a period of 17 years from the date of application.

Assuming that our patent system had been completely Americanised five years back, we should to-day have an army of inventors, measured on the basis of the United States ratio of the number declared valid and sealed to those applied for, of no less than 79,843 patents, a great proportion of which would most probably be found to be of sterling value, and owing to the protection that would be provided by the Americanised British grant of patent, capital would be provided, and new enterprises would rapidly revivify the stagnant condition of many of our industries.

Tens of thousands of ingenious Britishers have expended time and hard-earned savings, with the mistaken belief that the granting of the 1883 patent provided them with a cachet or title of value. When they attempted to realise on their patent, or tried to secure some recompense for their efforts and sacrifices, they learned with disgust that their patent-deed (save the mark!) was a mere registration of date and nothing more, and after paying perhaps the fifth year's renewal fee, they allowed the patent to lapse, and our inventor either said, "enough of this," or started on his journey to the United States, where his inventive ability received proper recognition.*

In this policy of discouraging inventive talent the most vital interests of a nation relying for its prosperity principally on manufacturing enterprise are sacrificed. Let us try to realise the significance of the depressing fact that of the 138,517 British patents applied for in the last five years, probably 104,000 will have become void in the fifth or sixth year from the date of application, and the Government will have drawn fees from the deluded and unfortunate inventors amounting to some seven hundred thousand pounds sterling. It might almost be said with truth that it were better for a man to be born in this country with a defective, than with a creative, or inventive mind. For the care and protection of the former the country provides an asylum; for the latter it provides taxation.

The author some years since pointed out that the fiscal necessity that compels our Government to absorb the surplus from the Patent Department to make up for loss of Imperial revenue, making a budget deficiency, may actually originate from the evil effects of the policy of checking inventive energy by the imposition of unfair taxes, because, whatever some strangely constituted persons may say against any patent protection whatsoever being granted to inventors, it is a fact that some 80 per cent. of the prosperity of the United States, purely manufacturing industry (a prosperity without equal in the history of industry) is in a great measure due to the encourage-

* There has been a steady fall in the applications for British Letters patent since 1897. The figures are as follows: 1897, 30,952; 1898, 27,650; 1899, 25,800; 1900, 23,922.

ment given by the American patent laws. Besides, so very costly is the practical evolution of any important invention, that no one, save a madman, will run the risk of pioneer work, i.e. if he is not certain of some return for the risk he runs. And without adequate protection by law how is he to obtain the recompense for the risk in time, money, and often health, expended? It should be known that often there are almost as many risks to health, and even life, in pioneering and practically exploiting an untried (and especially engineering) invention as there are in many of the engagements of modern warfare.

Obviously, therefore, the first and most essential incentive to industrial enterprise and pioneer work associated with the practical exploitation of an invention, is the adequate protection by the Government of the inventor. If the title deeds of property were merely registration dates of transfer and statements as to the area of the property, how many transfers would be effected, and who would dare to build on the property of such uncertain tenure? In brief, our patent laws are a decided deterrent to manufacturing enterprise.

The cost and risk associated with the practical evolution of an important invention, whether process or apparatus, is really serious, and if this fact were realised in its full significance, there would, in this country, be greater readiness to admit the justice, and to adopt the truly economic policy of generously acknowledging that the industrial pioneer of an invention should be well recompensed, and the patent acts would be framed and invariably interpreted to do this, as not only do the American, but also the German patent laws.

One explanation of the remarkable expansion of the United States, is the readiness to adopt new industrial developments and inventions. An up-to-date American manufacturer argues that if the adoption of a patented invention will provide in, say, ten years, the depreciation to cover the extinction of his existing plant, and in addition provide him with an increased profit, he is prepared to adopt the new invention and throw his obsolete plant on to the scrap-heap. This policy is economically sound, and it will have to be adopted in this country. The manufacturer who refuses to fall into line with the policy that means maximum all round efficiency must inevitably go to the wall.

The work of perfecting and launching an invention of useful service on to the market, even assuming all the financial conditions existing to permit this to be done, invariably requires for revolutionary inventions a long period of time, extending sometimes to eight or ten years, or even longer. In the meantime, if the improvement has involved the taking out of several patents, the yearly payments at the end of the ten years attain a considerable financial proportion. This recurring tax becomes positively cruel, and it has in many instances meant the last straw to the burden, and the inventor and the almost perfected invention are sacrificed on the altar of a mistaken national policy.

In the United States this lapsed British patent might have provided the basis of a great industry. Certainly, if proof were required that the successive Governments of this country were out of touch with the requirements of the people, and did not realise the character of the reforms that are urgently needed to enable British manufacturers to successfully stem the waves of American competition, the fact of the marvellous stimulus possessed by the American patent system in the development of American manufacturing, and especially engineering, industry, would long ago have been recognised and more or less exactly imitated by our Government. The British Government has however, recently attempted to ascertain the opinion of the public on the defects of the British patent system. It would naturally be thought that inventors themselves would have been called upon to give their views. It will, however, probably not surprise anyone to learn that the committee which sat to hear the evidence was not composed of even manufacturers selected by our Chambers of Commerce, or by all the great technical institutes or engineering societies of this country, but merely gentlemen of great legal eminence, and probably two Members of Parliament, neither of them engineers, with representative patent agents.

Of the witnesses called to give evidence, only a fraction (one-tenth) could by any stretch of argument be considered to represent the inventors' class.* Resolutions of some of the technical institutions were, however, put in as evidence. The evidence given before the Commission leads to the conclusion that an approach to the American examination of patents is desirable. The opinion of the Committee, as stated in the report, is as follows:

"Clause 9.—Both from the evidence before us and from the knowledge possessed by members of our Committee, we are of opinion that the grant of invalid patents is a serious evil, inasmuch as it tends to the restraint of trade and to the embarrassment of honest traders and inventors; and this fact, coupled with the result of the foregoing inquiry, is in our opinion a cogent argument in favour of some inquiry as to anticipation by prior Letters Patent.

* Of the twenty witnesses that appeared before the Commission of Enquiry, two were from the Patent Office, one solicitor from the Board of Trade, two referees, two delegates, seven patent agents, three consumers of patented articles, one legal expert, two inventors.

"Clause 10.—We are therefore of opinion that in addition to the existing inquiries, an examination ought to be made in the Patent Office into the question whether any invention claimed in a deposited specification has been claimed or described in any, and what specifications of Letters Patent granted in the United Kingdom dated less than 50 years previous to the date of the application: that this inquiry should not be extended to provisional specifications which have been published but not followed by a complete specification, and that consequent upon the limitations of this inquiry, an enactment should be passed to the effect that the publication of an invention in specifications of Letters Patent granted in the United Kingdom dated 50 years or more previous to the date of the application, or in a provisional specification of any date of the kind before mentioned, shall not of itself be deemed an anticipation of the invention."

The Committee then proceed to detail a scheme suitable for giving effect to their recommendations; after which they say: "Clause 12. We are of opinion that if this scheme receive the sanction of the Legislature it will be highly beneficial to honest and *bonâ fide* applicants, whilst it will tend to discourage the taking out of patents for objects other than the protection of a real invention."

It is satisfactory to the author to be able to state that in the House of Commons on February 10, 1902, Mr. Gerald Balfour introduced a Bill to amend the law with reference to application for patents and compulsory licences and other matters connected therewith, the provisions of which followed very closely the recommendations of the Committee of 1900, which, he said, were unusually detailed and precise. He stated that the Government intended to allow some considerable period to elapse between the first and second readings, and in the meantime they undertook to consider most carefully any recommendations that might be made to them. After Mr Lewis had thanked the Government for introducing that measure, the Bill was read a first time.

This important reform is most welcome, but it must be accompanied by the adoption *en bloc* of the American scale of fees, so that a British inventor shall be able, for the sum of about 7*l*., to secure a more or less valid patent for a period of 17 years from the date of application. No doubt the examination of the applications as recommended by the 1901 Committee would reduce the proportion of the number of patents granted by some 40 per cent. But the 60 per cent. remaining would be more or less good and valid patents and the chances of their being overthrown, all things being equal, would probably be in a ratio to the number of patents tested similar to that in the United States and to the ratio declared invalid. One authority stated that out of 540 cases tested in the American courts 198 were declared invalid—a proportion equivalent to a ratio of 26 per cent., whereas in a similar period in England only 26 cases were tested, and of these 11 were declared void, or equivalent to 42 per cent. These figures alone are eloquent testimony to the value of the American system of a searching examination. Moreover, the American ratio is based on more than twenty times the number of test cases brought before the English courts. In England only a small proportion of the patentees dare risk the enormous expense of a law-suit: they know only too well that patent cases are submitted before a tribunal more or less unfitted or incompetent to weigh the delicate quantities or measure the subtle qualities that often give an invention its intrinsic and important value.

Beyond this, in England the legal definition of an invention is obscure, and the legal interpretation is generally given against the inventor, whereas an inventor of Teutonic origin, who effect a striking technical improvement of industrial value by the use of well-known agents is, in Germany, granted a patent, and this policy simplifies the formula that defines what does and does not constitute an invention. The same principle is carried further in the United States. There the original inventor, or the man who first put the idea to actual test, receives the greatest consideration. In this country the law is often strained to protect the public against the inventor—a policy of disastrous folly. The public, instead of being benefited, is prevented from enjoying an invention which, if perfected, might have proved of great public utility, because an inventor, however ingenious he may be, is not so foolish as to persevere in the perfection of his invention when once he knows the facts relating to our patent system, or unless he can influence immense capital, which in this country is the only protection. Of course there are the small Birmingham class of articles that can easily be perfected and without much financial risk, but these are usually outside the category of the inventions that make a country great and prosperous. As long as England had the trade and industry of the world practically to herself, she could handicap the representatives of her industries with almost intolerable burdens with impunity, but these burdens must now be removed, or England's champions will be compelled to cry, "Peccavi," and retire discomfited from the arena.

The author trusts that the arguments and statistical data advanced by him will convince British engineers that the complete Americanisation of the British patent system is most urgently desirable.

APPENDIX A.

PATENTS SEALED under Act of 1883, and kept in force by Payment of Renewal Fees.

Year.	Patents Sealed and in force to end of 4th Year. Number.	Proportion per cent. of Applications.	5th Year. Number.	5th Year. Proportion per cent. of Patents.	6th Year. Number.	6th Year. Proportion per cent. of Patents.	7th Year. Number.	7th Year. Proportion per cent. of Patents.	8th Year. Number.	8th Year. Proportion per cent. of Patents.	9th Year. Number.	9th Year. Proportion per cent. of Patents.	10th Year. Number.	10th Year. Proportion per cent. of Patents.	11th Year. Number.	11th Year. Proportion per cent. of Patents.	12th Year. Number.	12th Year. Proportion per cent. of Patents.	13th Year. Number.	13th Year. Proportion per cent. of Patents.	14th Year. Number.	14th Year. Proportion per cent. of Patents.
1884	9,983	58.3	2921	29.3	2137	21.4	1746	17.5	1453	14.6	1173	11.7	1002	10.0	857	8.6	716	7.2	599	6.0	452	4.5
1885	8,775	54.4	2685	30.6	2049	23.4	1650	18.8	1381	15.7	1162	13.2	981	11.2	838	9.5	732	8.3	613	7.0	484	5.5
1886	9,099	53.0	2724	29.9	2070	22.7	1676	18.4	1400	15.4	1152	12.7	987	10.8	829	9.1	730	8.0	615	6.8	497	5.5
1887	9,466	52.4	2822	29.8	2111	22.3	1715	18.1	1403	14.8	1171	12.4	1034	10.9	917	9.7	805	8.5	677	7.2	529	5.6
1888	9,817	51.4	2840	28.9	2182	22.2	1752	17.8	1430	14.6	1194	12.2	1025	10.4	874	8.9	760	7.7	637	6.5	..
1889	10,664	50.8	3369	31.6	2381	22.3	1827	17.1	1533	14.4	1284	12.0	1089	10.2	917	8.6	772	7.2	..	..
1890	10,598	49.7	3162	29.8	2253	21.3	1796	16.9	1479	14.0	1251	11.8	1072	10.1	909	8.6	..	..	..
1891	10,922	47.7	3351	30.7	2445	22.4	1974	18.1	1653	15.1	1416	13.0	1182	10.8	..	..	..	..
1892	11,599	48.0	3752	32.3	2757	23.8	2173	18.7	1801	15.5	1490	12.8	..	..	..	..	..
1893	11,779	46.9	4003	34.0	2964	25.2	2354	20.0	1958	16.6	..	..	..	..	..	..
1894	12,042	47.4	4124	34.2	3080	25.6	2455	20.4	..	..	..	..	..	..	..
1895	12,346	49.3	4159	33.7	3025	24.5	..	..	..	..	..	..	..	..
1896	14,170	46.9	4560	32.2	..	..	..	..	..	..	..	..	..
1897	14,465	46.7	..	..	..	..	..	..	..	..	..	..
1898	13,452	48.7	..	..	..	..	..	..	..	..	..	..
1899	13,471	52.2	..	..	..	..	..	..	..	..	..	..

Patents kept in force by payment of Renewal Fees at commencement of

APPENDIX B.

PATENTS SEALED under Act of 1883, which have expired through Non-payment of Renewal Fees.

Year.	Patents Sealed and in force to end of 4th Year. Number.	Patents Sealed and in force to end of 4th Year. Proportion per cent. of Applications.	5th Year. Number.	5th Year. Proportion per cent. abandoned.	6th Year. Number.	6th Year. Proportion per cent. abandoned.	7th Year. Number.	7th Year. Proportion per cent. abandoned.	8th Year. Number.	8th Year. Proportion per cent. abandoned.	9th Year. Number.	9th Year. Proportion per cent. abandoned.	10th Year. Number.	10th Year. Proportion per cent. abandoned.	11th Year. Number.	11th Year. Proportion per cent. abandoned.	12th Year. Number.	12th Year. Proportion per cent. abandoned.	13th Year. Number.	13th Year. Proportion per cent. abandoned.	14th Year. Number.	14th Year. Proportion per cent. abandoned.
1884	9,989	58·3	7062	70·7	784	7·9	391	3·9	293	2·9	280	2·9	171	1·7	145	1·4	141	1·4	117	1·2	147	1·5
1885	8,775	54·4	6090	69·4	636	7·2	399	4·6	269	3·1	219	2·5	181	2·0	143	1·7	106	1·2	119	1·3	129	1·5
1886	9,099	53·0	6375	70·1	654	7·2	394	4·3	276	3·0	248	2·7	165	1·9	158	1·7	99	1·1	115	1·2	118	1·3
1887	9,466	52·4	6644	70·2	711	7·5	396	4·2	312	3·3	232	2·4	137	1·5	117	1·2	112	1·2	128	1·3	148	1·6
1888	9,817	51·4	6977	71·1	658	6·7	430	4·4	322	3·2	236	2·4	169	1·8	151	1·5	114	1·2	123	1·2	...	...
1889	10,664	50·8	7295	68·4	988	9·3	554	5·2	294	2·7	249	2·4	195	1·8	172	1·6	145	1·4	...	...	...	...
1890	10,598	49·7	7436	70·2	909	8·5	457	4·4	317	2·9	228	2·2	179	1·7	163	1·5	...	...	...	...	...	...
1891	10,922	47·7	7571	69·3	906	8·3	471	4·3	321	3·0	237	2·1	234	2·2	...	...	...	...	...	...	...	...
1892	11,599	48·0	7847	67·7	995	8·5	584	5·1	372	3·2	311	2·7	...	...	...	...	...	...	...	...	...	...
1893	11,779	46·9	7776	66·0	1039	8·8	610	5·2	396	3·4	...	...	...	...	...	...	...	...	...	...	...	...
1894	12,042	47·4	7918	65·8	1044	8·6	625	5·2	...	...	...	...	...	...	...	...	...	...	...	...	...	...
1895	12,346	49·3	8187	66·3	1134	9·2	...	...	...	...	...	...	...	...	...	...	...	...	...	...	...	...
1896	14,170	46·9	9610	67·8	...	...	...	...	...	...	...	...	...	...	...	...	...	...	...	...	...	...
1897	14,465	46·7	...	...	...	...	...	...	...	...	...	...	...	...	...	...	...	...	...	...	...	...
1898	13,452	48·7	...	...	...	...	...	...	...	...	...	...	...	...	...	...	...	...	...	...	...	...
1899	13,471	52·2	...	...	...	...	...	...	...	...	...	...	...	...	...	...	...	...	...	...	...	...

APPENDIX C.

APPLICATIONS FOR PATENTS, PATENTS SEALED, ETC., under Act of 1883.

Year.	Number of Applications.					No. of Patents Sealed and remaining in force to end of Fourth Year.
	For Patents.	Abandoned. § 8, s.s. 2.	Void. § 9, s.s. 4.	Open to nder § 10, ut not proceeded with to the Sealing Stage.	On which Patents were refused. § 11.	
1884	17,110	7,013	63	40	11	9,983
1885	16,101	7,236	58	22	10	8,775
1886	17,176	7,952	79	36	10	9,099
1887	18,059	8,434	76	62	21	9,466
1888	19,089	9,141	77	38	16	9,817
1889	21,004	10,221	90	18	11	10,664
1890	21,309	10,570	100	23	18	10,598
1891	22,878	11,791	122	18	25	10,922
1892	24,179	12,417	122	15	26	11,599
1893	25,107	13,162	121	15	30	11,779
1894	25,386	13,180	116	14	34	12,042
1895	25,062	12,530	146	7	33	12,346
1896	30,193	15,829	157	9	28	14,170
1897	30,952	16,251	204	10	22	14,465
1898	27,650	13,959	206	8	25	13,452
1899	25,800	12,075	181	57	16	13,471
1900	23,922	..	..	..	..	..

Number of Patents, Designs and Trade Marks applied for and registered, the Receipts and Expenditure, and the Number of Staff employed in the Patent Office, in The United Kingdom and The United States of America in the Years 1896–1899.

THE UNITED KINGDOM.

Year.	Number of Applications for Patents.	Designs.	Trade Marks.	Number of Patents issued.	Designs registered.	Trade Marks registered.	Receipts. Patents.	Designs.	Trade Marks.	Sale of Publications.	Total.	Expenditure. Salaries. Technical Officers.	Other Officers.	Total.	Pensions.	Establishment.	Buildings.	Printing.	Total.	Number of Staff. Technical Officers.	Other Officers.	Total.
							£	£	£	£	£	£	£	£	£	£	£	£	£			
1896	30,193	22,849	9,466	12,473	21,727	2917	185,619	3835	9,723	7685	206,862	23,885	31,808	55,693	2725	6643	4,536	26,757	96,354	57	194	251
1897	30,952	20,417	10,624	14,210	19,301	3358	200,850	3725	10,397	8564	223,536	24,731	32,291	57,022	2590	7194	14,280	31,150	112,236	61	201	262
1898	27,650	20,049	9,767	14,063	18,830	3137	200,419	3574	10,891	8535	223,419	25,740	32,298	58,038	3270	9400	1,991	28,350	101,049	62	205	267
1899	25,786	19,495	8,927	14,160	18,470	3777	202,977	3287	11,354	8082	225,700	26,794	32,850	59,644	4227	7978	24,117	27,250	123,216	62	205	267

THE UNITED STATES OF AMERICA.

Year.	Patents.	Designs.	Trade Marks.	Patents issued.	Designs registered.	Trade Marks registered.	Receipts. Patents.	Designs.	Trade Marks.	Sale of Publications.	Total.	Expenditure. Salaries. Technical Officers.	Other Officers.	Total.	Pensions.	Establishment.	Buildings.	Printing.	Total.	Examination Staff. Experts.	Assistant Experts, Etc.	Total.
							£			£	£			£		£		£	£			
1896	42,154	1828	2100	21,928	1445	1846	262,308			2504	264,812	No information.	No information.	137,473		9822		75,387	222,682	200	405	605
1897	45,765	2150	2038	22,163	1631	1701	272,513			2615	275,128			137,901		8726		77,941	224,568	200	405	605
1898	33,999	1813	2162	20,464	1803	1473	225,344			2203	227,547			144,059		5921		77,259	227,239	228	434	662
1899	39,043	2400	2831	23,388	2139	2260	222,684			2407	225,091			150,881		4911		86,564	242,356	228	434	662

DISCUSSION.

The PRESIDENT said that he was sure that all those present had shared with him the pleasure of listening to a paper read by one who was obviously an enthusiast, as well as an expert, in the subject before them. Speaking for himself, he (the President) had no knowledge of the intricacies of the subject, and therefore could look at it impartially. That it was of pressing importance at the present time, was proved by the fact that the Government had thought it worth while to bring in a bill this session, making some very vital changes in the patent law. The author suggested that that bill did not go far enough, and that was a point upon which the discussion would no doubt throw some light. However, whether the author's views were agreed to or not, he was sure the meeting would be unanimous in appreciating the merits of the paper Mr. Thwaite had given them. He had compressed his opinions, as was so desirable in papers of this sort, within the narrowest possible compass; and he (the President) would suggest that those who took part in the discussion should endeavour to follow the author's excellent example. He had very great pleasure in proposing a hearty vote of thanks to Mr. Thwaite for his paper.

The vote was carried with acclamation.

SIR HIRAM S. MAXIM said that in discussing the paper he did not know that he could do better than to give some of his own experiences, as they would show the practical working of both the British and the American systems. He could fully endorse the greater part of what the author had said. The advantages of the American system over the English system, however, were not so great as one might be led to believe from the paper. It might be quite true that the American system, considered as a whole, was considerably better than the English system. Still, there were glaring defects in both systems. It certainly was a great hardship to the struggling inventor to be forced to pay a standing tax for simply having his drawings and specification registered and dated at the Patent Office, for this, in reality, was all that an English patent amounted to. If they registered a paper, which might be of much greater value than a patent, they were only called upon to pay once, and not to pay a stated tax every year. It was quite true that any article could be patented over and over again in England. Several of his (the speaker's) patents had been patented over and over again. The automatic gun had been re-patented certainly a hundred times.

It would certainly be a good plan, and one very much to

the advantage of the inventor, if they had a system of examining the applications before patents were granted, similar to the one employed in the States. In England the date of a patent was the date of filing—that was, the date of filing the provisional specification—and the inventor could not, at any subsequent time, go beyond what was clearly shown in his provisional specification. Moreover, no one could put in an application for a patent in England at a later date, and make it antedate the original application, and by perjury beat the original inventor who was the first to file his application—and that was an advantage in favour of the English patent system.

Suppose an inventor applied in the United States for a patent which he imagined might be very valuable—for instance, a deposition of carbon on the filaments of the incandescent lamp. He (Sir Hiram) thought that out, and made an application for a patent for it, and laid it before the capitalist who was paying the bills. The other electrician said that he did not wish any better proof to show what a fool he (Sir Hiram) was to make such a suggestion, and that everyone knew that if a carbon was heated by electricity in petroleum vapours nothing but a disastrous explosion would result. However, he applied for the patent; and the other party was discharged, and when he found that the process was a success, he applied for it himself on the top of him (Sir Hiram). He (Sir Hiram) was limited to the truth, whereas, the other man was not: that man employed his father and his brother to swear that he had invented it a long time before he (Sir Hiram) did, and the priority of invention was decided in the man's favour. He (Sir Hiram) was told that if the patent had been properly taken, it would have yielded a million dollars a year, as no lamp could be made without it. Mr. Edison was forced to use the patent, and a regular triangular duel went on for some time. In the end the party who had caused the trouble went to Washington, abstracted a patent, made alterations in it, and put it back again into the office; however, the fraud was detected. Shortly after that the party shot a man, and was sentenced to the penitentiary, and died. Then the patent was made common property, so that Edison could use it without paying a royalty. By that means he (Sir Hiram) lost the most valuable invention he ever made in his life. That would not have happened under English law, for his patent was in the office some six months ahead of all others. Here was one point in which the English law certainly had advantages over the American law.

The great fault of the American system was, that it permitted a patent to be attacked after it had been filed, or even after it had been granted. Suppose one applied for a patent which promised to be very valuable; suppose that the inventor had conscientious scruples against committing perjury, and was in the very nature of things confined to giving accurate dates; but another party who was not troubled with these scruples, and wished to possess himself of another man's property, swore himself, and got others to swear, that he made the same invention several years before, that all this was conducted privately, and that he had delayed in getting his patent from some cause or other, duly stated; it became an extremely difficult matter for the real inventor to prove a negative, and in that case, he, like him (Sir Hiram), might lose a very valuable patent. He thought, therefore, it must be very evident that there were some defects in the American patent law; and if the law in England was to be changed, he was certainly of the opinion that it should be so framed as not to permit of the frauds which he had indicated. The gentlemen who made a living by attacking the patents of others in the States were known as "patent sharks," and it was not the object of these sharks to prevent the issue of a patent, or to break it down after it had been issued with a view of making it public property, but to get the patent themselves and reap the benefit. Suppose, for example, that it should be impossible for anyone to receive a patent on an invention which had already been filed at the Patent Office by another, the motive for attack would be completely done away with. Suppose, however, that one should have been engaged in manufacturing a certain article, and had failed to patent it, the fact that he had not patented it would be sufficient indication that he did not care to do so. However, should another person put in an application for a patent, all the manufacturer would have to do would be to show that he had been working the invention himself before the date of this application, and this would make the patent common property, but would not enable the manufacturer to take out a valid patent. And that was about the only improvement that he could suggest in the American patent law.

It must be understood that the object of the dishonest party in taking out a patent in the States was not to make the invention public property, but to get possession of it, or to dispose of it, or to blackmail the real inventor out of some part of it. As long as it was known that a patent could be attacked, there was always someone ready to do so if it was worth anything.

As regarded patent litigation, that was exceedingly slow, unsatisfactory, and expensive, in both countries. He had been engaged in many patent cases, both in England and the United States, and in all instances the lawyers had attempted to impress upon him the stupidity and almost imbecility of the judge, before whom the case was to be decided. He remembered one case in which he was told that the judge did not go by the quality of the evidence at all, because he did not understand the subject, but he looked rather at the volume of the evidence. The solicitor admitted that he (Sir Hiram) had an extremely good case, and that the facts were all in his favour; and he (Sir Hiram) naturally supposed, as those had been very clearly set forth by a few competent witnesses, and as there were plenty of documents to show the truth of their contention, that that would be all that was necessary. However, the other side had put in an immense volume of testimony, of an extremely weak nature, much of it not bearing on the case at all, but the lawyer said that the judge would go very largely by the amount of testimony and not by its character. So other witnesses were employed, and he (Sir Hiram) suggested that if it was true that the judge did not understand the question at all, and if it was quite impossible to make him understand it, the evidence should be put on the scales and weighed, without being read at all, and that, as far as he (Sir Hiram) was concerned, his case should be largely written on one side of very thick paper. The scales of justice being completely impartial, he would then be sure to win his case. Perhaps his suggestions were carried out to some extent, for the judge, who was said to know nothing of the case, decided it in his favour.

He did not remember ever having been in a patent suit in which it was not impressed upon him by his legal advisers that the very first thing they had to do was to educate the judge; and in many cases he had heard most eminent counsel, in opening a case, speak for many consecutive hours in delivering a speech, which, in reality, was nothing more nor less than a primary lecture on science and mechanics—a lecture delivered for the purpose of giving to the judge just sufficient knowledge of the subject to enable him to understand the evidence. Why was that? They certainly had many scientific men among them who were quite competent to decide almost any scientific or mechanical question without any special lecture at the time. He certainly could pick out a dozen men in London, any one of whom would be perfectly able to understand any patent case that might arise. Why, then, should they not have a committee of experts, recognised by the Government, who would be able to decide patent cases in a few hours for 200*l.* or 300*l.* expenses, which now required many years and cost fabulous sums? In the United States of America it was no uncommon thing for a patent case to drag on for six or seven years, and to cost from 50,000*l.* to 100,000*l.*; and in many cases the patent expired long before the case was decided. He spoke feelingly on the point, for he "had been there." He had a case now that had been going on for seven years, and it would go on for more than another seven years—if the patent lasted as long as that.

Although the American patent law might be quite equal to any other patent law in the world, everything considered, still he was decidedly of the opinion that it was not advantageous, as a rule, for European inventors to apply for patents in the United States of America. He thought he should be able to count on the fingers of one hand all the European patents that had ever paid in the States. As a rule, Americans did not recognise the validity of a patent until after it had been contested in the courts; and as that was altogether too long and expensive an affair for any European, it was quite an easy matter for the American manufacturer to make the patent article without rendering any equivalent whatever to the original inventor, and that species of sharp practice was not discouraged. His advice, therefore, to Europeans was to be very chary about spending their money in taking out American patents, which, as a rule, instead of being a source of profit, could only be a source of annoyance and vexation.

In regard to the lifetime of a patent, he thought it should be at least seventeen years. In many military and naval inventions, it took nearly fourteen years to get them understood and introduced into an unwilling service. In his opinion, this class of patent should not only be very cheap, but should also have an exceptionally long life.

Mr. HORACE ALLEN said that he found, from the first table given in the paper, that the fees for an English patent together with interest really amounted to 115*l*., compared with seven guineas, the cost of a United States patent. From figures given later on, it might be assumed that the really remunerative patents were those carried on to the full extent of fourteen years; the others would be more or less unremunerative, and these unremunerative patents only amounted to from 5½ to 10 per cent. of the original applications. One other point which would be of interest had struck him, upon looking at the diagram, and that was, that of the patents applied for in 1888, filed and paid up to between the fourth and fifth years, the patentees would have paid no less a sum than 48,540*l*. without having received any remuneration whatever. That sum should produce in return, at the low figure of 3 per cent., 14,562*l*. in nine years, which with capital amounted to 63,102*l*.; that amount should reasonably benefit the inventors, and they had to obtain that return, in the short space of nine years, by the exploitation of the remunerative patents.

Mr. H. SHERLEY-PRICE said that he had noticed something

which appeared to him to be rather striking. He deduced from the statement made in the early part of Mr. Thwaite's able paper, that an American inventor knew that his patent fees would either secure him a good title or disclose to him the disagreeable fact that his invention was not new. But he (the speaker) had yet to learn that America, or any other country, guaranteed a patent because they issued it. If they did so, surely there would be no litigation. The American Patent Office certainly did make a search, and they declined far more patents than were granted in England, but they did not guarantee a patent, so far as he understood. He scarcely thought that an American patent gave a title which was unassailable. There was no doubt that we did require rousing from our slumbers in England. But he thought that the author took a too pessimistic view of England's position; and he believed that one great cause of the decadence of England (if there be decadence) was that England kept to old methods. It was often a matter of grief to him, in going over works, to see the hundreds and thousands of tons of old plant which ought to have been on the scrap-heap twenty years ago. He was delighted to find that in large works, such as Armstrong's, Vickers', Whitworth's, and many others, they are throwing out the old plant and bringing in new. That was not because the old did not work, but because it did not work so well, or so cheaply, or so quickly as the new. If the advantages of the new could be drilled into the heads of all manufacturers, England would take a leap forward, patent laws or no patent laws.

He was thoroughly in sympathy with the author as to the penalising effect of the English patent laws upon the poor man. He believed that the majority of our inventions were the result of the observations of working men. But, if a workman made his invention known to his employer, in most cases he feared he would get few thanks for it. Therefore the man kept the knowledge to himself, until he had got a little money, and until he could take out a provisional specification. Perhaps, when the second payment became due the man would be without money, and his invention would be lost. The working man in England was penalised by our patent laws. It was a well-known fact that wealthy manufacturers took out patents, which they had no intention of working, so that no one else should take them out.

He hoped that the new patent law which was to be introduced this session would afford more security to the poor inventor, and would reduce the rate to about that of the American patent. At all events, he thought that the cost of a patent should be reduced in its preliminary stages, for in the earlier stages the workman could not afford to pay much for his patent.

Mr. ALEXANDER SIEMENS said that he had a very large amount of sympathy with the paper. He observed the statement that in the United States "the original inventor, or the man who has put the idea to actual test, received the greatest consideration." He thought that that was the proper principle of a patent law. He differed from Mr. Thwaite when he said in the next sentence, "in this country the law is often strained to protect the public against the inventor." He thought that that was very necessary. Sir Hiram Maxim had told them about the "patent sharks" in America. There were similar people in this country. There were people who simply took up a subject without knowing anything about it, and rushed to the Patent Office and tried to patent it, and of course the patent was absolutely useless. It was the useful people who ought to be encouraged. He thought that the paper was absolutely misleading as to the value of the American patent law. There was not a single thing in the American patent law to recommend it. Like Sir Hiram Maxim, he had law suits pending in the United States, in France, in Germany, and in England. The author stated that a financial comparison alone was an adequate testimony of the superiority of the American system. Anyone who took out a patent in America which was of any value at all, knew that the supposed cheapness was an absolute fallacy. He could quite confirm Sir Hiram Maxim's warning to European inventors not to try to patent their inventions in America. The idea that an inventor could get an American patent for 7*l*. 5*s*. 10*d*. was absolutely misleading. Anybody who had obtained a patent in America knew that that was absolutely impossible.

As to the validity of an American patent, he could at once confirm what the previous speaker had said. In France there was no examination whatever, and an inventor could patent whatever he liked. In Germany there was a strict examination. If Mr. Thwaite wanted to prove the advantage of the American system compared with the English, by quoting Mr. Abel's paper and stating that of 100 patents brought into court 42 failed in England and only 26 in the United States, he (Mr. Siemens) could prove the advantage of the French system, which had absolutely no examination beforehand, by quoting from the same paper that, out of 100 patents brought into court 41 failed in Germany, and only 23 in France.

The proper way for an inventor to treat his invention was to be moderate in his demand for royalties, and then it would not be to the interest of anybody to upset the patent. The inventor should not try to make his fortune in three days. Let him be moderate in his demands, and make friends of the manufacturers and give them also the benefit of the price, and then nobody would try to upset his patent. All the talk about American machinery being brought into England was only a proof that the British manufacturers had so much to do that they could not possibly do more.

As to the introduction of new machinery, there were two sides to the question. Of course, as the author said, useless machinery should be thrown away and replaced with better. That was quite right. But, on the other hand, they might do too much in the way of getting rid of old machinery. British manufacturers introduced new machinery with a certain amount of slowness. British conservatism he admired, to a great extent; and it was much better than rushing into new things and extending too much.

Mr. C. ROUS-MARTEN said that when he returned to England about nine years ago, after having been in various colonies studying colonial methods, particularly in New Zealand, he had the pleasure of spending a day at the works of one of the principal manufacturers of agricultural machinery; and that gentleman told him that just previously his firm had sent out their best agent, with the object of introducing to the colonies some highly improved agricultural machinery which they had just invented, but as soon as their man reached the colony they received a cable message to the effect that all the newly invented machinery which he had come to introduce had been in use there for several years, having been imported from America. The agent also found several things there that he had never seen in this country. Of course that was a shock, but the result was that the manufacturers simply set to work to revolutionise their old system, and they very soon regained a first place.

He ventured to think that what one of the speakers had said was very much to the point, and that was, that the differences between England and America did not consist so much in the patent laws as in the readiness to adopt improvements where they were desirable. There was no doubt that it might be judicious, in some cases, even to throw away old machinery which was no longer capable of doing good, up-to-date work.

He was rather sorry to hear the author of the paper take such a pessimistic view of England with regard to its commercial supremacy having been already lost. He hoped that the author meant rather that the supremacy would be in danger of being lost, unless we were prepared to move with the times. If the

E

author meant that, he thought that the Members would all be prepared to agree with him. The principle which underlay the whole subject appeared to him to be that broadly stated by the Prince of Wales, in the very able speech which he delivered in the City, after his return from his tour round the world. That was perhaps the most remarkable speech ever delivered by a British prince. His Royal Highness said that, if England was to maintain its industrial supremacy, it must "wake up." He (Mr. Rous-Marten) thought that that expressed the whole matter; and, the more we could improve our patent laws by really encouraging inventors, the better chance we should have of maintaining that supremacy which we had enjoyed in the past, and which we hoped still to have in the future.

Mr. B. D. HEALEY said that he had taken out about fifty patents in England on his own account. About 10 per cent. had proved remunerative, and 5 per cent. had proved fairly remunerative. He thought that they must all give credit to the author for the able manner in which he had raised the different points of the subject, and shown the weakness of the British patent system and certainly some slight advantages of the American. He hoped that in the amendment of the British patent law some of the hints which were contained in the paper would be embodied. He was quite certain that, if only the Government could see its way clear to bring in at once a law which embodied the greatest part of the suggestions contained in the paper, the trade of the country generally would flourish almost immediately. He thought that the members would be surprised if they knew of some of the older plants that were lying waste. In some of the old works the proprietors were beginning to wake up.

Mr. G. F. EMERY wished to say a few words from the point of view, not of the inventor, but of the lawyer. The real difference between the American system and the English system seemed to be that, in the English system, the Patent Office honestly said that the inventor must take all the risk; while, if the American system was what Mr. Thwaite would make it out to be, it looked as if the American Patent Office took the risk, but that in fact it did not do so. The American Patent Office sometimes put the inventor, as they had heard, to very great expense by putting him on interference, that caused tremendous delays and expense, and when the patent was at last obtained it was no better than an English one, although the English one was got with practically no trouble. The English method was cheap and quick; it was true that there might be opposition, but opposition was not very common and was comparatively inexpensive. Did they want to substitute for the English method a system which did not secure what it pretended to secure, and which was certain to increase the cost of getting a patent? When an inventor applied for his patent, not only would he have to pay extra for the patent investigation, but the Patent Office would probably bring to light somebody who could come in and oppose him, and so increase the cost of the patent. It seemed to him that everything in the nature of a Government inquiry at the commencement of the patent tended to make the patent more expensive. Inventors and everybody connected with English patents ought to be very careful about incurring the risk of extra preliminary expenses for something which might or might not turn out to be worth anything at all, for it was quite clear that official searches in the Patent Office were of very little use in securing the validity of a patent. However much they searched, as in Germany and in America, the validity of the patent had to be decided in the Law Courts; and he must say that it seemed to him, upon looking at the different systems all round, and considering especially the fact that in France the number of the patents upset was comparatively small, the English system, which had gone on so long with extremely satisfactory results, was on the whole quite as good as the American system.

Sir Hiram Maxim had spoken of the expense of litigation. No doubt patent litigation was very expensive, but that was not the fault of the legislature. He believed that the legislature intended and expected to cheapen patent litigation. The Act of 1883 provided that a litigant, whether plaintiff or defendant, could claim as of right to have the stupidity of the judge assisted by a competent assessor to be paid out of public moneys. That Act had been in force for a great many years, but as far as he knew there had never yet in England been an assessor appointed to sit with the judge. That looked as if Sir Hiram Maxim's plan of having a scientific judge did not meet with the approval of the litigants. He (Mr. Emery) should very much like to see it tried, for he must say that he was a great believer in cheap litigation. He would rather have a lot of cheap suits than a few big ones, and he should like to see a scientific assessor sitting with the judge, and disposing of patent actions for the benefit of the poor inventor and not for the benefit of a number of expert witnesses. He thought that it was a great pity that the system which provided for assessors had never yet been tried; he hoped that it would be tried, and he was sure that it would be found to be a great boon for the inventor. That would enable him to test his patent cheaply, and would be far more satisfactory than putting him to the expense of an official search and costly oppositions, which would only tend to make him think that his patent was better than it was.

Mr. CROYDON MARKS said that he thought the author must have been putting the spy-glass to the blind eye when he looked at British industry. If there was one thing that his paper established more than another, it was that it was not the inventor who had been at fault, but rather the manufacturer and the indifference of people in taking up the inventions of the country. From the beginning there had been a difficulty in getting inventions introduced into this country; from the time that Trevithick proposed the use of high pressure steam, when an Act of Parliament was brought into the House of Commons to prevent steam of a pressure of above 10 lb. being used, until now. There was an unwillingness on the part of manufacturers to take up what was called new notions. It was that which prevented progress. When one saw old firms working with square shafting, and drilling machines bolted to the wall, for turning out marine engines to-day, one wondered that they existed at all, rather than that they were able to get a little job now and then. That was not the fault of the inventor. There had been any amount of opportunity for manufacturers to adopt inventions, but they had not done so, and it was their inaction that prevented further progress. If Mr. Thwaite, in giving the number of patents that had been taken out, had put the population of the country as well as the number of patents, he would have found that England was the most inventive country in the world. In England, in one ordinary year selected, viz. 1899, there was one patent taken out for every 1462 people in the population of about 40 millions; in America, at the same time, there was a population of about 64 millions, and one patent taken out for every 1583 people— so that the proportion of patents was greater here than in America.

With regard to the fees, an English inventor could get his invention into the Patent Office for a payment of 1*l.*; in America, he only got his invention into the Patent Office by paying 3*l.*—so that the American inventor at the start had to pay 2*l.* more. When the examination had been concluded, he had a further $20 dollars to pay, which was practically equal to about 4*l.* sterling. Thus he had 7*l.* to pay for the patent before it was issued at all, whereas the English inventor paid 4*l.* for his patent, which lasted him for four years, and at the end of that time, if he had done nothing with it, it was better that it should drop. An invention that was standing on the books at the end of four years, and had proved to be of no benefit as far as the inventor was concerned, would be a dog-in-the-manger to better inventions that might come afterwards. As a rule, inventions were not taken further when they got to the fourth year, because there had been proved one of two things: either that they were not wanted, or that the inventor had opened his mouth too widely. It sometimes took a man three or four years to find out what he ought to ask for his invention.

With regard to the official searching, if that system were introduced in England there would be more litigation than ever. The figures given in the paper showed that there were 540 cases decided in America during a certain period, and that in the same time there were in England only 26 cases; and in the 540 cases arising in America, every one of the patents had been examined, and every one was supposed, therefore, to have some degree of validity about it. When a man obtained a patent in America, there was produced a feeling which was a false kind of security. He (Mr. Marks) had obtained patents in America that could not have been upheld in England, and he had overthrown, or had to amend many American patents that had been granted before they could be brought into an English court. Of the scores of patents that came from America, more than 75 per cent. of them had claims that had to be eliminated before they could be considered valid in this country. He had seen a patent come from America with 98 claims, 96 of which required to be struck out before starting an action. When a man obtained his patent in America, he was generally not sufficiently versed in the law to know the value of his claim. A large proportion of the patents granted in America were practically worthless, and would never stand half a day in an English court because they would have been anticipated by prior publication or want of invention.

In the bill now before Parliament, the suggestion was that the patents should be examined, and that the examiners should search back for fifty years and not beyond, and that the patent would then be valid; but how could it? There were other records besides the English specifications. There were the records of America, and these were not to be searched, and there might be specifications from Germany and France which would entirely destroy the validity of the patent. But the Patent Office under the Act would only search the English records. If they were to search at all, they should make the search free from restriction. There was no suggestion that a book more than fifty years old should not anticipate a patent, though a specification published forty-eight years before would make a patent invalid. The new Act would be a leap in the dark. The inventor would think that he had an advantage. The only advantage would come to the patent agent, and the new Act would be a glorious thing for him. He (Mr. Croydon Marks) was an American patent attorney as well as an English one, and he knew the difference between the systems of the two countries. His number on the American attorneys' roll of patent agents was over 3000; and there were over 4000 in America altogether. In England there were under 300 patent agents. Thus in England there were only about half the applications made, and 300 patent agents over here were all that were needed, while in America they had over 4000 agents practising.

Mr. JOSIAH ODDY said that he was one of the people that Sir Hiram Maxim had made such good-humoured play about. The difficulties of the judge were, perhaps, frequently increased by the expert witnesses. A case occurred to his mind in which Lord Kelvin was being examined as to the electrolysis of sodium caustic, and it was extremely difficult for the examining counsel to get anything like a direct answer to any of the questions put. One could imagine the extreme difficulties of a judge, even when he was not in the hopeless condition that Sir Hiram Maxim had described, if a distinguished scientist like Lord Kelvin failed to help the judge by giving direct answers.

Mr. Thwaite seemed to be dissatisfied with the constitution of the committee of inquiry into the working of the Patent Laws, which sat at the Board of Trade, but it was doubtful whether any better results would have been arrived at if a greater number of inventors had been on the committee or even called as witnesses before the committee. The presence of such men as Lord Alverstone, Sir Edward Fry, and Mr. Fletcher Moulton, was a standing testimony to the strength of the committee.

Mr. Thwaite seemed to imagine that the state of the British patent law was the root-cause of any degeneracy in British trade. He seemed to forget that there were many other causes at work, and that England was a very much older country than the United States, and that the United States had much more leeway to make up than England had. The late Lord Playfair pointed out in an article in the *Contemporary Review*, in 1888, that from 1870 to 1884 the manufacture of pig-iron in England rose 131 per cent., whereas that of the rest of the world rose

237 per cent. during the same period, and he said that the reason was that the other countries had so much leeway to make up. England was doing as much as ever it possibly could do in the way of manufactures and commerce, and it was not true that it was going backwards, as Mr. Thwaite suggested. That was shown in an article which appeared in the *Contemporary Review* last year. Therefore, it would seem there was not anything in the patent laws which was standing in the way of our progress.

With regard to the question of Patent Office fees, it was not true that the American patentee had done everything when he had paid the small amount which appeared in Mr. Thwaite's table. In practice, it was a general experience that many other payments had to be made.

As to the large number of patents that lapsed in England at the end of the fourth year, the obvious answer was that it was not worth while for the inventor to pay the very small sum which was then required to continue the patent.

Further, Mr. Thwaite seemed to think that the degeneracy of English trade was due in some way to our fiscal system, and one would imagine that he was advocating some method of protection, as against a free trade system. But Lord Playfair, to whom he (Mr. Oddy) had just referred, gave it as his opinion that the free trade system was probably the best one for this country.

The reason why so much machinery which, as some of the speakers had said, ought to be on the scrap-heap, was still being used was probably due to the extraordinary backwardness of the English industrial classes in the technical education which had been much talked about in the last few years. The working classes needed to be educated, in order to appreciate the value of invention and of new ideas, and so be induced to throw away their old machinery. That was, he thought, more at the root of the question than the character of the English patent law. The Patent Bill now before Parliament was a step in the right direction, and it was as much as could be hoped for from the legislature at the moment. It would be regrettable for the bill to be wrecked, just because the changes it proposed were not sufficienlly radical.

Mr. WILLIAM H. BOOTH said that with regard to the question of fees, it was perhaps desirable that the fees should be less, and that there should be a deferred system of payment. The present payment, which now stood for four years, should cover the payment for seven years, and whatever further payment might be necessary should be made after that time.

As regarded litigation, would it not be a good thing if

expert witnesses were put a stop to altogether, and the court had power to appoint technical men to give evidence before it in an entirely independent manner, and quite independently of the litigants? He believed that if that plan were adopted litigation would be much cheaper, and the court would get the truth. They would not have witnesses in the box trying to get the better of counsel, and counsel trying to get the better of the witnesses. It was questionable whether there should be either counsel or expert witnesses called by the parties to the suit. The expert witnesses should be in the employ of the court, and the litigants should be charged for their use. In that case men who gave evidence in such cases should be registered as expert witnesses. They might pay an annual fee for the privilege.

Mr. THWAITE replied to some of the points of the discussion, stating that he would subsequently make a fuller reply in writing. He highly appreciated the remarks that had been made, but some of them would require a little digesting before he could give a sufficient answer to them.

The remarks of Sir Hiram Maxim had practically confirmed the paper. The main difference between him and Sir Hiram was the character of the constitution of the courts for testing the validity of patents. He quite agreed that England should have a system of testing patent disputes something like that which existed in Germany. Under that system, technical experts, men of technically scientific attainments, composed what was essentially a jury of experts. The questions at issue in Germany had to be discussed, first by written arguments and then before the court *viva voce*. That system was, in his opinion, far cheaper than the present English law court system. Of course it might be that the American patent system was not perfect. But what was really perfect in this world? We all admitted the magnificent progress of America. Let the meeting listen to the following recently expressed opinion of a famous American expert, Patent Commissioner C. H. Duell: "I assert, without fear of contradiction, that we Americans owe to our patent system such foothold as we have gained in forty-six years in foreign lands for our manufactured products." Could anyone speak like that about the English patent system? The graphic diagram accompanying the paper showed instanter the deplorable position of the bulk of English patents. Unless we had some different system to help the workman inventor, England would continue on the down grade. Another American authority said that all classes had need to be thankful for what inventive genius had done for the United States: it had furnished the primary agencies by the employment of which

American manufacturing interests had attained their present magnitude.

Mr. THWAITE contributed the following further replies to the discussion on his paper by correspondence:—
The author agrees with Sir Hiram Maxim, who says, concerning the British patent system, that it is hard that the struggling inventor should be forced to pay a statutory tax for mere registration of date and publication of a patent specification that can with impunity be patented over and over again. His opinion is confirmed by the Hon. Charles Parsons, who has written to the author to say that he thinks that the British patent system is hard on the inventor and that the term of the British patent is too short. Sir Hiram admitted the advantages of the United States system of examination for novelty, so that he coincides with the author in the main conclusions of his paper.

He quite agrees with Sir Hiram Maxim's remarks respecting the importance of retaining the British system of preventing the obvious injustice that may spring from the privilege of antedating a patent article, and that this is a defect of the American system. The author agrees in that the new laws should make it impossible for patent sharks to maintain a profitable existence in this country.

The author considers that, seeing that the American patent system provides a search for novelty for a fee payment of 3l., and that this search is remarkably thorough, it is therefore advisable, until the new patent bill comes into legislative force, for all British inventors to apply for an American patent before applying for a British one. The author is glad to know that Sir Hiram agrees with him that an extension of the life of the British patent is desirable, and that fourteen years is an inadequate period if the average inventor is to be reasonably remunerated.

The author thanks Mr. Allen for drawing attention to the fact that the British patent fees with added interest really amount to 115l.

In reply to Mr. Sherley-Price, the author would observe that if an American patent is granted it is certain that its claims for novelty are well justified. The author notices that Mr. Price agrees with him in his remarks relating to the necessity of scrapping obsolete or inefficient plant; but Mr. Price does not sufficiently realise that financiers or manufacturers object to risking capital in developing or pioneering a new invention without they believe that they will obtain an adequate return for their financial risk; and, unless a British patent represents as good a title as that of an American, financiers or manufacturers will not be ready to adopt new processes. It is therefore obvious that England cannot leap forward unless the desired character is given to the British patent, and one that at present it does not possess. Mr. Price's remarks on the penalising of the British inventor are to the point; but, as he says, not only does the British inventor keep his invention to himself, but he not rarely crosses the Atlantic, and his invention is there protected, and may be [secures for the Americans an added source of national prosperity.

The author is glad to know that Mr. Siemens agrees with the principle of the American patent law, i.e. the patents are always to be so interpreted as to uphold and not to destroy the right of the inventor. But Mr. Siemens spoils his appreciation of this American principle by his statement that it is very necessary that the law should be strained to protect the public against the inventor—the two opinions are inconsistent. The application of the policy of crushing the inventor, for an assumed public benefit, has frequently proved disastrous to the best interests of our country, because, unless the invention, however good, is perfected for, and exploited in the market, it is of no public use or benefit. The author agrees with Mr. Siemens in his remarks relating to patent sharks. No doubt the suggested council of scientific and technical experts would be able to detect the true from the false inventor, and the law should be made so severe that the pirates and patent sharks could only have a short life of activity for evil.

The views of Mr. Siemens and Sir Hiram Maxim are antagonistic, as to the merits of the American patent system. The expression "misleading," used by Mr. Siemens, is not justified. A perusal of the author's paper, and an examination of the graphic diagram, show that all the figures given are based on official statistics, and the conclusions to be drawn are definite, and prove, that with some minor alterations, the American patent system is as practically perfect as a system can well be. Mr. Siemens' opinion of the value of the American patent system is erroneous. The cost of the American patent is as stated in the author's paper, and the fact that considerably over 20,000 United States patents are issued yearly, is the best proof that the author's statements are justified.

From the author's statistical data, Mr. Emery will see that the ratio of the patents actually allowed and issued to those applied for in the United States is about one in two, a very satisfactory ratio, considering the stringent character of the search. Of course, if our inventors ask for impossible claims from the United States patent department, and persist in attempts to obtain them, there is a considerable delay; and because certain inventors are avaricious, that fact should certainly not be brought forward as a defect of the United States patent system. The author's experience shows the objections raised by the United States patent examiners are in 90 per cent. of the cases thoroughly justified. The author's opinion, as will be seen from his paper, is that the chances of a contested patent being upset are less in the United States than in England, besides the fact that the American system, which exposes anticipations and grants a title of novelty, is valuable to the inventor; and had this system been adopted by the British Government, the author, like thousands of British inventors, would have been saved needless expense, time and worry.

The objections raised against the examination system do not seriously apply to the United States system, as statistics prove; but the author readily admits that the trail of disaster, produced by the present British system, will seriously interfere with the granting of British patents after the examination system is adopted, and this without providing the public with any benefit whatever. The author is glad to hear that Mr. Emery, as a lawyer, is in favour of cheap legislation, and also to learn that he is in favour of the application of the proposal to employ expert assessors to assist the legal element of a court in arriving at a decision in patent cases.

Mr. Croydon Marks' reference to the conservatism of British manufacturers is interesting, but the British patent system has always discouraged inventors. Even if the statement made by Mr. Marks, that the United Kingdom is the most inventive country in the world, is true, which the author greatly doubts —because it is only fair in making a comparison to deduct the black population of the United States—this statement only serves to show how important it is that this latent talent should be protected and encouraged to the utmost. The author refers to the graphic diagrams to show how crushing is the effect of the British patent system on the life of British patents.

The author does not agree with Mr. Marks that a patent should be allowed to lapse if the inventor had done nothing with it in the short period of four years. Very often from seven to eight years are required to develop a patent for the market, and until this is done the inventor cannot obtain any return. If it requires four years for an inventor to realise what amount he should ask for his invention, then that is a powerful argument in favour of the American system, which gives him still thirteen years from the time the inventor arrives at a reasonable opinion of the value of his patent.

Referring to Mr. Marks' allusion to the number of cases decided in the United States compared with those contested in the United Kingdom, it is well known that patents are rarely contested until they are well advanced in age—eight or nine years at least. Now, as statistics prove, only about 1 in every 20 of the British applications is alive at the end of eight years, the renewal fees have destroyed 19 out of every 20 British patents; whereas at least 1 in every 2 of the American applications is in force, and for seventeen years as against the fourteen years of the British maintained patents. The author's experience of the value of the expert search of the Washington office is the reverse of that of Mr. Marks.

The author agrees with Mr. Marks' views of the restrictions of the proposed new Patent Act, but only as far as they refer to American and other patent specifications published in the English language. There must be some limitation; otherwise, as Mr. Marks rightly says, the new experiment will prove to be a leap in the dark. If the examination and search covers American specifications, we shall have the benefit of the American search into novelty for the period of fifty years, the limitation proposed to be set by the new Act.

Mr. Oddy is mistaken in assuming that the author considers that the imperfections of our patent system constitute the root-cause of the statistically demonstrable falling-off of our progress compared with that of America. With respect to the importance of technical education, he fully agrees with Mr. Oddy, but an important part of this question is the education of the younger sons of British manufacturers, who are destined to control British factories. Instead of the university training of the usual British kind, they should have a training like that followed by the sons of German and American manufacturers, in which applied science constitutes a considerable proportion of the curriculum.

The suggestion made by Mr. W. H. Booth, to the effect that expert witnesses should be permanently retained by the court, is valuable, and its adoption would certainly tend to check the regrettable practice between such witnesses and opposing counsel referred to by Mr. Booth.

The following communication was received from Mr. GRIFFITH BREWER subsequently to the reading of Mr. Thwaite's paper:—

I am quite in accord with the author's principal contention that the American system of examination for novelty is of great value, as it saves many inventors from living in a fool's paradise, thinking they have invented something when they have only reproduced some obsolete thing that has been previously patented and then thrown aside as useless. In this country there is no such safeguard at the present time, and it occurs every day that patentees have patents granted to them for inventions that are obviously old, and this often leads the inventor into expense in manufacture that could have been saved had a search for novelty been made in the first instance.

The bill now before Parliament aims at effecting a remedy for this obvious ill; and, in return for only 1*l*. extra fee, it is proposed to search back into prior specifications for a period of 50 years, and give the applicant the entire benefit of the information thus gained. It is ridiculous to say that this Government search will benefit the patent agent, for it will take away one of his chief sources of profit, namely, searching on behalf of his client; but it will certainly benefit the patentee, for he will get a search by an examiner, skilled in the particular subject of the invention, for one-fifth of the cost he would incur if a patent agent made a similar search, and if his invention proves to be old the applicant can even save the 1*l*. by abandoning his application.

Up to this point I believe the author and myself are in accord; but I now come to the question of abolition of the Government renewal fees, against which I will give two serious reasons. The first reason is, that whether they are right or wrong, in view of the present state of the national exchequer, no Government at any near date will have the courage to abolish such an easy source of revenue as that derived from the tax on successful inventions. Therefore it is impracticable to try at present to accomplish this, and we may as well direct our energies to those reforms more within our reach.

The second reason against discontinuing these renewal fees, lies in the doubtful benefit such a measure would have upon the trade and enterprise of the nation. The renewal fees do not commence until the patent is four years old, and if the invention is not worth paying a 5*l*. tax on at the end of four years—which is generally the case with an invention in an impracticable form, or with an invention owned by an incapable man—it is time the patentee made way for others with more enterprise. If it were not for these renewal fees we should have numbers of patents kept alive for the full 14 years, for the sole purpose of clogging the trade of the country. The system of renewal fees should not be abolished, but I see no reason why they should not be reduced when the finances of the country will permit it, though even then they should never be so far reduced as to stultify the effect of weeding out patents covering disused inventions.

In conclusion, I would point out that the bill at present before Parliament introduces some of the beneficial features of the United States practice, whilst it steers clear of its objections; and none but those who at present benefit by the issue of patents for old inventions, can possibly suffer by the Patent Office privately informing the inventor of the prior state of the art, which in most cases will be of the greatest value to the patentee.

16

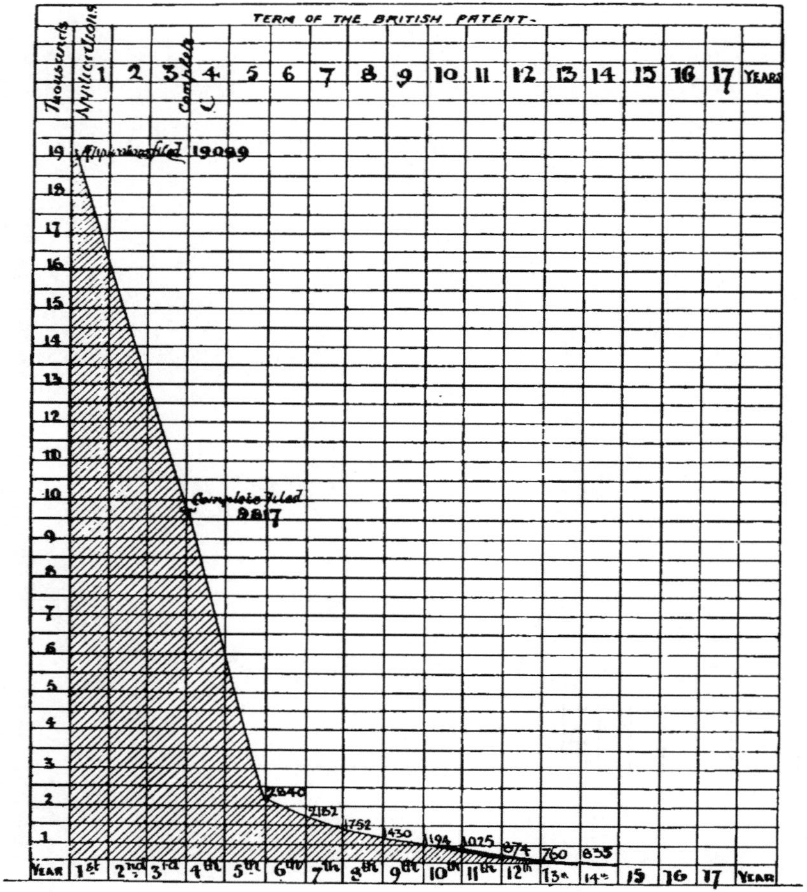

Fig. 1.

April 7th, 1902.

PERCY GRIFFITH, President, in the Chair.

AUSTRALIAN TIMBER BRIDGES, AND THE WOODS USED IN THEIR CONSTRUCTION.

By Herbert E. Bellamy.

Introductory.

In presenting the present paper for the consideration of the Society, the author would first observe that the subject of carpentry as applied to bridge and wharf construction is one to which he has devoted much time and attention. He trusts, therefore, that the present communication from him, as a foreign member, may be of some slight use to young engineers who contemplate going to the Colonies, and also to others who are interested in the employment of timber for bridges or other structures. Bridges for heavy traffic, composed entirely of timber, are now seldom constructed in England, but in the Colonies they are very extensively erected. The life of such bridges is very varied, depending upon the situation, the nature of the wood employed, and its treatment. Generally speaking, experience proves that from 35 to 55 years would be a likely average, although repairs are expected to be required in about 10 years after erection. The author has every reason to believe, that as soon as the Australian timbers become properly known, and their superior qualities recognised, they will be extensively used, not only in Australia, but also in Europe and America. The author proposes, in the first place, to describe and illustrate a timber bridge which he designed and erected in Queensland, and afterwards to describe, by the aid of specimens, the various Australian timbers used in such structures.

Description of Bridge.

The bridge, which the author submits as a practical illustration of that class of structure in the Colonies, spans a creek

10 feet deep at high water, but the waters of which, in flood times, rise to the level of the top of the fence, or about 25 feet above ordinary high-water mark. Figs. 1 and 2, Plate I. are respectively an elevation and a plan of the bridge; Fig. 3, Plate II. being a cross section, and Fig. 4 a longitudinal section of one span.

The total length along the central line, including the wing, is 320 feet, and the width between the gravel beams is 18 feet 6 inches. The piles are all of ironbark timber well creosoted, 16 inches being the minimum diameter of the butt, and 11 inches at the top or toe of the pile, while the length of any one pile does not exceed 40 feet. All piles over this measurement were scarfed (2 feet 3 inches scarf), plated by 3 feet 3 inches by 6 inches by $\frac{1}{2}$-inch wrought-iron plates, and bolted by five 1-inch bolts. The piles were all shod with wrought-iron shoes weighing 45 lb. each, and were driven with the top end of the natural tree downwards. In driving, the heads of the piles were protected with 3-inch by 1-inch wrought-iron circular hoops. The size of the tenon on top of the pile which fitted into the headstock was cut to 9 inches by 5 inches by 4 inches. The piles of the two piers in the creek were protected and covered by Muntz metal of 20-oz. gauge, bedded upon felt to 2 feet above high-water mark, and to 12 feet below high-water mark. This was secured with Muntz metal nails driven in 2 inches apart. The Muntz metal was fixed complete for 4s. 6d. per lineal foot. The test of the piles having been sufficiently driven was, that they should not be driven more than $\frac{1}{8}$th of an inch by 6 blows of a ram weighing 25 cwt., and falling a distance of 6 feet at each blow. The lengths of all the piles driven are clearly shown in Fig. 1.

It may be mentioned that there was a depth of not less than 20 feet of black mud below the bed of the creek, which had been washed down in flood times. This therefore explains the large discrepancies of the lengths of the piles. It might also be stated that borings were taken on each row of the piers before the work was commenced, and that all the piles pulled up well at the specified depths. Each pile was capable of safely carrying a load of 18 tons. The actual cost of driving the piles complete, including materials, labour, plant, etc., was 4s. 6d. per lineal foot.

Upon the head of each pier of piles, headstocks of rough-hewn ironbark timber, well creosoted, were fixed. They were 23 feet long, and were dressed to 14 inches square; rough-hewn timbers being preferable to sawn for large scantlings. Upon these headstocks, ironbark timber corbels, properly dressed (as shown in Fig. 4) were fixed, each 8 feet long by 10 inches wide, for the purpose of distributing the weight evenly on to the pile piers. Upon these corbels, the longitudinal girders or stringers of spotted gum were fixed at a distance of 6 feet 5 inches apart, centre to centre. Each stringer was 25 feet long (excepting those spanning the creek, which were 28 feet long), 14 inches deep by 10 inches wide. All were well creosoted. The joints of these stringers were of the ordinary half-scarfed joint and 2 feet long. These stringers were bolted on to the tops of the piles by anchor bolts. The length of the bolts over all was 6 feet 4 inches. The bolt was 1 inch round and welded on to a strap 2 feet by $2\frac{3}{8}$ inches wide by $\frac{7}{8}$ inch thick. The bolts passed through the stringers, corbels, and the sides of the headstocks. These girders, of course, carried the decking-platform and the post and rail fence. The length of these girders should never exceed 24 times their depth.

The formula for the strength of spotted gum girders as usually adopted by the author is $W = 7\frac{1}{2} \frac{b d^2}{L}$; where $W =$ breaking weight in cwt., $L =$ span in feet, $b =$ breadth in inches, $d =$ depth of girder in inches. For ironbark substitute 8 for $7\frac{1}{2}$. The maximum strength of a timber girder is obtained when the breadth is to the depth as 5 is to 7.

The decking was fixed at right angles to the stringers or girders and was of 6-inch by 4-inch spotted gum timber well creosoted, in lengths of 20 feet 2 inches, except at the railings and struts, where it was of 8-inch by 4-inch timber and in lengths of 25 feet, which allowed a projection on each side of the stringer of 2 feet 5 inches, so as to carry the struts which supported the fence. There was an open $\frac{1}{4}$-inch joint between each deck plank, and the planks were secured by two 6-inch by $\frac{1}{2}$-inch round wrought-iron spikes at each intersection of the four stringers. The cost of this decking, including all materials, labour, etc. was 99s. per square. The gravel beams or curbs of spotted gum well creosoted, fixed on each side of the deck platform, were 25 feet long by 10 inches by 6 inches. They were bolted by 1-inch bolts to the side stringers between the corbels, at distances of about 6 feet 5 inches. The joints were 12 inches long and of the ordinary half scarf. These beams or curbs were also mortised for the upright posts of the fence.

With regard to the post and rail fence, which was of spotted gum, the uprights were 5 inches by 5 inches square, fixed 4 feet 4 inches above the gravel beams and tenoned both into the upper rail and gravel beam. The top rail was 6 inches by 4 inches chamfered on upper edges, mortised to receive the uprights, and spiked with 6-inch by $\frac{1}{2}$-inch spikes into all uprights. The middle rail was 5 inches by 2 inches notched and bolted by

$\frac{1}{2}$-inch bolts into all uprights. The struts supporting the fence were 5 inches by 2$\frac{1}{2}$ inches, tenoned and bolted by $\frac{1}{2}$-inch bolts to the uprights and also to the 8-inch by 4-inch deck planks. The cost of this fence was 33s. per rod. In cases where the lengths of the piles standing out of the ground exceeded 7 feet but were under 9 feet, the piers were stiffened by a single 9-inch by 6-inch brace of ironbark, well creosoted, fixed transversely by 1-inch bolts to each pile and extending from the headstock across to the ground line. When the piles were more than 9 feet out of the ground, the piers were stiffened by double diagonal braces, the feet of which were tenoned into ironbark walings 23 feet long by 10 inches by 6 inches and fixed on the ground line.

All ironwork fixed was painted with four coats, as follows: before erection 1 coat of boiled linseed oil, two coats of red oxide paint, and after erection one coat. The post and rail fence, struts and wing fence were painted with three coats of the best linseed oil and white lead.

The total weight of all the timber in the bridge as fixed, was about 200 tons, while the weight of the ironwork fixed was 4$\frac{1}{2}$ tons.

The cost (materials and labour, etc. in fixing) of the ironbark headstocks, walings, braces, spotted gum corbels and stringers was 4s. per lineal foot, and of the spotted gum gravel beams was 3s. 6d. per lineal foot. The total cost of the structure including a small portion of the approach roadway was 1900l.

Australian Timbers.

The author will now describe the varieties of Australian timbers mostly used for bridges, reference being made to the specimens which accompany this paper. Of different kinds of timbers suitable for bridges, ironbark, spotted gum, blue gum, bloodwood, blackbutt, box, mahogany, karri, and swamp mahogany are amongst the most durable. Ironbark, mahogany, blue gum, bloodwood, swamp mahogany, turpentine or peppermint, tea, she-pine and cypress-pine are very durable when constantly immersed in water or wet ground, and are therefore well adapted for the piles, etc. for the foundations.

Blackbutt (native names "Tcheergun" and "Toi").—This is a very large tree, the bark of which is persistent at the base, but falling off in strips from the upper part of the trunk and branches. The leaves are thick, tapering from the base towards the point, and more or less curved. It is more or less a mountain tree, but it is also found on level country near rivers in South

Queensland, also in New South Wales and Victoria. It is a tree of rapid growth, and derives its name from its dark bark, which is blackest near the butt. The wood is of a light grey colour, hard, tough and durable, and is much used for piles and where strength and durability are required. When subjected to sudden strain or concussion it is even tougher than ironbark. Its crushing weight per square inch lengthwise is about 8440 lb. and its specific gravity 0·99. The weight required to break a bar transversely 1 inch square and 1 foot long is 857 lb. The mean diameter of the trunk of this tree is 48 inches, and the average length of the trunk or bàrrel 60 feet. (Specimen A.)

Bloodwood (*native name "Boona"*).—Found in tropical Queensland, North Australia, Victoria and New South Wales. The wood is of a red colour, containing large cavities full of gum, and is very durable if used whole, as for piles, posts, etc. It is a tall tree with persistent, spongy, somewhat fibrous bark, which is more than 1 inch thick; the leaves rather thick, tapering towards the apex, 3 to 7 inches long. Its specific gravity is 0·918. The gum contains 43·71 per cent. of kino tannin; it contains also a great deal of insoluble matter kinoised, especially catechu and kinoin. The mean diameter of the trunk is 27 inches, and the length 40 feet. (Specimen B.)

Blue Gum (*native name "Mungara"*).—Common on good land in Queensland north and south, also in New South Wales, Victoria, South Australia and Tasmania. A tall handsome tree, the bark deciduous, leaving here and there patches of a bluish colour. Leaves 4 to 6 inches long, the flowers with a lid often more than ½ inch long and of a nearly white colour. Fruit nearly globular. It is widely and extensively used for piles, bridges, planking wharves, construction of railway wagons, and for building purposes generally. The wood is of a red colour, close-grained, and its specific gravity before seasoning is about 1·04, and 0·95 after seasoning. The blue gums are the highest and largest trees in the world, exceeding even the enormous trees of California. The mean diameter of the trunk of this tree is 45 inches, and the average length of trunk is 80 feet. The weight required to break a bar transversely 1 inch square and 1 foot long is 759 lb. Experiments prove that the transverse strength increases with the drying or seasoning. (Specimen C.)

Brisbane Box (*native name "Tubbil Pulla"*).—A fine handsome tree, the bark persistent on the base of the stem, deciduous above, and on the branches of a brownish colour. Leaves crowded at the ends of the branches and very large. Found north and south on the Queensland ranges, also in New South

Wales and Victoria. Wood of a dark grey colour, hard, tough, close in the grain and very durable when kept dry. It shrinks very much in drying, and is used largely for girders, joists, sleepers and posts. The mean diameter of the trunk of this tree is 40 inches and the average length of the trunk is 80 feet. The weight required to break a bar transversely 1 inch square and 1 foot long is 948 lb. (Specimen D.)

Cypress Pine.—A coast pine, head very dense and dark green, found on the Queensland coast and also in New South Wales. The wood is of a light colour, close-grained, fragrant and durable, and capable of a high polish. It is used for piles of wharves, and is supposed to resist the attacks of the teredo. The mean diameter of the trunk is 36 inches and the height is 70 feet. (Specimen E.)

Ironbark (variety " Rostrata ").—A large tree, the bark black and thick, deeply furrowed but still separable into layers. Leaves very large, often from 2 to 6 inches wide on young trees. Found in Queensland, New South Wales, Victoria and Tasmania. For strength and durability it is unrivalled. The wood is red, close in the grain and is the heaviest wood in general use. Its specific gravity when just felled is 1·42. Its strength is 1·55 as compared with English oak at 1·00. It is very largely employed for piles, girders, sleepers, and for engineering work generally. The mean diameter of the trunk is 36 inches and the height 70 feet. The weight required to break a bar transversely 1 inch square and 1 foot long is 888 lb. (Specimen F.)

Karri.—This is a member of the eucalyptus family, as also is the jarrah, and is found in West Australia and on the coast range of Victoria; height of trunk up to 80 feet and mean diameter of the trunk 70 inches. The timber is red, hard, dense, elastic and tough, and is used extensively for piles, bridges, wharves, jetties, street paving, etc. Specific gravity 0·980; tensile strength 7070 lb. per square inch. Crushing weight per square inch lengthwise about 12,500 lb. It is almost a non-absorbent wood, and as it contains eucalyptus oil, the antiseptic qualities of which are well known, the author considers it to be the most hygienic material for street paving that can be used. The maximum absorption of karri is only 7 per cent.; that of jarrah 10 per cent.; of other ordinary soft paving woods, 23 per cent. Although this wood is not equal to granite it is more lasting than other material, and its surface does not become so slippery to horses' feet as jarrah. It is a very rapid grower, and reaches a sufficient size for cutting in 30 years. (Specimen G.)

Kauri Pine or Dundathu Pine.—A tall tree with the

branches in whorls. Leaves of a deep green colour, ovate, 2 to 5 inches long, and from 1 to 2 inches broad. Found on Queensland coast country, usually on ranges, and also in New Zealand. The mean diameter of the trunk is 30 inches and the length 60 feet. This wood is of a light yellow colour close-grained, entirely free from knots, soft and easy to work, largely used in house building, and sometimes for piles. Its breaking weight is 460 lb. per square inch. (Specimen H.)

Mahogany " Jarrah," Red or Flooded Gum.—Usually a very large tree with a rough reddish fibrous bark, the leaves large, straight or curved, and tapering towards the point. It is found in New South Wales, Victoria, Western Australia, Tasmania, and a few places in South Queensland. The gum contains 65·57 per cent. of kino tannin. The specific gravity is 0·90. The wood is of a rich red colour, close-grained, strong and durable, shrinks but little and is most useful for piles, girders for bridges, and wood block paving for streets. The mean diameter of the trunk of this tree is 50 inches and the length 80 feet. The weight required to break a bar transversely 1 inch square and 1 foot long is 780 lb. Jarrah is more deficient in fibre than karri. (Specimen I.)

Red Cedar, or Toon Tree.—A very large tree with spreading head, losing its leaves in the winter. A common tree of the Queensland and New South Wales scrub bordering rivers and creeks. The timber of this tree is that known in the English market as Moulmein cedar. The wood is beautifully grained, of a rich red colour, easy to work and very durable. It is principally used by cabinet-makers, although sometimes it is used in bridge work. Its breaking weight is 440 lb. per square inch. The mean diameter of the trunk is 60 inches and the length of the trunk 70 feet. (Specimen J.)

She Pine (native name *" Kidneywallum "*).—A tall erect tree, with a thin somewhat stringy bark and long linear glossy green leaves from 2 to 6 inches long. A common tree of the coast scrubs in Queensland and New South Wales. The wood is of a light yellow colour, close-grained, strong and durable, and is largely used for piles as it fairly resists the attacks of the teredo. (Specimen K.)

Spotted Gum (native name *Urara*).—This is also known as the bastard box, or mountain ash. It is found in southern Queensland, usually on stony ridges, also in New South Wales and Victoria. A fine handsome tree, the bark deciduous, falling off in patches, leaving an indentation where each piece was peltately attached, thus giving a spotted appearance to the trunk. The gum contains 34·97 per cent. of kino tannin. The wood is of a light grey colour, free from veins, little liable to warp, very elastic and durable, and is largely used for piles, and building purposes generally. The specific gravity is 0·981. The mean diameter of the trunk is 36 inches, and the length 70 feet. (Specimen L.)

Stringy Bark.—A moderate-sized tree, with a fibrous, persistent bark. Found in New South Wales, Victoria, South Australia and Queensland. The wood is of a grey colour, close in grain, hard, and much used for building purposes. It is occasionally used for girders, but it is inferior for this purpose to blue gum. This wood shrinks 4 per cent. of its thickness, and is sometimes marred by large gum veins. It is very quickly attacked by white ants, and when used in sea water it succumbs to the attack of sea-worms in about eight years. Its specific gravity is 1·020, or 63½ lb. per cubic foot. The mean diameter of the trunk is 30 inches, and the length 60 feet. (Specimen M.)

Swamp Mahogany (native name " Boolerchu ").—A moderate-sized tree, the bark somewhat fibrous and persistent, the leaves oval, 3 or more inches long, and more or less downy or hoary, as well as the young shoots. It is found all over the Queensland coast lands, often in swamps, also in North Australia and New South Wales. The wood is of a red colour, resembling Spanish mahogany, hard and close-grained, but best fitted for underground work, extensively used for piles, as it is found to resist the ravages of the teredo longer than any other wood yet tried. The mean diameter of the trunk is 36 inches, and the length 50 feet. (Specimen N.)

Paper-Barked Tea Tree (native name " Atchoourgo").—A large tree, the bark white, spongy, in thin paper-like layers, the leaves alternate, 2 to 4 inches long, broad or narrow, with from 3 to 7 nerves. Found in tropical Queensland. The wood is of a pinkish colour, hard, and close-grained, and very valuable for underground work and wharf piles. (Specimen O.)

Turpentine or Peppermint.—A large tree, with a reddish, fibrous, persistent bark. Leaves narrow, ovate, about 4 inches long, dark green. A common tree on the hills of southern Queensland, New South Wales, and South Australia. The wood is of a grey colour, close-grained, very tough and durable, difficult to burn, and used for piles, planking, girders, and ship and house-building purposes. The gum contains 53·32 per cent. of kino tannin. (Specimen P.)

Woollybutt.—A very large tree, the bark rugged and persistent for a distance up the stem, after which it is white and deciduous; leaves long, thick, the base broad, but tapering towards the point. It is found in Victoria, New South Wales, Tasmania, and in mountain gullies and river flats in southern

Queensland. The wood is of a red colour, close in grain, hard, tough, and durable, and is used for girders and joists. Its specific gravity is 1·034, crushing weight, lengthwise, 7997 lb., transverse, 2968 lb. per square inch. The mean diameter of the trunk is 36 inches, and the length 60 feet. (Specimen Q.)

Characteristics of Good Timber.

There are certain appearances which are characteristic of strong and durable timber, no matter to which class it belongs. In the same species of timber that specimen will in general be the strongest and the most durable which has grown the slowest, as shown by the narrowness of the annual rings. The cellular tissue, as seen in the medullary rays (when visible) should be hard and compact. The vascular or fibrous tissue should adhere firmly together, and should show no signs of softness at a freshly cut surface, nor should it clog the teeth of the saw with loose fibres. Too much confidence in passing dark-coloured timbers as possessing strength and durability should not be indulged in. In wood of a given species the heavier specimens are in general the stronger and more lasting. The freshly cut surface of the wood should be firm and shining. A dull, chalky appearance is a sign of bad timber. Amongst resinous woods those which have least resin in their pores, and amongst non-resinous woods, those which have least sap or gum in them are in general the strongest and most lasting.

Timber should be free from such blemishes as "heart-shakes," or splits radiating from the centre of the tree; "star-shakes," in which several splits radiate from the centre; "cup-shakes," or cracks which partially separate one layer from another; rind-galls or wounds in a layer of the wood which have been covered and concealed by the growth of subsequent layers over them; upsets where the fibres have been injured by compression; foxiness, which is a red or yellow tinge caused by incipient decay; twisted fibres, caused by the action of a prevalent wind, and which has turned the tree constantly in one direction; and hollows or spongy places in the centre or elsewhere, indicating the commencement of decay.

Seasoning Timber.

The best time for felling timber in tropical climates is the dry season, when the sap is at rest, and in temperate climates in the midsummer or mid-winter. It is desirable after timber has been felled, that whatever sap there is remaining be removed as quickly as circumstances will permit. This is termed seasoning the timber, and is effected either by stripping off its bark and exposing the tree to the air, taking care to protect it from any heavy rains until all sap is evaporated; or by a process which consists in immersing the freshly-cut timber for fourteen days in a pond of water, by which the sap becomes dissolved, and afterwards by exposing the timber to dry in the sun. Care must be taken that the timber is entirely submerged, otherwise the log will be injured along the water-line.

Hot-air seasoning or desiccation consists in exposing the timber in an oven to a current of hot air, which dries up the sap. This process, however, is seldom if ever carried out for engineering purposes. The disadvantage of this artificial drying or hot-air seasoning is, that the method of drying is too rapid, and seems to take away the stability of the timber, leaving it less firm, more brittle, and duller in appearance.

Decay of Timber.

The general causes of decay in timber are the presence of sap, exposure to alternate wet and dryness, or to moisture accompanied by heat and want of ventilation. Rot in timber is decomposition, generally occasioned by damp, and which proceeds by the emission of gases. There are two kinds of rot which are known—dry rot and wet rot. Both of these arise from the same origin, the fermentation and consequent decomposition of the sapwood, caused in one by turns of first wet and then dry, and in the other by the want of a free circulation of air round the timber. Both these kinds of decay arising from the presence of sap, it is of the utmost importance to lessen its quantity as much as possible.

Destruction of Timber.

Timber both in its growing and converted states is subject to the attacks of worms and insects, and when these exist in large numbers they remove so much of the wood as to impair the strength of any structure depending upon the timber. The white ant is a cream-coloured insect, not quite a quarter of an inch long, with a black head and lobster claws. It is found sometimes in Europe, but more especially in Australia, Africa and tropical climates. Its nests are in the timber or in the ground. White ants will eat the whole timber work of a house without noise, and they play frightful havoc with bridge scantlings. They bore close to the surface of the wood but without absolutely destroying it. They will even bore through piles and girders, leaving them as mere shells. Experiments prove that the ironbark resists white ants for a considerable period when seasoned; that the peppermint is the least affected of any timbers; that the red gum and sugar gum if cut while the sap is flowing, are very much interfered with, but if well seasoned are not so liable to the attacks; and that generally, dry timbers resist white ants better than when fresh cut.

Creosoting with bone oil is the best preservative against white ants for bridge structures. Kerosene is often used, but is only effective while its smell remains. The use of arsenic is sometimes resorted to, but it may be considered as entirely ineffectual.

The *Teredo navalis* is the most common enemy to timber used in submarine works. It is found largely in Australian waters, and penetrates nearly all kinds of timber. It is first deposited upon the timber in the form of an egg, from which in time emerges a small worm; this worm soon becomes larger and then commences its destructive work. It is equipped with a shelly substance in its head shaped like an auger which bores into the wood. At the same time it coats the hole which it makes with a very thin coating of carbonate of lime, and closes the opening with two small lids. The general opinion seems to be that the boring is assisted by the aid of an acid secretion. As the work of the teredo advances its size also increases.

The *Lycoris fucata* is a little worm with legs, something like a centipede, and which lives in the mud. It crawls up the pile inhabited by the teredo, enters the hole in which the teredo is located, eats the teredo, enlarges the entrance to the hole, and then lives in it.

A large number of methods have been tried in order to protect timber in marine works from the ravages of worms. Creosoting when properly carried out is more successful than any other process, and is the only one to be recommended. Copper sheathing is not permanently effectual by any means, although it is an additional safeguard. In time it is found that the worm gets in between the copper and the timber.

Preservation of Timber.

The best means for preserving timber from decay are to have it thoroughly seasoned and well ventilated. Several processes have, however been introduced with a view of preventing decay in timber by excluding moisture, or by drying up or expelling the sap within it. A few of these processes will now be described.

Painting preserves timber if well seasoned previous to the paint being applied. The same may be said with regard to tarring. All cracks and shakes should be filled up with good white lead in oil before the wood is painted over.

Charring timber is a process extensively adopted, and is fairly successful. The lower ends of posts which are put into the ground are charred with a view of preventing dry rot and the attacks of worms. Care should be taken to see that the timber is thoroughly seasoned, otherwise the process confines the moisture and rapidly induces decay. In this connection it may be well to mention that all posts should be put in upside down, that is with regard to the position in which they originally grew.

Creosoting, known also as Bethell's process, consists in extracting the moisture and air from the tubes of the timber, and then forcing in oil of tar at a high pressure. The timber after being dried is placed in a wrought-iron cylinder, after which the air is extracted from the cylinder and the pores of the wood by a pump. Creosote at a temperature of about 120° F. is then forced into the cylinder and penetrates the wood at a pressure of from 100 to 150 lb. per square inch according to the length of timber, for a period varying from three-quarters of an hour to two-and-a-half hours. The amount of creosote pumped in depends of course upon the nature of the timber and the purpose for which it is intended. It is usual to specify for the injection of 5 lb. of creosote oil per cubic foot of hardwood timbers for marine works. Of all the preservative processes at present known, creosoting is the most successful. It coagulates the albumen of the wood, fills its pores with the oily liquid, destroys insects and fungi, repels worms, excludes moisture, and prevents dry rot.

The result of impregnation with metallic salts has not in all cases been satisfactory. The following are the systems sometimes used; they are, however, seldom if ever, carried out in Australia: Kyan's bi-chloride of mercury, diluted with 150 parts of water, applied cold without pressure. Margery's sulphate of copper, diluted with 40 or 50 parts of water, applied with pressure varying from 15 to 30 lb. per square inch for six or eight hours. Paine's sulphate of iron and sulphate of barium. Burnett's chloride of zinc, with 25 to 40 per cent. of metallic zinc (3 parts of hydrochloric acid to 1 of zinc). The mixture is diluted with from 30 parts of water, and applied under a pressure of 100 lb. to 120 lb. per square inch for about two hours.

Conclusion.

The author has now completed the agreeable task he set himself of preparing for his fellow members at home what he hopes may prove to be, as an example of Colonial practice, a useful contribution to the Transactions of the Society. His only regret is that, holding a Colonial appointment, he is unable to present his paper in person.

The following communication received from Mr. HENRY ADAMS, Past President, was then read by the Secretary.

The bridge illustrated in Mr. Bellamy's paper is of a very simple character, but the contribution is particularly welcome as being from one of our Colonial members, who might favour us more often in this way than they have done in the past.

There seems to be no objection to halving the timbers horizontally for the scarf where shown in the drawings, but frequently in timber construction a vertical scarf is better. In Fig. 4, I should have used bed bolts for bolting down the hand rail over the piles, instead of the eye bolts shown. The formula for strength of timber W is presumably the load in centre, but is not so stated, and the breaking weight for H and J is presumably transverse load in centre on a beam 1 foot long and 1 inch square, but this is not clearly stated. It is not quite evident why the Muntz metal sheathing to the piles was bedded on felt. It would almost seem that a closer fit could have been made at the joints if there had been no felt. Creosote does not enter well in hard timber; 5 lb. per cubic foot seems rather a large allowance. Many foreign woods are very apt to split in nailing; it would have been interesting if the author had referred to this quality in describing the various kinds of Australian timber. The remarks on characteristics of good timber apply equally to timber from other countries.

Only one example (Q) is given of the strength of timber to resist pressure transversely. It is very important to know the strength of timber in this direction, as in beams resting on columns, but the records of experiments are almost universally silent on this point. Some years ago I used 20,000 cubic feet of creosoted Memel in constructing a large coal hopper and had occasion to search the records. I found some experiments on a cubic inch of timber under transverse compression, and also some experiments on the depression produced by a load upon one square inch area pressing upon the surface of a large beam. Neither of these was of any practical use, and I had to wait for experience in the finished structure, when I found that a maximum working load of 250 lb. per square inch might be put upon the bearing surface of creosoted Memel beams, upon the heads of cast-iron columns, and this was sufficient to squeeze out some of the pickle.

DISCUSSION.

The PRESIDENT said that he was sure that everyone who had heard the paper would appreciate the fact that the author was a Colonial member, and being so far away, he could not participate in the debate. The paper gave the Society the advantage of experience which was beyond the reach of most of the members. They would always, he was sure, appreciate records of actual experience given by members and others who were situated in other lands, and he trusted that for that reason alone the members would join with him in expressing very strongly their appreciation of the paper. There was, at the present time, a very general desire to bring the Colonies into closer communication with the mother country, and exchanging experiences in practical matters of engineering construction was certainly one way in which that communion might be strengthened. He was sure that every one of them would feel inspired, especially after the experience which this country had passed through recently, to strengthen to the utmost the bond of sympathy already existing between the Colonies and the mother country.

There was one point upon which criticism might possibly arise with regard to the paper, and that was the title, which was "Australian Timber Bridges and the Woods used in their Construction." He believed that to the majority of persons the main interest of the paper would lie in the description of the Australian woods. He must, however, ask them to give special attention to the diagrams which rendered the descriptive portion of the paper so valuable and interesting. As to the different kinds of wood, he trusted that the paper might tend to induce engineers in this country to use those woods to a greater extent in the future than they had in the past. The details of the bridge would also prove of very great value to the junior members of the Society.

Unfortunately, as Mr. Bellamy was in the Colonies they could not have the pleasure of hearing his reply to the discussion that evening. He (the President) had very great pleasure in proposing a hearty vote of thanks to Mr. Bellamy for his paper.

The vote of thanks was accorded with acclamation.

Mr. E. T. SCAMMELL said that he attended as the representative of one of the Australian Government Agencies, Western Australia, and, on behalf of the Agent General, the Hon. H. B. Lefroy, expressed his satisfaction that so important a subject as the suitability of Australian timbers for engineering purposes had been introduced for discussion at a meeting of the Society of Engineers. He disclaimed more than a general interest in, or knowledge of, engineering work, but he had some acquaintance with the subject of Australian timbers. He said that it was important, if those timbers were to be introduced to any large extent into this country, that the engineer should know exactly what he was buying. For that reason he deemed it necessary that not only the characteristics but the distinctive names of the various woods should be clearly defined. In many cases the common or vernacular names were utterly misleading, so that it was desirable in dealing with the subject to give the botanical names as well. In Mr. Bellamy's paper the common names only had been used. He had, therefore, been at some pains to ascertain their botanical equivalents, and these he would proceed to give in the order in which they stood in the paper.

Blackbutt, Eucalyptus pilularis, Queensland, etc. *Eucalyptus patens,* Western Australia.

Bloodwood, Eucalyptus corymbosa, is considered an altogether inferior tree.

Blue Gum, Eucalyptus tereticornis, Queensland, New South Wales, where it is also known as grey gum or bastard oak; and Victoria, where it is known as grey, red, or flooded gum.

The *Eucalyptus globulus* is the best known Australian blue gum, and is the wood most in demand for dock and harbour works. Its principal habitat is Victoria and Tasmania.

Brisbane Box, Tristania conferta, of which there are several varieties.

Cypress Pine or White Pine, Callitris robusta, of which there are two varieties in Queensland, both of which it is stated resist the teredo.

Iron Bark, Eucalyptus siderophloia (var. *Rostrata*), which has a number of varieties and is widely distributed. It is also known as red, white, and flooded gum.

Karri, Eucalyptus diversicolor, is strictly confined to Western Australia, and does not grow on the coast range of Victoria. It is one of the largest trees on the continent.

Kauri, or Dundathu Pine, Agathis robusta, is similar to the well-known kauri pine of New Zealand, *Agathis australis.*

Jarrah, Eucalyptus marginata, not mahogany, Jarrah, red or Flooded Gum, is found only in Western Australia.

Red Cedar, Cedrela Toona, is specially suited for cabinet work.

She Pine, Podocarpus elata, is a scrub tree, known also in New South Wales as the native plum and damson.

Spotted Gum, Eucalyptus maculata, is a useful timber, but not of the character of some other Eucalypts.

Eucalyptus goniocalyx is the spotted gum of Victoria, where it is also known as white gum and bastard blue gum, and bastard box.

Stringy Bark, Eucalyptus acmenioides, is the Queensland variety. The ordinary stringy bark of Victoria is *Eucalyptus macrorrhyncha.* But the best known and most valuable wood of this name is the *Eucalyptus obliqua* of Victoria and Tasmania, which is also called Messmate in Victoria, where, in Gippsland there is a yellow stringy bark, *Eucalyptus muelleriana,* which is highly regarded.

Swamp Mahogany, Tristania suaveolens, of which there are several varieties. The *Eucalyptus robusta* of New South Wales is also called Swamp Mahogany.

Paper Bark Tea Tree, Melaleuca Leucadendron. Many varieties of this tree are found in Australia. In the north it grows to a good height and is superior to the Tea Tree of the south.

Turpentine or Peppermint, Eucalyptus microcorys. This is the well-known Tallow-Wood of New South Wales, where there are two or three varieties of turpentine.

Woollybutt, Eucalyptus botryoides, is called also Bastard Mahogany in New South Wales and Gippsland, Victoria, where the name Woollybutt is applied to the *Eucalyptus longifolia.*

Mr. Scammell said that he presumed that the tests given by Mr. Bellamy were taken from Laslett's well-known book. But those tests were subject to revision by subsequent and fuller engineering experience. The measurements in the paper he failed to understand.

In concluding his remarks he said that most of the woods referred to by Mr. Bellamy, and other Australian woods that might be mentioned, were of great use to engineers, but the subject required fuller treatment than was possible in a paper of the kind to which they had had the pleasure of listening.

Mr. JAMES STIRLING, Government Geologist of Victoria, said that he cordially endorsed Mr. Bellamy's statement that, when Australian timbers became properly known, and their superior qualities recognised, they would be more extensively used in Europe and America.

He noted that many of the woods referred to by the author belonged to the timber trees growing in Queensland and in

Northern Australian territory, while others had a wider territorial range in Australasia. The character of the timber used for construction purposes by the municipal engineers, public works and railways department in the South-east Australian States presented considerable variety. That came into view when they studied the distribution of the species of Eucalyptus, under different climatic conditions of temperature, humidity, rainfall, etc., and the geological formations over which they were found.

He might state that there were no less than 131 well-defined species of Eucalyptus in Australasia. Many of them were shrubby or stunted arboraceous forms, of no special use for construction purposes, but, as in the case of the *Mallee, Eucalyptus dumosa, E. oleosa,* etc., were largely used for the distillation of oils. Of the total number of species, Queensland and Northern Australia possessed 53, New South Wales and Victoria 50, Western Australia 46, South Australia 28, and Tasmania 14 species. There was a notable affinity in the specific forms growing in the mountains of Victoria and Tasmania.

He noted that Mr. Bellamy had given the vernacular and also the aboriginal names of the species described in the paper. Such vernaculars were frequently misleading when any attempt was made to co-ordinate the facts of observation of the timber to be used for construction purposes, over large areas; while the aboriginal names obviously differed with the local tribe using them. He would give some illustrations, taking the woods in the order given by Mr. Bellamy.

Blackbutt.—The species of Eucalyptus used in New South Wales and parts of Victoria were varieties of *Eucalyptus pilularis,* but the trees in that region attained a height of 300 feet, girth 45 feet. The timber was used for bridges, piles, railway sleepers, and carriage wood. Professor Warren, of Sydney University, gave the weight of a cubic foot as from 58 lb., while Baron von Mueller, the late Government Botanist of Victoria, gave for the weight of the Victorian varieties, from 68 to 77 lb. per cubic foot. In Victoria, the name "blackbutt" was often applied to a variety of the sub-alpine species *Eucalyptus siebieriana,* the "Yanut" of the East Gippsland aboriginal tribes of "Kurnai," which attained a maximum growth at between 3500 and 4000 feet altitudes. In Western Australia the species *E. patens* was known as blackbutt.

Bloodwood.—He did not recognise this species from the vernacular.

Blue Gum.—If by this was meant the *E. globulus,* the "Baluk" or "Wang Gnara" of the aborigines (*wang* means bark, and *gnara* a string; the bark hung from the tree in long strings, and was shed from season to season), then the species was most extensively used in Victoria and Tasmania, where it attained its maximum development. The toughness and elasticity of the wood, the hardness of the grain, power of resisting great transverse strain (its transverse strength was about equal to British oak), and its vertical or crushing strain, were remarkable, averaging 3·078 tons per square inch, in experiments on 2-inch cubes. Its elasticity enabled it to withstand the heaviest blow of the pile-driver, without curling or rending. It was, however, not always easy to distinguish its wood, when cut and stacked away for some time, from that of the messmate or several of the stringy barks, as *E. obliqua, E. capitellata, E. macroryncha*, or *E. siebieriana*, etc. Its weight, however, generally enabled a distinction to be made in cases of doubt. It ascended to altitudes of 4000 feet in the Australian Alps. In Victoria, it was specially in demand for railway sleepers, bridges, poppet heads of mines, joists, studs, rafters, or heavy scantlings. It was rather pale in colour, sometimes with a light bluish-pink tint; the texture was often more twisted than *E. obliqua* (messmate), *E. amygdalina* (white gum), and many other fissile kinds, but was not so interlocked in the grain as *E. rostrata* (red gum), and *E. polyanthema* (red box). The specific gravity of the Victorian samples varied from 0·698 to 1·108. The species *E. goniocalyx* (spotted or bastard blue gum) which was of inferior quality, was often mistaken for the true blue gum. In Western Australia, the species *E. megacarpos* was called blue gum.

Brisbane Box.—The vernacular did not enable him to say to what species that corresponded, as there were no less than five distinct species called box in South-east Australia, viz.:—

1. *E. hemiphloia* (white or grey box, "den" of the aborigines), growing in South Australia, Victoria, New South Wales, and Queensland.

2. *E. melliodora* (yellow box, "dargan"), Victoria, New South Wales.

3. *E. botryoides* (red box, or snowy river mahogany, "binak"), Victoria, New South Wales and Queensland.

4. *E. polyanthema* (red box, "den" or "dern"), Victoria and New South Wales.

5. *E. odorata* (yellow or grey box, "dargan") South Australia and Victoria.

The Queensland wood referred to by Mr. Bellamy might possibly be *E. hemiphloia*, which was used for piles, marine work, or railway purposes, also for carriage building and all purposes for which West Australian jarrah (*E. marginata*) was used.

E. botryoides was a softer wood, but was valuable for ornamental purposes.

E. polyanthema, as it occurred in the sub-alpine areas of the Australian Alps, was an extremely tough and durable timber for many purposes.

Cypress Pine.—There were no pines, strictly speaking, in South-east Australia. Certain representatives of the order Coniferæ, such as the genera Dammara and Callitris, of which there were several species forming small timber trees, were used principally for flooring or lining boards in schools or other public buildings, owing to the strong aromatic scent proving a preventive to the ravages of the white ant. The trees ranged from 40 to 60 feet in height, and had an average diameter of from 15 to 24 inches. The wood was hard, brittle and knotty. When polished, it had a close, marble-like grain. Freneld (*Callitris verrucosa*) was called "cypress pine" in Western Australia.

Ironbark var. *rostrata.*—Again the vernacular name was confusing, as *E. rostrata* in Victoria was a red gum. There were several distinct species of ironbark, including: *E. resinifera*, New South Wales and Queensland, of value for ship-building; *E. siderophloia* (dark ironbark in Victoria; stringy bark in New South Wales), New South Wales, Victoria and Queensland; *E. cereba* (grey ironbark), New South Wales, Queensland and North Australia; *E. leucoxylon* (white iron-bark, "yirik" or "bwurawi") Victoria, South Australia, New South Wales, Queensland; *E. paniculata* var. *fasciculosa*, South Australia, Victoria, Tasmania and New South Wales.

The ironbarks were well known by their very rugose bark and the special hardness of the wood, the tensile strength of which was about half that of good wrought iron; the weight varied from 69 to 76 lb. per cubic foot. They were used for bridges, sleepers and general purposes where hardness and durability were required.

Karri.—Unless this was the *E. diversicolor* of Western Australia, or a variety of the red gum trees of South-east Australia, of which *E. rostrata* and *E. tereticornis* are examples, he (the speaker) was unable to identify it by the vernacular. In Gippsland, *E. tereticornis* is a littoral species, much prized for railway sleepers. It was known from Victoria, New South Wales and Queensland under the names red, blue and slaty gum, and by the aborigines as "yuro." *E. rostrata* occurred in Western Australia, Victoria, South Australia, New South Wales, Queensland and North Australia, and was well adapted for street paving—it was not the red gum with which Westminster had recently been paved, but ought to have been. *E. calophila* of Western Australia was also a red gum.

Kauri Pine did not occur in South-east Australia.

Mahogany, Jarrah, Red or Flooded Gum.—Here again the vernacular was indefinite; as previously stated, the mahogany of Victoria was *E. botryoides.* The jarrah of Western Australia was *E. marginata* (the latter a hard, close wood with interlocked grain, and of great strength and durability, lasting well in water), while the flooded gums of Western Australia were the species *E. rostrata, E. rudis* and *E. decipiens.*

Red Cedar or Toon Tree.—Not known by the vernacular.

She Pine.—Also not known by the vernacular.

Spotted Gum, Bastard Box or Mountain Ash.—The vernacular was very misleading, unless the *E. maculata* of New South Wales and Queensland, which was known as the "spotted gum," was meant. The spotted gum of Victoria was *E. goniocalyx,* and the mountain ash is *E. siebieriana* (also sometimes called woollybutt). *E. amygdalina* was also called mountain ash, although that species was better known as white gum or peppermint gum. *E. amygdalina* was the giant gum of Southeast Australia, the variety *regnans* attaining a height of nearly 400 feet. It was used for floors, lining, etc., and for culverts. In Gippsland, Mr. A. W. Howitt had described three varieties, under the native names of Chunchuka, Yertchuk, and Wang Gnara. Another white gum was the species *E. pauciflora,* which ascended as a stunted scrub to the higher elevations of the Australian Alps, at 6500 feet altitudes; its timber was of no particular value.

Stringy Bark.—This was the most difficult of all the eucalyptus trees to define by the vernacular, there being no less than six distinct species in South-east Australia, viz.:—*E. piperata* (white stringy bark, "yangura"), *E. capitellata* (mountain or red stringy bark, "dumung"), *E. muelleriana* (yellow stringy bark, "yuroka"), *E. pulverulenta* (silver-leafed stringy bark, "bindirk"), *E. eugenoides* (white stringy bark, "yangura"). The species *E. obliqua* or messmate was also sometimes included in the stringy barks. Most of the stringy barks were used for building purposes, fencing, etc. The species *E. muelleriana,* of Gippsland, was specially valuable for its durability; posts were known to have been in the ground for over forty years without much decay. It was fissile, free from gum veins or shakes, and clear in the grain.

Swamp Mahogany.—The vernacular gave him no means of identifying it with South-east Australian species. The swamp gum of Victoria, *E. Gunni,* ascended to alpine habitats of 6000 feet along with *E. pauciflora.* In the lowlands it formed a fine timber tree, and was used for a variety of purposes.

Paper-barked Tea Tree.—This was also difficult to correlate, as the Victorian tea trees were all members of the genera

Melaleuca, Leptospermum and Callistemon. The *Melaleuca leucadendron* of Western Australia is known as a paper bark.

Turpentine or Peppermint.—The species known as peppermint in South-east Australia was generally a variety of *E. amygdalina.*

Woollybutt.—The woollybutt of South-east Australia was a variety of *E. siebieriana.*

He thought he had said enough to show how necessary it was to give the proper botanical names when descriptions of woods were given, which were intended to supply useful data for engineering purposes, and that the employment of vernaculars or purely local names led to confusion, in consequence of the great variability in the qualities of different species occurring over widely separated areas under different climatic and soil conditions.

With regard to the variation in the quality of the wood, he had noted that of the trees of the same species growing in a district, those on the northern sunny slopes were more durable than those on the moist southern slopes, although the latter were frequently more fissile.

With regard to seasoning, a process known as the Rieser had been extensively used in Victoria, as at Wandong on the North Eastern line of railway.

The destruction of timber in the growing state was effected in Victoria, not only by the white ants, which had a predilection for certain species, but also by certain larvæ of nocturnal Lepidoptera, such as the *Uruba lugens*, which had caused great destruction to the red gum trees. This insect, however, did not seem to have affected the white gum (*E. viminalis*), swamp gum (*E. Gunni*), and yellow box (*E. melliodora*), growing in the same locality, although other Lepidoptera might affect those species.

Mr. R. Y. ARDAGH said that it seemed hardly fair to criticise the paper as the author, Mr. Bellamy, was not present. Having been to Australia he knew something about its timbers, especially those of Western Australia. There were practically three timbers, jarrah, karri and blackbutt, in Western Australia that were commercially known in this country and largely exported for engineering purposes; the others, though good, were not grown in sufficient quantities for exportation.

With regard to the statement that all the piles and bridge timbers " were well creosoted," the impracticability of impregnating Australian hardwood timber with creosote was well known. It was impossible to properly creosote soft woods unless they were thoroughly seasoned.

The author spoke of blue gum as being largely used. From

an Australian point of view it was perhaps true, but as an exported timber it bore no comparison with jarrah or karri which were extensively shipped to every continent.

It was singular that Mr. Bellamy should state that the blue gum was practically the largest tree in the world and specify such as 80 feet high by 45 inches diameter, and then describe a karri tree as 80 feet high and 70 inches in diameter. Supposing the girth were taken at half the length, viz. 40 feet, the karri tree was nearly two and a half times larger than the blue gum. He (Mr. Ardagh) had measured a karri tree that was 34 feet in circumference as high as he could reach, and the tree was reckoned to be 250 feet high and 150 feet up to the first branch. This was the largest tree he had seen in Western Australia.

Concerning the numerous varieties dealt with in the paper, engineers had only to deal with the commercial aspect of the timbers and not the botanical one, as, unless the woods could be readily obtained, it would be of no use to specify them.

The author gave the weight required to break a bar transversely 1 inch square and 1 foot long, as 857 lb. for blackbutt, as 759 lb. for blue gum, and as 780 lb. for jarrah, but the breaking strain of karri was omitted.

The certified breaking strain of karri by Mr. W. H. Stanger, the average of eleven specimens measuring in section ·973 by ·947 of an inch, span 12 inches, load at centre, was 1087 lb. The highest was 1321 lb., and the lowest 918 lb.—a sufficient proof that karri was much the strongest wood.

Under the heading of seasoning timber Mr. Bellamy remarked, "it was desirable after timber had been felled that the sap be removed as quickly as possible by immersing it in water for fourteen days." He (Mr. Ardagh) did not think that period sufficiently long to be effectual. If the logs were immersed for from three to six months possibly it might have some effect, but he did not know whether even then, except in the case of saplings, it would be of any use with Australian hardwood. It was stated that the seasoning of timber was effected by stripping off its bark and exposing the tree to the air, taking care to protect it from heavy rains until all sap was evaporated. He did not know how the author proposed to carry this out, unless he was going to build sheds to protect the timber.

Mr. Bellamy gave the weight of the timber in the bridge as about 200 tons, which he (Mr. Ardagh) calculated would be about 12,000 cubic feet at the inclusive price of 3s. per foot. Karri, as well as being used for piles, bridges, jetties and street-paving, was more extensively employed than any other Australian timber by all the principal English railway engineers for the construction of railway rolling stock, longitudinal and crossing timbers, sleepers, decking for platforms, warehouses, piers, telegraph arms, and the like.

Mr. C. STANLEY CHURTON said that with regard to that part of the paper which related to the preservation of timber, Kyan's method of preserving timber was out of date. The copper process also was not used now. It was very good and cheap, but it had died out. Only a few months ago he recommended the use of sulphate of copper for dipping the plugs for American sleepers. He suggested that they should put the plugs in a strong solution of sulphate of copper and then drive them in. With regard to Paine's process of sulphate of iron, sulphate of iron was largely used on German railways, but it was not used in this country except for experimental purposes. The sulphate of iron process was fairly good, and it hardened the wood. Burnett's burnettising solution of chloride of zinc was still very much used by the Government and by the Post Office and others where they wanted a white surface, because it did not discolour, and it rendered the wood absolutely proof against the white ant. In India, Singapore and places like those in the tropics, where the white ant abounded, there was nothing better than the chloride of zinc process. The worm had a very great objection to the taste of chloride of zinc, and consequently would not touch it. When wooden houses, barriers and walls were burnettised with chloride of zinc, neither ivy nor any other parasite would cling to them. He had seen ivy trained up for many years against wood which had been so treated, but the ivy would not fix itself into the wood, for the ivy, like the worm, objected to the chloride of zinc, and would not touch it.

Jarrah was a most beautiful wood, but for road paving there was nothing like the usual yellow block, properly creosoted. Mr. Bellamy stated in his paper that jarrah wood would absorb 10 per cent. of moisture, and that yellow deal would take in 23 per cent. of moisture. But when yellow deal was properly creosoted it would not take in any moisture at all, and it was waterproof. The creosote not only went into the wood, but it formed a coating outside which was nature's own way of preserving the timber and acted like the bark of a tree. The properly creosoted block was perfectly waterproof. Again, the creosoted blocks gave, whereas the hardwood, like jarrah wood, tore off, and gradually became shaved away. Therefore the longevity of the one was greater than that of the other.

Mr. A. T. WALMISLEY said that the description of the timber bridge was instructive, and likely to prove of service to the younger members, being accompanied by excellent diagrams. The value of the diagrams was that they helped to explain points which the author might have made a little more clear in the letterpress. For instance, he mentioned timber of 8 inches by 4 inches used in connection with decking, but he did not say whether the 8 inches or the 4 inches was the depth. That was important as the failure of the beams in the recent sad accident at Glasgow was attributed to insufficient depth for the unsupported span. The author's diagram, however, made the point clear. But the most valuable effect of the paper was the information which had been elicited upon the subject of Australian timber. That was a matter which engineers wanted to know a great deal more about than they did at the present time. The remarks which had been so far added to the paper in the discussion had certainly resulted in contributing importance to the subject, but he was sorry that they had not the advantage of the presence of the author, because he should have liked to ask him some questions about the important subject connected with the shrinkage of this scale of timber. The author mentioned this detail in the case of one or two of the samples which he had described in his paper, but he might have profitably told them more about it, and commented upon its cause and effect. They wanted to know to what extent the Australian timber would shrink, especially when it was not felled at the time and in the way that the author described. It appeared that there were some persons who would fell it at a disadvantageous season, and Mr. Ardagh, an expert in the matter, had agreed with the author in his remarks as to the time of year for the process of felling the trees.

Some of the heavy timber obtained from Australia had been used for seaside works where, owing to the fact that it sank in water, the engineers had to employ buoys to place it into position before they could turn it up vertically to drive it in the shape of piles. Some of the blue gum piles had been employed in that way, and the advantages claimed for the use of that wood in piling for sea works had been that, if any of the timbers broke, they would sink at once, instead of floating about as wreckage and doing damage to surrounding property. He had inspected a pier constructed of jarrah wood at Purfleet, and he was at the present time considering the adoption of Australian timber for a landing stage at the head of a pier in the river Medway.

Mr. T. B. GRIERSON said that he could not help being struck with one or two botanical observations made by Mr. Scammell and by Mr. Stirling. He knew Mr. Stirling as a very eminent geologist, but botany was a new rôle for him to play, and both he and Mr. Scammell had shown that they possessed an extensive knowledge of the subject. It should however, be borne in mind that the paper was written by an engineer, not by a botanist, and if the author had made a slip in any of the Latin names he should not be taken too severely to task as his evident intention was to mention those Australian timbers which he had found suitable for engineering works.

He thought that one of the great points with regard to the paper was that it let English engineers know what their Australian brethren were doing. The Australians had come forward with blood and money to help this country, and the least that English engineers could do was to specify those materials which the Australians could supply, in order, if only to show that Englishmen were grateful for the help which the Australians had rendered.

With regard to the timbers mentioned in the paper, the only Australian timber which he had used personally was blackbutt. About five years ago he ordered 200 pieces of blackbutt to take the place of a timber called Sally. That particular timber took nearly a year to come from Australia. It was very cheap as compared with the timber which it was to replace, and it was exceedingly suitable because it was of a tough character, and had great resistance to detrusion or pressure, and it was a marvellous timber for the particular class of work it had to do. That timber was also good for sleepers, owing to its resistance to decay, its hardness and toughness, but it and other Australian timbers of the same quality were at present too expensive as compared with those to be obtained from Russia, the Australian prices as obtained by him being 6s. per sleeper against say 3s. for the same size of the Baltic timber. It was entirely out of the question for British railway engineers to be expected to adopt such expensive sleepers. Of course all the Australian timbers were useful in their own way, but he wanted British engineers to try and find out what the Colonies could supply them with. Could the Colonies supply England with the timber that was wanted in this country as cheaply as Russia could?

He wanted to know what that experimental piece of timber, 12 inches long, 1 inch square, of each class mentioned in the paper, would carry. They were not told whether the weight was applied in the centre or distributed. Of course everybody knew that the piece of timber would carry twice as much weight if the weight was distributed, as it would carry in the centre. His object in asking the question was that it might be used with the formulæ given in the paper by Mr. Bellamy, when English engineers were calculating the dimensions of timber bridges, using the proper coefficient for each timber respectively. There was a timber, called American white oak, which had been largely used in this country for piling, and he wanted to know what was the cost of the Australian timber recommended for piles by the author, as compared with it per cubic foot, because the bridge was not a cheap one as it stood. He presumed, however, that the bridge was a great deal cheaper than a wrought-iron bridge, or any other bridge which Mr. Bellamy could get under the circumstances. He wanted to know roughly how the whole thing worked out per cubic foot.

Mr. J. W. WILSON said that he was interested to see their old friend the *Teredo navalis* turn up again in the paper, but he wanted to protest against it being called a worm. On Professor Huxley's authority, he would point out that the *Teredo navalis* was not a worm but a bivalve mollusc, more of the nature of a mussel or an oyster. The author mentioned another interesting animal which was described as following the *Teredo navalis* into his hole and then killing him. He (Mr. Wilson) should have thought that that was rather a good thing than otherwise. The author stated that copper sheathing was not permanently effectual, although it was an additional safeguard. At Portsmouth, where they suffered from the teredo, Colonel Raban, now Chief Director of Works to the Admiralty, got over the difficulty by putting piles down and cutting them off short rather below high-water mark, and then slipping over a 1½-inch cast-iron square case which penetrated into the ground some 18 inches and projected a slight distance above high-water mark. Then he placed the continuation of the pile in the upper part, making an ordinary butt joint, and carrying it up to the superstructure. No teredo had ever penetrated into those piles. The author said that the charring of the lower end of the post was to prevent the attacks of worms in the ground. But worms did not get into the ground. They depended upon the water. According to his (Mr. Wilson's) experience, they did not get down far below the ground line, and they did not go far above high-water mark. Though Colonel Raban's remedy was not cheap, still, if it kept the teredo out, it was a good plan to adopt.

Creosoting he did not regard as a perfect panacea. There had been much criticism from time to time about many of the timbers which had been mentioned, and which were represented by samples now on the table. He had been looking at them, and it did not appear to him that any sensible teredo would think of touching several of those timbers, but further experience would be of value to the engineer.

The author spoke of putting his planking down with quarter-inch spaces between the planks. He should have thought that it would have been better to make it rather wider. Otherwise, he imagined, the bridge would soon have no space between the timbers at all, and circulation of air between the joints was necessary in order to keep it in good condition.

Mr. PERRY F. NURSEY said that if any of those present had not been to the Colonial Exhibition which was being held in the Royal Exchange, he would advise them to go there. They would there find, in the West Australian court, a most interesting collection of Colonial timbers, both rough and worked into finished shapes, from paving blocks to carved and polished cabinet work. Amongst the former were paving blocks which had been in use for a number of years in various towns and cities in the United Kingdom, and which showed a remarkably small amount of wear under heavy traffic. Thus among the jarrah blocks were some which had been down for four years, the wear being only $\frac{3}{32}$ of an inch, whilst others showed a wear of only $\frac{1}{8}$ of an inch after five years use. Some karri blocks which had been down in Manchester for seven years had only lost half an inch under heavy traffic. In all cases the blocks were said to have been down under heavy commercial traffic.

The longevity of some of the Colonial timbers under water was illustrated by a jarrah pile which had been in use in Bunbury harbour for thirty years. Another jarrah pile shown had been submerged thirty-six years, and a jarrah fencing post was exhibited which had been drawn after being in the ground for thirty-nine years, whilst yet another had been in the ground for sixty years. Another specimen was a jarrah joist which had been in use in the dry for fifty-one years. All the samples were remarkable for their excellent state of preservation, most of them being sawn through to show their interior soundness.

Mr. H. SHERLEY-PRICE, referring to the cost of the bridge, said that Mr. Bellamy in his paper put the total weight at about 200 tons roughly, and the total cost might be taken in round numbers as 2000*l.* That would be equal to 10*l.* a ton. He presumed that timber was much cheaper in Australia than it was in this country, but engineers could build a bridge of wrought iron or steel at 10*l.* a ton. He did not mean that they could take the iron out to Australia and erect the bridge there for that money, but they could certainly construct a plain simple iron or steel bridge weighing 200 tons at 10*l.* a ton.

The PRESIDENT said that as Mr. Bellamy was in Queensland and therefore unable to be present at the meeting, the discussion would be forwarded to him for his reply which would appear in the Transactions of the Society.

Mr. BELLAMY's reply.

In replying upon the discussion, I desire first to thank those present at the meeting for the cordial vote of thanks which they gave me, and for the way in which my paper was received and so well discussed. I am confident that the valuable discussion which followed the reading of the paper cannot but prove to be of considerable advantage to all those interested in the subject of Australian woods. One or two of the speakers said it was desirable that the botanical names of the trees should always be used, but as they are rarely or never employed by the majority of engineers in Australia, I purposely omitted them. Indeed, all the timber merchants I have met, have never as yet studied that part of the question.

In reply to Mr. Adams, I would state that the strength of timber W as mentioned, was the load applied at the centre of a beam 1 foot long and 1 inch square. This applies to specimens H and J. In the case of the Muntz metal sheathing being bedded on tarred felt, experience proves that a closer fit can be obtained by its use especially where it passes over the bolted wrought-iron plates. With regard to the splitting of Australian woods, due to nailing, my experience is that nearly all hardwoods of small scantlings are subject to splitting, and that boring has to be done almost in every case. As a rule with nearly all hardwoods, even a 6-inch by 1-inch board will split without first being bored. The experience of many carpenters also proves that even with the same class of wood, it is impossible to lay down any rule which would be applicable to accurately determining the amount of splitting by nailing, as the effect must naturally depend upon the grain of each particular piece.

With reference to the valuable information so kindly imparted to the meeting by Mr. Scammell and Mr. Stirling, I feel personally indebted to them for having brought the botanical names of the trees before the members. I have, however, to state that the common names of the trees are well known to practically all the timber merchants in the Colonies, and that if the botanical names are employed, great confusion and disappointment will be likely to result. Many specifications have also been perused by me, and not in any one case has the botanical name even been mentioned. If there had been any specific advantage to be gained by its use, surely it would have been inserted. It is well known to engineers in England that if they specify oak to be of the very best quality, free from knots and all other defects, they either accept or reject it, and this is the practice in Australia. A particular class of wood is specified, and if not up to specification requirements it is rejected.

Mr. Stirling mentions that many of the trees described were not recognisable to him by the vernacular. As an engineer I perhaps can more easily distinguish them. It seems strange that the red cedar, so well known throughout the building trades, should not be known by the vernacular to Mr. Stirling. And it is still more strange to me that the well-known Brisbane box should not also be recognisable.

When I was in England I never saw any of the following oaks specified by their botanical names, viz. *Quercus robur*, or *Quercus pedunculata*, or the *Quercus sessiliflora*; and I am almost sure that very few of the merchants there work to those names.

However, where there is any reason for doubt, the classes to which the trees belong are described as follows:—

Blackbutt, A. *Eucalyptus patens*.
Bloodwood, B. *Eucalyptus corymbosa*.
Blue Gum, C. *Eucalyptus globulus*, and is found where stated by me.
Brisbane Box, D. *Tristania conferta*, and is well known in Australia.
Cypress Pine, E. *Callitris robusta*, and is also well known.
Ironbark, F. *Eucalyptus siderophloia* (var. *rostrata*).
Karri, G. *Eucalyptus diversicolor*. I am informed by an eminent botanist that this has been found on the coast range of Victoria, but not to a large extent.
Kauri Pine, H. *Agathis robusta*. Well known and extensively used in house building.
Mahogany, Jarrah, Red or Flooded Gum, I. Under this heading, I experienced great difficulty in coming to a decision, because I found that the word "jarrah" was used by many engineers indiscriminately in different States. This is the chief reason why the title was adopted. It is, however, very misleading, as jarrah (*E. marginata*) is confined to Western Australia.
Red Cedar, J. *Cedrela Toona*. Well known, and used by cabinet makers.
She Pine, K. *Podocarpus elata*. Not known as a native plum or damson as mentioned by Mr. Scammell.
Spotted Gum, L. *Eucalyptus maculata*.
Stringy Bark, M. *Eucalyptus pilularis*. Well known.
Swamp Mahogany, N. *Tristania sauveolens*. Well known and much used.
Paper-barked Tea Tree, O. *Malaleuca leucadendron*.
Turpentine or Peppermint, P. *Eucalyptus microcorys*.
Woollybutt, Q. *Eucalyptus botryoides*.

The tests mentioned by Mr. Scammell were for the greater part carried out by me. Mr. Stirling is perfectly correct in stating that the character of timbers varies considerably according to different climatic conditions.

With regard to Mr. Ardagh's remarks, comparing the sizes of the blue gums with the sizes of the karri trees, I would observe that the sizes given by me are averages, and that to quote isolated cases would be rather misleading. I know of a blue gum cut in Victoria only last year, which measured 480 feet from the root to the topmost branch. Mr. Ardagh is right in saying that engineers have only to deal with the commercial aspect of the timbers, and not the botanical one. If Australian merchants cannot supply timber to engineers' specifications, they cannot possibly look forward to securing orders.

With hardwoods, experience proves that for seasoning timber, its immersion in water for fourteen days is sufficient, and that no benefit would accrue by adopting a six months period.

I am indebted to Mr. C. Stanley Churton for the information put before the members, but am surprised to hear that the yellow deal blocks wore for a longer time than the hardwoods.

With reference to the point raised by Mr. A. T. Walmisley as to the shrinkage of timbers, from information obtained from several large timber exporters, and from my own experience, I find that practically all timbers shrink to a large extent if not properly seasoned, or if they are cut at the wrong season of the year. I can cite a few cases observed by me. The first is as follows :—In the construction of a small bridge, spotted gum was used for the decking which was cut at the wrong time of the year, and placed in position without having been previously seasoned. The scantlings were 8 inches by 4 inches and these had no spaces left between them. Within the short space of two weeks there was an opening between the planks of $\frac{3}{16}$ of an inch. At the end of four weeks it had enlarged by an additional $\frac{2}{16}$ of an inch. At the end of three months it showed an additional $\frac{2}{16}$ of an inch. At the end of six months the opening was exactly $\frac{1}{2}$ an inch, and at the end of twelve months the opening was just the same as at the six months period, thus showing that shrinkage had ceased.

Take another case of spotted gum which had been cut at the proper time and was properly seasoned, the scantlings of which were 8 inches by 4 inches. The planks were also used for bridge decking, and no spaces whatever were left between them. At the end of twelve months the greatest space between any two planks was only $\frac{1}{8}$ inch.

Ironbark (*E. siderophloia*) appears to be the least affected of all the timbers for shrinkage, whether seasoned or not. Blue gum (*E. globulus*) and red gum appear to be about the worst of all the timbers for shrinking when not properly seasoned. Stringy bark (M) is hardly, if ever, used for bridge decking now, on account of its shrinkage to an indefinite extent, even after seasoning. Kauri pine, when seasoned, is little liable to shrink.

With reference to Mr. Grierson's remarks, I would observe that with the experimental piece of timber 12 inches long and 1 inch square, the weight was applied in the centre in every case, and was therefore not a distributed load. During the past few months, I have made some more experiments with timbers as regards their breaking strain, and the results obtained are as follows. The experiments were made with a test piece 1 foot long and 1 inch square. The results obtained were as follows:—*Bloodwood*, breaking weight, 720 lb.; *cypress pine*, 640 lb.; *she-pine*, 750 lb.; *spotted gum*, 820 lb.; *stringy bark*, 785 lb.; *swamp mahogany*, 770 lb.; *paper-barked tea tree*, 728 lb.; *turpentine or peppermint*, 846 lb.

With regard to the cost per cubic foot of the whole work, I estimate that there is 13,000 cubic feet of timber fixed in the bridge, and this therefore works out at a cost of 2s. 11d. per cubic foot.

Mr. J. W. Wilson describes a system which was adopted at Portsmouth by Colonel Raban for resisting the ravages of the teredo. I would briefly refer to two systems carried out in San Francisco some years ago in connection with harbour works. (1) The piles were jacketed with large sewer pipes, and the space between the pipes and the pile filled in with cement concrete. The result was that the piles were not attacked, except in places where the pipe had broken. (2) The piles were coated with different kinds of compositions, one, a mixture of asphaltum and burlap, others with varieties of marine cement. These were applied to different kinds of American timbers, which were exposed under water for a little over five years. The result was that all the piles were destroyed by the *Teredo navalis* which is so prevalent in the harbour of San Francisco.

Mr. P. F. Nursey refers to a jarrah pile which was on exhibition at the Royal Exchange, and which had been submerged for thirty years in Bunbury Harbour. I quote the following confirmatory passage from a South Australian Parliamentary report: "Jarrah makes excellent piles in both salt and fresh water. Some karri piles have stood well in bridge work, but karri is unsuitable for piles in sea water, being much more subject than jarrah to the attacks of the teredo." I also find from the same report that in Western Australia, experiments have been tried with karri and jarrah by submerging the timbers in the sea at Fremantle for one year. The result is stated as follows: "The karri is completely riddled by the *Teredo navalis*, while the jarrah is perfectly sound."

Mr. H. Sherley-Price, in speaking upon the cost of the bridge, assumed it at 10l. per ton. The bridge certainly appears to be an expensive one, but the minimum wage for labourers in Queensland is 7s. per day, and for bridge carpenters 10s. per day, which is considerably more than the wages paid for labour at home. This therefore accounts for a great deal of the apparently large cost of the bridge.

In conclusion, I would mention that the various specimens of wood sent by me were obtained from the respective States in which the trees have their habitats. In cases where any one tree is found in more than one State, specimens were obtained from each State, and were most carefully compared.

The following communication was received from Mr. J. O. NEUMANN, subsequently to the reading and discussion of Mr. Bellamy's paper.

During the discussion on Mr. Bellamy's paper, a question was asked as to the transverse strength of Australian timbers. Some light is thrown on this subject by an excellent table giving the results of tests made by the Railway Commissioners of Victoria, which appeared in a report made by the Carriage Timber Board of that Colony and published by Walter May and Co., of Melbourne, and from which I extract the following information.

The samples tested were 7 feet long by $1\frac{1}{8}$ inch square. The distance between the bearings was 6 feet, and the weight was gradually applied in the centre until the sample broke. Taking the blue gum (*Eucalyptus globulus*) of Victoria, the following particulars are given. Three kinds were tried: No. 1 was grown at Mirboo; No. 2 at Range south of Waterloo, Gippsland; and No. 3 at Corner Inlet. Three samples of No. 1 were cut December 23, 1882, and were tested January 28 and 31, and February 2, 1884. Three samples of No. 2 were cut April 1883, and were tested on January 31 and February 4, 1884. One sample of No. 3, date of cutting unknown, was tested on January 24, 1884. The tree from which samples of No. 1 were cut had a diameter of 4 feet $4\frac{2}{3}$ inches, and the average weight was 10·75 lb. per sample. The diameter of No. 2 was 3 feet 6 inches, and the average weight 11·42 lb. per sample. The diameter of No. 3 is not given, but the average weight per sample was 11·25 lb.

The average weight of the whole of these samples per cubic foot was 65·18 lb., and the average specific gravity was: No. 1, 1·008; No. 2, 1·071; No. 3, 1·055; giving a total average of 1·045. The breaking weight of each sample was as follows:—

	cwt.	qr.	lb.	Average in lb.		cwt.	qr.	lb.	Average in lb.
No. 1.	9	1	25		No. 2.	9	3	24	
	9	3	13	1111·6		8	3	6	1023·6
	10	1	21			8	2	17	

	cwt.	qr.	lb.	Average in lb.
No. 3.	10	2	25	1201·0

The deflections at point of rupture were: No. 1, $3\frac{1}{2}$ inches, $4\frac{1}{4}$ inches and $4\frac{3}{4}$ inches respectively, giving an average of 4·16 inches, and No. 2, $4\frac{1}{2}$ inches, $3\frac{1}{2}$ inches, and $3\frac{1}{2}$ inches respectively, giving an average of 3·83 inches, and an average specific strength of 3036. The formula used is $\dfrac{WL}{4BD^2}$ where W = breaking weight in lb., L = length in inches, B = breadth in inches, D = depth in inches. Nos. 1 and 2 were grown in the mesozoic strata, whilst No. 3 was out of the granite formation and of which no further particulars are given. No. 1 was grown between 700 feet and 800 feet, and No. 2 about 1250 feet above sea-level.

The average specific strength of blue gum, tested by Baron Sir Ferdinand von Mueller, is given as 2666; ironbark of New South Wales (*Eucalyptus siderophloia*) coming next in his tests with 2859; blackbutt (*Eucalyptus piperita*) tests gave 1495; Queensland canary wood (*Eucalyptus hemiphloia*), 2678; and ironbark, 2854, when tested by the Commissioners; the latter figures are nearly the same as von Mueller's. Extremely interesting wood paving tests, ranging over a period of twelve years, are detailed in a report made by a former city engineer of Sydney, Mr. R. W. Richards, and presented to the Minister of Mines and Agriculture of New South Wales in October 1893.

Besides the information to be found in the various publications of Baron von Mueller, there is a little book entitled 'Australian Timbers,' by Mr. W. K. Warren, Challis Professor of Civil and Mechanical Engineering at the Sydney University, which is the best and most comprehensive work of its kind, and which was published by the Government printer in Sydney in 1892.

I have given detailed figures of the blue gum, because it was the paragraph in Mr. Bellamy's paper dealing with this wood which led to the question referred to, and also because this wood is more suitable than others for timber construction, the supply being plentiful and the price lower in consequence. Where price is no object, blackwood will be considered. Baron von Mueller's test fixed the average specific strength of this wood at 1709, but he obviously dealt with logs coming from comparatively very low sea-levels. The question of selection of soil and of altitude was carefully considered by the Railway Commissioners, and in the report to which I have referred, details are given of four logs, two of which were cut at 1300 feet above sea-level, and two at 700 feet and 1200 feet respectively. It was found that their average specific strength was 2610, the average breaking weight of the best sample rising to 1236 lb. The recommendations of the Commissioners —thoroughly practical men—bear so much (especially in their concluding remarks) upon all Australian woods, that I quote the paragraph in full: " Blackwood (*Acacia melanoxylon*) under test presented results superior to any other timber, whether of indigenous or foreign growth. Of all the indigenous timbers, this one may be said to possess in a pre-eminent degree every quality desirable for the purpose stated (i.e. building of railway stock). It seasons without losing bulk, is of good colour and figure, and resisted the heaviest strain before fracture, whilst in cost it is less than half that of Indian teak, to which it is superior in most respects and inferior in none. The most favourable conditions under which this timber obtains is in ranges displaying rock of the mesozoic period, as at Mirboo, and the sources of the Moe, Narracan and Morwell rivers. As this timber is being largely used in building railway vehicles, its excellent qualities may be taken to be recognised, and these tests may be held to confirm, beyond all question, the wisdom of the selection of this timber by the Railway Department, who have pleasure in very strongly recommending its continued use, care being taken, however, to obtain supplies from the localities where the natural conditions are favourable, and that the trees be felled in season when the sap is down."

With regard to the West Australian woods, a former Governor of the Colony, Lieut.-Colonel Sir Gerard Smith, K.C.M.G., in a paper recently read before the Royal Colonial Institute, said:—

"It is only in the last five years that the world in general has awoke to the appreciation of Western Australian hard woods, especially of karri and jarrah. With 8,000,000 acres of forest trees in the Colony, of which only about 1,700,000 acres are under lease, it may be safely prophesied that the supply of these hard woods is likely to be equal to any demand made for them.

"The uses to which these hard woods are increasingly put are many in number. It may be confidently asserted that an immense future lies before jarrah in railway workshops and in permanent ways, in the latter case for the excellent reason that the life of jarrah as a railway sleeper is about twenty years, as against the seven and a half years life of the ordinary creosoted soft wood article, thus more than counterbalancing increased initial cost.

"I believe jarrah and South Australian red gum to be the best sleepers yet known to the railway world. I doubt whether there is much to choose between the two. The karri wood in the particular purposes for which it is suited cannot be excelled, and we have some excellent examples daily before us in the paving of our streets. For building, or for purposes where great weights have to be borne, it is as good as the best English oak.

"There is, however, one condition essential to the utility of all the hard woods known, the neglect of which would be certain to lead to disappointment, and that is that they must all be thoroughly seasoned before using."

STRALIAN

20

"AUSTRALIAN TIMBER BRIDGES" BY HERBERT E. BELLAMY. PLATE I.

ELEVATION. FIG. 1.

PLAN. FIG. 2.

ERBERT E. BELL

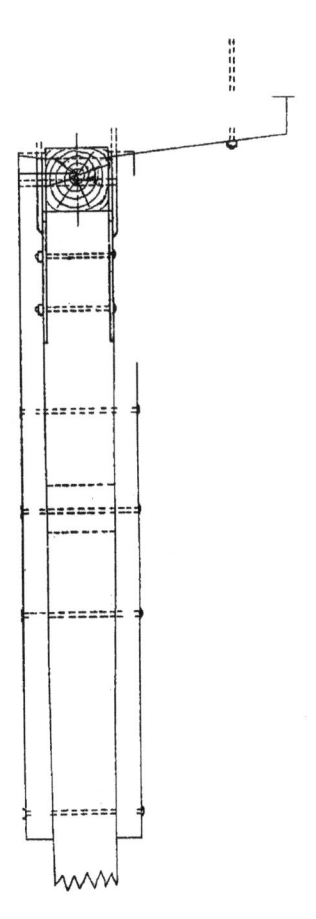

CROSS SECTION.
FIG. 3.

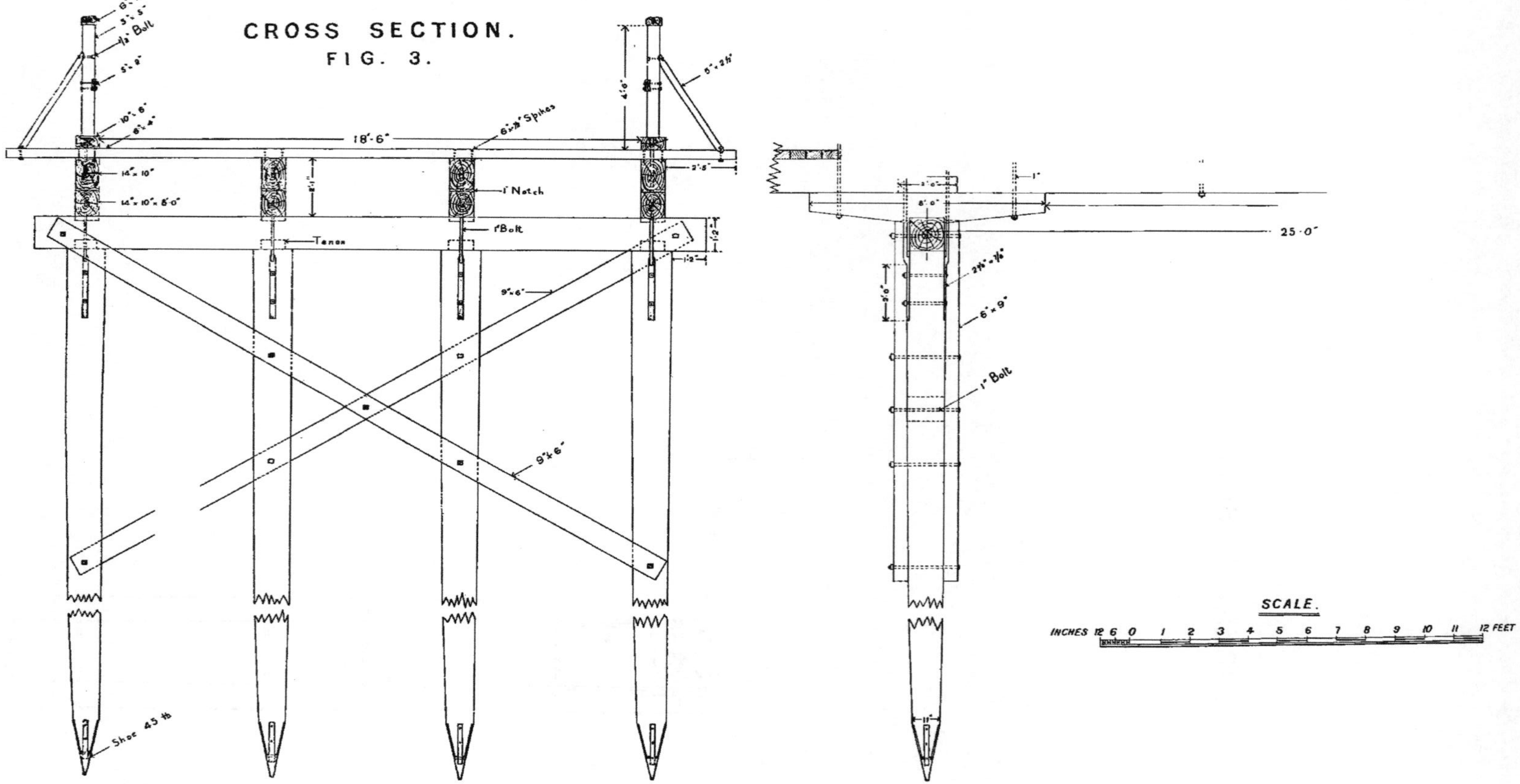

May 5th, 1902.

PERCY GRIFFITH, PRESIDENT, IN THE CHAIR.

RECENT BLAST FURNACE PRACTICE.*

BY BRIERLEY DENHAM HEALEY.

INTRODUCTION.

WITHIN the past few years the question of blast furnace practice has attracted considerable attention, even outside metallurgical circles, and has been seriously discussed in the public press. At one time the British iron industry has been threatened with extinction, whilst at other times we have been assured that large combines were being formed to protect it.

Forty years ago the British ironmaster could take things comfortably, as, in most cases, a well-managed ironworks at that time was a veritable gold mine. Then the output of furnaces seldom exceeded an average of 200 tons per week, whereas at the present time we have several British furnaces which smelt 200 tons per day. The great leaders of this most important industry are now fully alive to the fact that if they do not move with the times, the times will move along without them, or to quote the words of our observant Prince, "they must wake up." The only danger which awaits them is that if they awake too suddenly they may adopt methods which have not been sufficiently proved.

For at least three decades very little has been done in this country to improve the wasteful methods of utilising its vast mineral wealth, and our total production of iron for last year was only 7,385,198 tons, whereas during the same period a total of 15,878,354 tons was produced in the United States of America. As a general rule, fuel has been so imperfectly consumed, that at least 60 per cent. has passed off in gaseous form to pollute the atmosphere, notwithstanding the numerous inventions for the prevention of smoke, and the penalties which may now be imposed by Parliamentary authority.

The by-products from iron smelting furnaces, and from gas producers for heating furnaces generally, have received very

* A Society's Premium was awarded to the author for this paper.

scant attention until within the last ten years or so, and even now very little actual work is being done, although at most of the works of modern design, by-product plants are certain to be profitable where the weight of coal used justifies the expenditure. In the following paper the author proposes to describe a typical blast-furnace plant comprising all the most recent improvements in practice.

A Modern Blast Furnace Plant.

The first thing to be considered is the selection of a site for the proposed smelting plant. Here the leading essentials are the localities of the fuel and mineral, and the delivery of the finished product with the view to minimise the cost of carriage. The site having been decided upon, the next thing is the general arrangement of the plant.

In grouping a smelting works, the following are the chief items for consideration, viz.—the smelting furnace; the hot-blast stoves; the equalisation of air-blast temperatures; the blowing engines; the power generators; tuyere and bosh cooling; stock of minerals and fuel; furnace charging apparatus; pig bed for iron output; slag removal and disposal; lighting the works; repairing and fitting shops; and chemical laboratory.

The author does not intend to discuss fully the details of the foregoing items, each of which would form the subject of a paper in itself. His description, however, will embody comparisons between old and new methods, whilst some results of recent working will be given which should be sufficient to convince ironmasters that there are means to hand whereby ordinary blast-furnace practice can be materially improved.

The Iron Smelting Furnace.—It has already been stated that forty years ago 200 tons per week was an average output, and the author well remembers the erection of new furnaces in 1863 which were to astonish the world. Their output reached 420 tons per week, and they had open tops around which was a flue to take off the gases. The pressure of the air blast was nominally 5 lb. but more generally 4 lb. per square inch, and the heat of the blast from the pipe stoves, then in common use, varied from 500° to 700° F. The furnaces referred to certainly produced very good iron, chiefly for conversion into Bessemer steel, and there was generally but little difficulty in the working.

Those were the days when various internal capacities were experimented upon, with varying sizes of the hearth, minute angulations of the bosh of furnace, and the question of maximum height or burden of charge that the fuel would carry. The questions of higher blast pressure and increased temperature were not even thought of in 1863, and to have mentioned a special charging apparatus for each furnace would have been considered sheer extravagance. The author has very vivid recollections of labour-saving designs which he made thirty years ago, only to be put aside for twenty years, but which are now in full use, saving in some works as much as 3000*l.* per year.

It is now well known that the internal capacity of a furnace should not be too great in proportion to the output. The furnace itself should be as high as the structure of the fuel will carry, and the blast pressure should be much greater than hitherto. The material within the furnace will then descend at a maximum rate, and the loss by radiation being nearly the same, regardless of quantity smelted, will be in inverse ratio to the output. The temperature of the air blast should be uniform and much higher than formerly, and adding to this an increased burden and extra power obtainable by means of suitably designed engines, the output should be in direct ratio to the weight of the air blast actually delivered to the furnace. All furnaces should be properly equipped with some form of bell top, in such a way as to prevent any serious loss of gas, as this has now become more valuable than ever owing to the introduction of apparatus for cleansing and using it by means of internal combustion gas engines, and for obtaining various by-products.

The advantage of driving a furnace at its greatest rate, can be determined with comparative certainty, by ascertaining the capital and renewal charges per ton of iron made during its life, as against the cost of production by another furnace at a lower rate, and the maximum rate of driving which may be possible, will be settled partly by the ability to feed the furnace with material and partly by the time required to rebuild one furnace whilst another is working.

The following tables of comparison between American and British blast furnaces provide very interesting and valuable data. The American furnaces are illustrated in outline in Figs. 1, 2, and 3, and the British in Figs. 4, 5 and 6.

The internal capacity of the furnaces, indicated by Figs. 4 and 5, is such, that with uniform weight of air-blast on each the output would be about equal, assuming them to be working on the same material, and the output of all the furnaces shown by Figs. 4, 5, and 6 may be considerably increased by an additional weight of air-blast.

To make a fair comparison between the tonnage of different furnaces, due allowance should be made for the character of the

Proportions of American Furnaces.

A Furnace, Height in feet.	B Hearth, Diameter in feet.	C Bosh, Diameter in feet.	D Internal Capacity in cubic feet.	E Capacity per ton per diem.	F Tons of Iron smelted per diem.	G Temperature of Blast, °F.	H Lbs. of Coke per ton of Iron.	J Pressure of Blast per square inch.	K Cubic feet of Blast per minute.	L Cubic feet of Blast per ton of Iron.
75	11	20	14,600	73	200	1100	1912	5	16,000	115,200
80	11	22	19,800	60	330	1100	1884	10	25,000	109,090
106	15	23	26,500	46	570	1100	1780	15	50,000	126,315

Proportions of British Furnaces.

A Furnace, Height in feet.	B Hearth, Diameter in feet.	C Bosh, Diameter in feet.	D Internal Capacity in cubic feet.	E Capacity per ton per diem.	F Tons of Iron smelted per diem.	G Temperature of Blast, °F.	H Lbs. of Coke per ton of Iron.	J Pressure of Blast per square inch.	K Cubic feet of Blast per minute.	L Cubic feet of Blast per ton of Iron.
60	10·0	18·5	10,012	105	95	1100	2352	5	9,540	144,606
65	10·5	19·0	12,610	87	145	1100	2268	6	13,500	134,068
85	11·0	20·0	18,495	89	206	1100	2206	7	17,263	120,673

mineral used, as it is well known that some minerals contain less than 28 per cent. of metallic iron, whilst some hematite ores contain nearly three times as much. The following are representative analyses of the essentials for the production of hematite iron, and of the liquid and gaseous effluents therefrom.

Analysis of Furness Hematite Ores.—Peroxide of iron 83·00; silica 15·50; moisture 1·50; carbonate of lime, trace; total 100·00.

Analysis of Furness Limestones.—Carbonate of lime 95·00; carbonate of magnesia 4·20; silica 0·50; alumina 0·30; total 100·00.*

Analysis of Durham Coke.—Carbon 92·38; ash 6·00; sulphur 0·70; moisture 0·92; total 100·00.

Analysis of Gases from Furnaces using Hematite Ore and Durham Coke.—Nitrogen (by volume) 58·51; carbonic oxide 30·97; carbonic acid 8·36; hydrogen 2·16; total 100·00.

Analysis of Hematite Pig Iron.—Iron 93·33; carbon 3·22; sulphur 0·02; phosphorus 0·04; silicon 2·94; manganese 0·45; total 100·00.

Analysis of Slag from the foregoing.—Silica 38·00; alumina 12·03; lime 42·19; magnesia 1·65; sulphuret of calcium 2·45; protoxide of iron 2·08; potash 1·60; total 100·00.

Hot Blast Stoves.—Formerly hot blast stoves were constructed of nests of cast-iron pipes, erected upon cast-iron base unions, the entire metallic structure being placed within rectangular brick chambers, which were lined with fire-brick, and provided with fire-grates. In later times, however, although the fire-grates were generally retained, along with the collection of the otherwise waste gas from the furnace tops, came the addition of gas firing apparatus for heating the air-blast. Owing partly to less cold air passing in, and because the firing doors and ashpits were then sealed, the temperature was then more uniform and generally increased to about 800° F. in the best constructed stoves of that type.

About thirty-three years ago the Whitwell recuperative hot blast stove was introduced. This system comprises an arrangement of fire-brick walls within an iron casing, and with a set of valves for directing the gaseous currents inwards and outwards as required. The associated brickwork of the interior did not lend itself properly to the rapid absorption of heat, and consequently was slow in giving it up again owing to the heating surface being insufficient, but the maximum temperature then obtained was 1200° F. Then came the Cowper stove with its chambers filled with fire-bricks, which when built up as shown in Fig. 7, presented numerous square cells leading from

* Small quantities of magnesian ore were added to the charge.

the top to the bottom of the stove, and the maximum temperature attainable was 1400° F. The dust lodging upon every horizontal ledge prevented the checkering absorbing the heat within the time theoretically allowed, and consequently this class of checkering always gives lower results than might be expected. The Cowper checkering was afterwards improved by the use of star-shaped bricks, as shown in Fig. 8, which are hexagonal in shape with six lugs projecting at 60° from each other. When set together the lugs join and form other hexagonal passages besides the one contained within each brick.

The Moncur stove checkering, with its projecting V lugs, is better than the Cowper filling, but in the opinion of the author it does not satisfy up-to-date requirements, which demand a freedom from all ledges where dust can lodge, and a maximum surface of heat-absorbing material, placed within a casing of minimum capacity. The most suitable bricks for the Ford and Moncur hot blast stoves, and the method of packing them, are shown by Figs. 9 and 10, from which it will be seen that the flat walls are well bonded together.

Next in importance is the checkering of Stevenson and Evans. This setting is so designed that the bricks cannot be displaced, either by scraping the vertical flues or by the action of the cleaning gun, as the checkering is all bonded together. By a proper manipulation of the chimney valve and hot blast valve, the dust may be quickly cleared out, instead of by using the cleaning gun, as there are no ledges on which the dust can lodge. Figs. 11 to 13 show a stove on this system. Fig. 14 shows an enlarged detail of a brick for the checkering, whilst Fig. 15 shows the method of setting the bricks so as to make good bond. When the stove is put on gas, the flame, after rising in the combustion chamber, passes down the whole of the checkering for about two hours, during which period the various sections A, B and C receive a proportionate amount, until about two-thirds of the filling is saturated with heat at the maximum temperature. The gas is then shut off and the stove changed in the usual manner, and the blast has a free passage through each section for the same period.

It has been the custom in many works to use three stoves per furnace, but in most of the modern works four are used. Some managers are of the opinion that more heat is lost by radiation and conduction with an equaliser, but with three stoves and an equaliser there would be no more loss in this direction than from four stoves. With the old cast-iron pipe-stoves, the temperature was low but regular. There was not nearly as much trouble with furnaces "hanging," and in the few works that still use pipe-stoves, such a thing as a furnace "hanging" is of very rare occurrence. Hence the importance, first of preventing the ingress of dust into the stoves, and second, of maintaining the air-blast at a uniform temperature.

Working with high heats generally means low coke; in other words, maximum heat, minimum coke, and although it has often been asserted, it has not been proved that high heats are worse for the furnace. The recuperative stove can be worked at any practical temperature, and if only a low heat is wanted it can be got just as well with a regenerative stove as with a cast-iron one.

Want of uniformity in temperatures is one of the chief causes of irregularity in working, and the author is of opinion that the benefit of the hotter blast given by recuperative stoves has not been fully obtained owing to the insufficient capacity of the apparatus. In a very short time after recuperative stoves came into use, it was discovered that when the heat-absorbing surfaces of the direct combustion recuperative stoves became covered with lime, carried into the stoves along with the blast furnace gases, the stoves very quickly lost their heat-recovery or recuperative efficiency. Now, if the stoves could be heated with producer gas made from cheap coal, as suggested in a paper* read at the Iron and Steel Institute meeting last September, not only would very much higher stove temperatures be obtainable, but the temperatures could be maintained at a suitable and uniform elevation, and the output of iron would be higher and more uniform in quality. The value of the blast furnace gas set free by this system would more than compensate for the cost of the coal and labour used at the producers, and the increased output of iron and its better quality would thus be gained without cost.

Equalisation of Air-Blast Temperatures.—In the ordinary working of the various forms of recuperative hot-blast stoves, there is inevitably a fluctuation of the temperature of the hot blast to the furnaces. This in some cases ranges from about 1000° F. to 1400° F., and interferes with the steady working of the furnaces. The equalising apparatus resembles somewhat an ordinary stove. It is built with a central division wall, and is entirely filled with checker-work, so that the hot blast, although entering at varying temperatures, becomes equalised, and it then enters the furnace at an average temperature. A little while after putting the apparatus into operation the blast, in passing through, arrives at a mean temperature, and if the recuperative medium is of sufficient capacity about one-third of the checker-work should always be at the maximum temperature. The difference in the maximum temperature, say

* 'The Profitable Utilisation of Blast Furnace Gas,' by B. H. Thwaite.

between a new stove and an old one, or between a clean stove and a dirty one, working on the same furnace, might be 400° F. as already stated.

In America, and perhaps in other places, it is the practice to level the heats by means of cold blast. This is done by admitting cold blast into the hot blast main, to keep the heat down to what may be termed the minimum temperature. Regulating the temperature in this way demands careful and constant attention, and it is obtained at the expense of reducing down to the minimum temperature, with the additional risk of bringing it too low, and moreover, thermally, it appears to the author to be a most uneconomical system. The difference between the quantity of heat carried into the furnace by the blast at 1200° and that at 1000° is equal to the combustion of about half a cwt. of coke per ton of iron made, and represents the difference between using the equaliser and the practice of diluting with cold blast.

The equaliser at the Normanby Iron-works, Middlesbrough, is the invention of Messrs. L. F. Gjers and J. H. Harrison; it is 20 feet diameter by 55 feet high. Figs. 16 and 17 show the construction of the equaliser. Fig. 18 is an autographic curve from a Uehling recording pyrometer showing the working of two stoves in good condition. Fig. 19 is an autographic record of a similar pyrometer showing the almost level line obtained by means of the equaliser. Fig. 20 is also from a similar recording pyrometer, and shows the working of two stoves, one in good condition and one in bad condition. It will be seen on reference to Fig. 20, that the average temperature when on the good stove is 1100° F., and when on the dirty stove it is only 975° F.

Blowing Engines.—A great many furnaces are still blown by engines of the very oldest type, running at from 16 to 20 strokes per minute, and with a maximum air-pressure of 5 lb. per square inch. A few of the most modern furnaces are working with a normal pressure of 10 lb. per square inch, and their output exceeds 1500 tons per week, as against the old average of 426 tons. Steam-driven blowing engines for this work have now reached immense proportions, and the speed of the largest steam engine has gone up to 50 revolutions per minute. The air cylinders at some of the older works are 12 feet in diameter, whilst 9 feet in diameter is a common dimension for modern blowing plants.

A new departure in blowing has been made recently, by using the otherwise waste gases of the blast furnaces for obtaining the blast pressure. The Thwaite-Gardner system of treating blast furnace gases and using them for driving internal combustion engines appears to offer the greatest advantages in this direction. The blowing cylinder is erected vertically over a central crank, and the gas engines are placed to the right and left, and work horizontally. A set of these engines has been erected at the Clay Cross Iron-works, Derbyshire. Their maximum speed is 150 revolutions per minute, and they develop over 500 I.H.P. The gas used may be taken at less than a quarter of the quantity formerly consumed for obtaining the same volume of blast, and the equivalent power developed. The exact volume of gas used, can be ascertained very readily at any time, by simply measuring the fall of the gas-holder and counting the engine strokes, say for five minutes. By taking indicator cards during the same interval, the gas used per horse-power is easily calculated. Figs. 21 and 22 show two views of the gas blowing engines at the Clay Cross works, whilst Fig. 23 shows the general arrangement of the complete plant at those works.

The enormous value of furnace gas when used in gas engines will be more fully recognised when dealing with poor gas, which will not burn in the boilers without the assistance of solid fuel. The author has often stated that such gas is quite strong enough for direct combustion in suitably designed gas engines, and the correctness of this statement was quite recently proved at the Clay Cross works. It is well known that, from a variety of causes, the gas varies considerably, even when taken from furnaces which are working pretty regularly upon the same charge, and the following practical illustration cannot be refuted. During the night, just about three months ago, the gas at the Clay Cross works was so poor that the steam supply gave out, and it was impossible to obtain the usual blast. The gas blowing engines were then started, and without any difficulty kept on, until the steam was at the normal pressure.

The gas from a 1500 ton per week furnace, working on coke, after allowing, say, one-third for heating the blast, 900 I.H.P. for gas-driven blowing engines, and 100 I.H.P. for furnace charging, pumping water, and clearing away the slag, leaves a surplus of 5000 I.H.P. for electrical energy generation or other power purposes. This would be sufficient to drive the whole of the machinery of an adjoining steel rolling mill of equivalent capacity to the blast furnace output, and also provide for lighting the works, offices and workshops. From the author's calculations, it may be fairly assumed that 6 I.H.P. per hour are obtainable for every ton of iron produced per week in coke-fed furnaces. From the total power thus ascertained provision has to be made for heating and compressing the blast, and for working the plant, as already shown in the 1500 ton example.

A new system of blowing engine was patented some five years ago by Mr. B. H. Thwaite. This consists of a series of two, three, four or more rotating blowers or fans. The first fan, developing, say, 2 lb. pressure at its outlet, has its pressure increased to 4 lb. by the second fan, and the progression is in a similar ratio at the outlets of the other fans. This rotating system may in the near future become a rival to reciprocating motion, just as certainly as the turbine may supersede screw engines for the propulsion of steamers.

Power Generators.—Up to the present time steam boilers of every variety in form and length have been used for generating power; and unfortunately, in a vast number of cases, they have been the greatest cause of anxiety to managers. The majority of boilers hitherto used have worked at comparatively low pressures, but a great number of boilers are now working at pressures of from 100 to 120 lb. per square inch. The increase in steam pressures has not reduced the risk which is inseparably connected with boilers at all times, and an arrangement for superseding steam boilers by a gas-treatment plant having no risks whatever should be welcomed at every works where gas can be used.

In the Thwaite-Gardner treatment plant for gas-power engines, which is illustrated in Fig. 24, the gas is drawn from the furnace syphons by means of a fan which is placed between the condensers and scrubbers. The gas first enters a series of washers, where the greater part of the dust is deposited, passing thence through the condensers to the coke and the sawdust scrubbers, and from the latter to the holder. The gases are then absolutely free from dust, and can be used without the slightest risk of injury to any of the working parts of engines. This is of the greatest importance in any plant designed to use furnace gases. By an ingenious arrangement the gas-holder causes part of the gas to be churned several times through the fan, and the latter is at once an exhauster and propeller. The whole process is practically independent of any continuous water supply, which is a very considerable advantage as compared with steam boilers.

At the present time an old iron-works in South Staffordshire is being entirely dismantled, and it is intended to rebuild it on modern lines generally. A very special fearure will be the absence of steam boilers, and the use of the power obtainable from the blast furnace gas. This plant will be supplied with electric motors, wherever power is required, and a very large surplus of electrical current will be available for electrical welding and motors generally at other establishments.

Tuyere and Bosh Cooling.—In former times the tuyeres of the blast furnace were formed of coils of pipes enclosed in a cast-iron casing. At first the pipes were ½ inch bore, but the size was afterwards increased to ¾ inch, and still later 1 inch pipes became the standard size. The tuyeres were, and are still, supplied with water of various degrees of purity, and at pressures ranging from 10 to 30 lb. per square inch. Owing to various inherent difficulties, and to others caused by irregular internal working of the furnaces, the average cost for tuyeres is generally a large one. In addition to the cost of tuyeres lost in working, many serious explosions have occurred owing to the water from defective tuyeres making its way into the hearth of the furnace; and although the greatest care may be taken to prevent accidents, the fact remains that under the pressure system there is always danger.

The latest departure in connection with tuyeres is the Foster vacuum system, which is illustrated at Fig. 25. This system has been at work for over twelve months at Darlaston Green furnaces. The water is drawn through the tuyeres instead of having to be forced through, and if a tuyere is burnt the water does not get into the hearth. In addition to this, each tuyere is fitted with an electrical tell-tale, and as soon as the vacuum begins to fall, a bell is sounded, and the number of the defective tuyere is indicated.

The vacuum is started and maintained by the pump illustrated at Fig. 26. To start the pump from the position shown, a pilot spindle A is pushed towards the right hand, the central collar B then opens the draining port C to the left-hand chamber D of the slide valve E. Then the live steam in right-hand chamber D^1 forces the slide valve E to the left, and thereby opens the port F, thus admitting steam to the right end of the cylinder, and causing the piston G to move to the left of the position shown. As the piston G approaches the left hand cover of the steam cylinder it closes the port F^1 and begins to slow down, as the steam thereby cushioned can only escape through the by-pass H, and at the same time the piston moves the pilot arm J to the left, and thereby puts the right-hand chamber D^1 in connection with the draining port C, and the return stroke then proceeds in the same way.

In many places a syphon may be relied on for starting a vacuum, although the water may have to be pumped up again to the service tank. A small tank, which is fitted with a ball-valve, is fed from the usual overhead pressure main. Two tuyeres may be supplied from each tank, and the suction main is fixed a little higher than the tanks.

The water-cooled bosh-plates which have been applied to some of the modern furnaces, have lengthened the life of the linings to four times the equivalent of the former tonnage. These bosh-plates are of iron, copper or bronze, built in courses in the bosh-walls every 12 inches or so in height, and the water gravitates from one course to another.

Stock of Minerals and Fuel.—The ancient methods of unloading stock and moving it about the stock-yard may still be seen in operation in some of the oldest works, but such a sight fills one with wonder, considering at what trifling outlay a few pence per ton could be saved on the gross inwards tonnage. In the stock-yards of modern works very capacious bins are provided for the different kinds of material, into which the wagons on overhead ways deliver their loads through drop doors. At some of the more recent works loading into the charging buckets is facilitated by having chutes down which the material is very easily drawn forward.

At works where a maximum output is aimed at, grab-buckets are used for loading up the stock, and this is absolutely necessary to get through the work, owing to the increased digestive capacity of the monster which has then to be fed. At each end of the main roads weighbridges are provided, so as to weigh the gross of every wagon as it passes inwards, and take the tare as it passes outwards. Smaller weighbridges are necessary for weighing the charges to the furnace, or the hoisting apparatus may be fitted up with automatic weight-registration machines.

Furnace-Charging Apparatus.—Up to within quite recently it has been the practice to have one incline or hoist for delivering the stock to as many as four furnaces, but this old method has latterly been superseded. Some modern plants have a special hoist for each furnace, which takes the charges from the stock-yard and delivers them automatically into the bell of the furnace. Under the old system there was a considerable amount of work to do on the upper stage, a great part of which is saved by the single direct charging apparatus. The method of dealing with the stock at the yard level under better supervision has also advantages, which may or may not outweigh a good top platform arrangement.

When the buckets or wagons are filled on the latest system, they are not only hoisted and delivered, but the bell top is so arranged that the material is distributed, and this assists in the more regular working of the furnace. A simple form of hydraulic ram, fitted up with differential speed sheaves and steel wire rope, would appear to be a very reliable type of driver; but in any truly modern plant where the gas is fully utilised and electrical energy is generated, the hoist of the future will be driven by electric motors, as it is already at some of the German furnaces.

After a careful examination of various appliances, backed by his experience with continuous elevators, which are giving complete satisfaction, the author has designed and patented a furnace hoist fitted up with somewhat similar gearing, but with essentially different details of construction. This hoist and charging hopper, which is applicable to existing furnaces, is shown by Fig. 27. There are only two buckets at equal distances apart, so that they balance each other, and these are carried by steel pitch chains, which work upon a pair of sprocket wheels at each end of the boom framing. The top and bottom members of the framing are employed as rails for the bucket wheels to run upon. The buckets are loaded as they pass upwards, and at about the same time that one is being loaded the other is passing over the head gear, and is discharging its contents into the top receiving boxes.

Above the usual bell of the furnace there is a receiver consisting of six boxes with a bottom door to each, the filling of five of which completes the usual stock charge, but the six boxes are always filled before the material is allowed to drop upon the bell. It therefore follows that as the sixth box takes the first part of another charge, the receiver distributes one and one-fifth of a charge every time it is emptied, and the material will be charged into the furnace in a remarkably regular manner. The hoisting chain also actuates another endless chain, by which the charging hopper is moved 60° after the discharge of each bucket, so that the whole system becomes an automatic feeder and distributor, which may be readily applied to existing furnaces. The lowering of the bell takes place after closing the flap doors, so that there is a very small loss of gas and no escape of flame.

Single direct feeding hoists might with advantage be fitted with indicating gear so as to record the time of sending up every charge. This would be a check on the charge-man, and it would be impossible for the burden to fall too low, without the time being told by a tell-tale, which is a checker that most men generally respect.

Some device of a thoroughly reliable character is absolutely necessary to prevent over-winding in all hoists of this description, as any hitch with the hoisting gear causes great inconvenience, and may upset the proper working of the furnace, especially where there is no platform for connecting the furnaces

A very effective arrangement of bell-top, which can be readily adapted to existing furnaces, has been designed by Mr. B. H. Thwaite, and is illustrated by Fig. 28. The usual hopper is provided with a cover which is made in four parts, and the usual bell and this cover can be so operated as to prevent the outflow of gas or the inflow of air.

Pig Bed for Iron Output.—The old established practice of moulding on pig beds is still commonly followed, and for an output not exceeding 700 tons per week it will doubtless continue in use. Those works which run a large proportion of the metal direct into pipes and other castings, and into mixers for steel making, should always have a pig bed for emergency requirements. However, for a tonnage exceeding 700 tons per week, some form of casting machine should be adopted for dealing with such an increased output of iron. The Uehling casting machine, or the apparatus designed by Mr. Hawdon, appear to meet the principal requirements of a modern plant, and their use effects a considerable reduction in the wages bill.

The casting machine invented by Mr. Hawdon, of Middlesbrough, has been selected by the author as the most simple and reliable apparatus, and it is illustrated by Figs. 29 and 30. The molten iron is poured from a ladle into spouts in which it flows to the moulds, and the latter are slowly carried upon inclined endless chains which are timed to allow the metal to set, before the pigs are delivered down a chute to a circular conveyor. The circular conveyor rotates in an annular trough of water, necessary for the final cooling of the pigs, and upon arrival at a fixed plough bar the pigs are thrust on to an elevator, the bars of which carry them to the top and shoot them into railway trucks. All the working parts of this machine can be lubricated, and the bearings are some distance from the moulds, so that the heat radiating from the apparatus in work, is somewhat dissipated, and the bearings remain comparatively cool.

Wherever the output is 100 tons or more per day some form of pig-breaker should be adopted, and it should deliver the broken pigs to an elevator, to be discharged into railway wagons ready for despatch. The pigs are lifted from the pig bed in batches by an overhead crane and taken to the pig-breaker table at a cost of $1\frac{1}{2}d.$ per ton as against $6d.$ per ton as previously paid.

Slag Removal and Disposal.—In many of the oldest works, the slag is allowed to run into tapered fixed cast-iron moulds, which are formed with a lifter in the centre of each. The iron lifting piece being well coated with loam, is easily driven out of the congealed lump of slag, after it is placed on the trolley which is to convey it to the tip, and it is capable of use for a considerable period.

In other works slag bogies are used with rectangular boxes which have folding sides, and these can be drawn away from the slag at the proper time, and so allow the bottom frame of the bogie to carry the lump to the tip.

At other places the slag boxes have conical sides with considerable taper to permit of their being lifted from the bottom of the bogies, which is generally done by a steam crane when the slag has set. This form of slag box is built up in sections which are held together by strong wrought-iron bands, and the continual change of temperature has very little effect upon its durability.

A considerable quantity of slag is now used for making paving blocks, and cement. The basic slag is selected and pulverised for sale to farmers, and the slag-wool trade and the slag-gravel trade are now taking a very large quantity. Last, but not least, the works bordering on our tidal rivers have been the means of causing great improvements to be made by tipping large banks of slag for the purpose of narrowing and consequently of deepening the streams.

For all high tonnage blast furnaces it is necessary to deal with the disposal of the slag in a very thorough manner in order to reduce the cost. With this end in view, a special form of cast-iron ladle has been introduced into practice by Messrs. Stevenson and Evans, a large number being in use. It consists of a cast-iron ladle carried in a plated casing, and is illustrated in Figs. 31 and 32. Fig. 31 shows these ladles in use, one having just been tipped. It will be seen that the slag leaves the ladle quite freely, whether it has set hard or remains molten. Two ladles, each of 10-ton capacity, are generally found sufficient for a 700-ton per week furnace. Fig. 32 shows a sectional elevation of the ladle on its truck.

Another form of slag ladle, the invention of Mr. John Dewhurst, of Sheffield, and which is also largely in use, is illustrated in Figs. 33, 34 and 35. It will be seen from Figs. 34 and 35 that the blocks of slag can be tipped clear of the ladle without any difficulty, by the works locomotive hauling on to the drag chain which is attached to the lower part of the ladle. The working cost of these different designs of ladles has so far been found to be very nearly the same, so that their choice depends mainly upon the views of the user.

The most expeditious method of dealing with blast furnace slag is to run it into a well which is partly filled with water. The slag is granulated as it comes into contact with the water, and it is then dredged up and loaded into trucks by elevators.

This is an up-to-date American system and requires less labour and space than any other, and there should be no difficulty in finding a market for a considerable quantity of such material.

Lighting of the Works.—Generally, the systematic lighting of smelting works has not been anything like what it should have been, the customary flare from the furnaces by night, and a few common lamps, being accepted as sufficient. In a few places gas lamps have been used, somewhat as a luxury, but of recent years such a great advance has been made with electricity as a lighting agent, that even the masters of smelting works have fallen into line with others, and have adopted this system. Additional light is absolutely necessary in any works where a maximum output is aimed at, and it can be obtained almost for the asking, by using a very small part of the otherwise waste gases of the blast furnace for internal combustion engines, which are now available on the Thwaite-Gardner system, and which are specially adapted for continuous running for power and electric lighting installations.

Fig. 36 is a perspective view of the first electric power installation driven by an ordinary gas engine using blast furnace gases, treated on the Thwaite-Gardner system, and is therefore historic. It was put up at the Wishaw Iron-works in 1896. Fig. 37 illustrates an electric power plant driven by a Thwaite-Gardner engine of 100 horse-power, working upon blast furnace gases. This installation, which is in duplicate, was put up at the Sheepbridge Iron-works, the first instalment in 1900 and the second in 1901. These engines have been working most satisfactorily, and it may be of interest to note that several other works are arranging to adopt the Thwaite-Gardner gas treatment system and engine, in one case to the extent of 900 I.H.P. for electrical current alone. These engines may be started on the shortest notice whenever there is gas in the flues, and at the end of five minutes they will be on full speed, which is a vastly different thing from having to get up steam.

Repairing and Fitting Shops.—Every smelting works is not of sufficient extent to require a fitting or a repairing shop, but they at least require a good smithy and carpenter's shop, and these should be spacious and situated at a convenient distance from the furnaces. Where the works are of sufficient size it is an economic policy to provide repairing and fitting shops, and to keep spare parts for renewal of things that are liable to great wear and tear, or to sudden collapse. At many smelting works where a considerable quantity of machine work is done, machines are still in use which should long ago have been on the scrap-heap, and at some of these works good modern-made machines are seldom found. It is of the greatest importance to the proprietors that they should adopt labour-saving machines, and be in a position to do any kind of work usually undertaken, quite as expeditiously and as cheaply as any of the up-to-date firms in the trade

Chemical Laboratory.—In the order of this paper the last item—the chemical laboratory—is by no means that of least importance. On the contrary, a good and regular system of analysing all mineral, flux and fuel is absolutely essential in every works of any magnitude. Upon the results obtained by analysis the contracts for stock are made, and the quality of the supply is kept up by constantly checking the material as it is delivered.

In the smaller works the manager is generally an efficient chemist, and things usually work very smoothly where such an arrangement exists. In larger works these duties are performed by different officials, and the laboratory is invariably ruled by the manager. In the largest works one or more assistants are required in the laboratory, besides ordinary labour for assisting in the preparation of the samples.

Conclusion.

It is now recognised that an increase in the size of the furnace-hearth provides space for extra tuyeres, which in a high-tonnage furnace is absolutely necessary for a proper distribution of the necessary weight of air.

The blowing engines, blast mains and heating stoves should be of sufficient strength for such an increase of pressure as will force its way through any "block" which is likely to occur in the furnace, although the steady delivery of the volume of blast at the normal working pressure is the main thing to aim at.

Reducing the size of the tuyeres to obtain a greater pressure has been tried, as an experiment, and it resulted in failure. The engines had more work thrown on them than they were designed to carry, and the result was actually a less volume of blast and a reduced output from the furnace.

The normal working pressure of blast must suit the nature of the minerals and fuel used, and consequently the height of the furnace and its capacity, all of which require to be carefully considered when laying down the interior lines of a blast furnace.

In order to increase the output of a substantially built existing plant, generally an extra weight of blast, together with a slight alteration of the furnace crucible and tuyeres, may be relied upon to effect the object desired, and this can be obtained most readily and economically by the addition of gas-driven blowing engines.

Unless the single direct charging apparatus is thoroughly reliable, it is far better to continue the use of barrows, along with separate hoists for all high-tonnage furnaces, retaining the upper platform as it is now, and when one hoist requires repair its work may be divided between the hoist on either hand.

ADDENDUM.

The following are the results of some tests of a 100-horse-power blast-furnace gas engine at the Sheepbridge Iron-works, near Chesterfield, constructed by the Blast Furnace Power Syndicate, Limited. Mr. Albutt and Mr. Mason represented the Sheepbridge Company, and Mr. Horace Allen had charge of the tests on behalf of the Syndicate.

The engine is of the single cylinder Otto cycle type; cylinder 22 inches diameter; stroke 30 inches. Connected with leather link belt 14 inches wide by $\frac{7}{8}$ inch thick to dynamo. Direct current dynamo, 230 volts, 300 amperes, 500 revolutions per minute.

The electrical current provides power for the pipe foundry, including one large Root's blower, one Sturtevant fan, one travelling crane, one turntable motor, one sand mill, and the lighting of the works.

A pressure governor holder 12 feet diameter, with 8 feet lift, was used for measuring the gas. The area being 113 square feet, 1 foot of fall is equal to 113 cubic feet.

The engine runs day and night except on Sundays, and the load is very variable. The gas washer is cleaned every week-end.

The tests were made on two different dates, namely December 7 and December 14, 1901.

	Dec. 7.	Dec. 14.
Furnaces in blast	One	Four
Fuel used in furnaces	Coal . . 1·45	1·45
	Gas coke . 1·00	1·00
	Hard coke . 1·00	1·00
Mineral used	Mixed	Mixed
Iron produced	Foundry and forge	Foundry and forge
Temperature of gas entering washer	241° F.	227° F.
,, ,, ,, condenser	103	73
,, ,, leaving condenser	68	41
,, ,, washer water	102	99
,, inside engine house	65	67
,, outside ,,	56	37
,, of water in cooling tank	70	88
,, of gas in holder	52	40
Volume of water in cooling tank	2300 gallons	3790 gallons

RECENT BLAST FURNACE PRACTICE. 115

Volumetric Analysis of Gas in Holder during Tests.

	Dec. 7.	Dec. 14.
Carbonic acid (CO_2)	8·3	7·8
Carbonic oxide (CO)	26·5	22·4
Oxygen (O)	1·2	2·2
Hydrogen (H)	5·0	3·8
Marsh gas (CH_4)	0·4	0·8
Nitrogen (N)	58·6	63·0
	100·0	100·0

	B.Th U.	B Th.U.
Thermal value per cubic foot at 60° F. { Higher	106·0	93·3
Lower	102·2	89·4
Coke and sawdust scrubbers	Worked 16 days	Worked 23 days
Pressure of gas at fan outlet	2·50 inches water	2·50 inches water
„ „ holder	0·75 „	1·375 „

From Indicator Cards.

Maximum pressure per square inch	208 lb.	199 lb.
Compression per square inch	116 lb.	110 lb.
Mean effective pressure per square inch	55 lb.	51 lb.
Revolutions of engine per minute	150·0	154·0
Ignitions during tests per minute	55·6	66·4
„ possible per minute	75·0	77·0
Ratio of load per cent.	74·0	86·2
Average indicated H.P.	84·8	97·5

N.B.—The charge was reduced to give high ratio of ignitions.

	Dec. 7.	Dec. 14.
Volts	230	230
Amperes	160	200
Kilowatts	36·8	46·0
Electrical H.P.	49·3	61·6
Efficiency I.H.P. per cent. to E.H.P., including all losses	58·0	63·1
Volume of gas per explosion	3·16 cub. ft.	2·64 cub. ft.
„ „ hour	10,558·00 „	10,526·00 „
„ „ E.H.P. per hour	214·10 „	170·80 „
„ „ I.H.P. „	125·40 „	108·00 „
B. Th. Units per E.H.P. „	21,881	12,487
„ „ I H.P. „	12,816	9,655
Thermal efficiency E.H.P., including all losses	11·6 per cent.	20·4 per cent.
Thermal efficiency I.H.P., including all losses	19·8 „	26·3 „

This table of results has been compiled from information supplied by Mr. Horace Allen, C.E.

DISCUSSION.

The PRESIDENT said that he was sure the pleasure he had experienced in hearing the paper had been shared by everyone present. Previously to the meeting, the author made two remarks to him which seemed to greatly enhance the value of the paper. One was that he had confined his paper, with a few exceptions, to works of British manufacture, the exceptions being examples of American practice which had been quoted only for the purpose of comparison. The other remark was that he had endeavoured to treat the subject as broadly as possible, with a view to interesting a large circle of engineers. He would remind the meeting that the author was not now before the Society for the first time. He favoured them with a very excellent paper on refuse destructors two years ago; and therefore the excellence of the present paper did not come altogether as a surprise.

The subject the author had taken up was one which lay at the very root of engineering, inasmuch as it dealt with what engineers must consider to be their "raw material." The general points affecting production of the raw material, and the description of new methods by which the cost of the production could be reduced to the lowest limit, must be of vital importance to every branch of engineering. In that sense, also, the paper was peculiarly suitable to a society which was almost unique in representing every branch of the profession.

From a national point of view, the subject with which the author had dealt—the economical utilisation of the natural products for which this country had hitherto held the first place—must also be considered a very important feature. The prosperity of this country had been very largely built up on its valuable natural products, and the economical utilisation of those products must obviously be a considerable factor in the maintenance of that prosperity. In these days no branch of engineering could possibly be left undeveloped. England had attained its present position in regard to engineering by the disposition of Englishmen to be always in the front. His Royal Highness the Duke of Cambridge had dealt with that point somewhat curiously in a speech which he made the other day. The Duke suggested that in these days there was a tendency to be too fast, and to go too extensively for modern inventions because they were modern, and to throw aside too lightly the habits and practice of our ancestors. He (the President) thought that it would be acknowledged by every Englishman that this was the most unwarranted of all the charges to which the engineering profession in this country could be subjected. He thought rather that it was in that direction only that their future success was to be found. If they were too conservative, and refused to take up modern improvements, their supremacy would soon be lost. He thought, however, that the paper which they had just heard was an effort in the right direction.

The smelting of ores was, he supposed, one of the most ancient branches of the engineering trade in this country, and for that reason it was, he feared, characterised by the most conservative practice. But the author had at least indicated, if not proved, that in order to keep themselves in the front rank among the engineers of the world, they must bring even the most ancient branches up to the most modern standard of development. That was the text from which the author had preached, and the lesson which he had commended to them. He (the President) might even go further, and say that this was the most important lesson which it remained for engineers in this country to learn. He had very great pleasure in acknowledging the completeness with which the author had accomplished his purpose, and also the comprehensive way in which he had brought the whole subject before them. He proposed a hearty vote of thanks to the author, and he was sure that it would be most cordially received.

The vote of thanks was carried by acclamation.

Professor BAUERMAN said that the paper was one which reflected the greatest credit on the author, for he had packed an immense amount of material into its pages, and it was not possible without considerable study to discuss such a paper. However, there were certain salient points upon which a few remarks might be made with advantage.

He observed that, all through, the author had tried to give them the modern practice in everything connected with the making of pig iron in the blast furnace. No doubt there were a good many novelties, and notably the greatest novelty amongst them all, he supposed, in connection with blast furnace practice, was the one which had been occupying them within the last few years—that of the direct application of gas. He agreed with the author that Mr. Thwaite was the pioneer in that matter, and he had indicated many interesting developments, notably in the original application to the driving of motors for electrical work, and more particularly the direct driving of blowing engines.

He was exceedingly pleased to see the figures of the English blowing engine driven by blast furnace gas. That he had never seen, but he had seen several foreign engines. He was glad to see that the author was bringing the English one forward in that way. If he might say so without disrespect to other parts of the paper, the most interesting feature the author had brought forward, was the direct use of blast furnace gas. It was a matter of controversy to some extent as to how far they could use blast furnace gas. It depended to a certain extent upon what they were smelting. If they were smelting hæmatite, there was no very great difficulty in getting the dust out of the gas, because the dust was a comparatively heavy material. It was to a very large extent finely divided hæmatite which had a high specific gravity. If, however, they were working ores containing a large amount of clay, and they had a large amount of flux, and so on, then taking the dust out of the gas was a very tedious process; but still it was to be done. Yet there was no doubt whatever that in the proper utilisation of blast furnace gas—that was when it was properly utilised apart from jeopardising the working of the furnace—the future of iron making had to be looked for.

He quite agreed with the author that, in a future time, with the proper utilisation of the gas in the furnace, it would be quite possible to drive heavy rolling mills by gas engines. He would rather drive them by gas engines than by putting in a dynamo and an electromotor between the prime mover and the mill. However, that was a matter of detail.

With regard to the granulating of blast furnace slag, the author very properly noticed Mr. Hawdon's method of disposing of this material, and he had alluded incidentally to its being used as a method for casting pig iron. But he had also told them that the most expeditious method of dealing with blast furnace slag was to run it into water, and then take it up and get rid of it somehow. That was characterised as an up-to-date American system which required less labour than any other. It might be an American novelty, but he (Mr. Bauerman) had known it in use for about thirty years in European blast furnaces.

Mr. HORACE ALLEN said that, having been associated more or less with blast furnaces for the last thirty years, he had listened to Mr. Healey's paper with very great interest and pleasure. Mr. Healey was to be congratulated upon the way in which he had put the paper together. He should very much like to see a works put down on the lines indicated. With regard to the selection of a site, it was quite possible, by getting a favourable locality, to make a difference of 4s. or 5s. a ton. If all the different points indicated by Mr. Healey were treated in the same way, there would be much more probability of a very good profit being made than there would be by means of works on the old plan. If a manufacturer took advantage of all the points which were mentioned in the paper, he would stand a very good chance of making a good profit while other makers were only just paying their way.

With regard to the utilisation of the gases, he visited a works a few years ago and drew attention to the value of the gases for the internal combustion of the engines, and he also drew attention to the fact that the gas was worth more than they thought it was worth, even for their boilers. At that time they were using at those works twelve wagon loads of coal a week under their boilers along with the gas, and by a careful modification of their boiler plant they had now dispensed with the coal, which they formerly had to use in consequence of the very poor method of combustion adopted.

He had that day come across an analysis of the flue dust from furnaces smelting hæmatite ore, with coke fuel. It consisted of 31 per cent. of ferric oxide, 25 per cent. of silica, and 24 per cent. of lime. From experience he knew that the cleaning of the gas was not such an easy matter as had been indicated by Professor Bauerman, because when it was burnt under the boilers in the usual way, and in the stoves, it was carried all along the stoves, and the white dust passed out in very large volumes; then the gas could be taken a very long distance without materially reducing the quantity of dust, unless some special apparatus was used.

Mr. B. H. THWAITE said that he quite agreed with Mr. Bauerman that the purification of the blast furnace gas was not so simple, and it was impossible to adequately clean the gas without special plant. The paper showed that the Americans had not done everything, and he thought that from that standpoint the paper was of a very interesting character. All that they wanted, as Britons, was fair play to inventors, such as was given in America.

Mr. R. F. STRONG said that he should have liked to have heard something from the author as to the quality of the metal produced in the high pressure blast furnaces. They heard a great deal about increasing the pressure, but there was nothing mentioned as to the quality of the product. In America they were increasing the pressure very considerably, and the Americans were sending pigs, bars, plates, etc. over to this country, but nothing was said about their quality. The American returns, as compared with the returns from the ship-building works in this country, were much higher. He had been told that the returns in the case of some of the American goods had been very high. The author did not say anything about that in his paper.

Dr. ROBERT LEE said that one remark he had heard that evening and which had struck him was, that England ought to encourage invention more, and give more support to it. They wanted more detail and more accurate work. It seemed from the paper that the value of laboratory work and chemical analysis in connection with the blast furnace was not being treated quite fairly. They wanted more experiment and more science than was now devoted to the subject. He thought that the author had let the masters off rather too easily. If England was going to hold her own, she would do so by means of brains and not by means of money.

Mr. W. E. PHILBROW said that the author had opened up a very interesting question. He thought that the great cry which they had heard going throughout the country about American practice had simply been raised by the newspapers. The Americans had not to start at the beginning of the iron industry. They started with all the knowledge of England on the subject, and they had everything that English brains could do to help them, and all that English experience could give them. In the cycle of time England would overtake America, and America would then have, comparatively speaking, old plant, and would not be able to scrap their huge furnaces and plants, without involving a large financial loss.

With regard to the question of utilising gas direct in internal combustion engines, he found that in America they did not seem to have tried it or to have any plant whatever for it, so that in that direction Europe was in advance.

With regard to modern furnaces and the increasing of the blast pressure, he thought it would be found that in English works they had not stoves which would stand increased pressure. If they increased pressure without rebuilding the stove to suit, they would blow the top of the stove away. In fact that had already occurred. There had always been a cry of England being behind; but he had no doubt that Mr. Healey could mention places where they were starting absolutely modern practice, quite up to any American practice. He thought that England could still hold its own against the much-talked-of American furnaces.

Mr. PERRY F. NURSEY said, with regard to the observations in the paper about the granulation of slag, he might mention that at an iron-works which he visited some years since, he saw a very good granulating arrangement. The molten slag was run on to a revolving circular platform, and water was sprayed copiously upon the slag. By the time the slag had reached a given point it had become cooled and granulated. At that point there was a fixed scraper by which the slag was scraped off the revolving platform into trucks below. It seemed to him a very simple arrangement, and he understood that it was a very economical and a very successful one. It appeared to him to be a better arrangement than that described by the author.

Mr. J. BERNAYS said that it was quite forty-five years since he had anything to do with blast furnace work, but there were one or two points in Mr. Healey's interesting paper to which he should like to refer. The system of drawing water through the tuyeres, instead of forcing it through, was a very excellent one, because it did away with the danger which used to be attached to the water-cooled tuyere. In the illustration, however, he saw that the small feeding tank was shown above the tuyeres, and it seemed to him that it should be placed below them, as he believed was done in practice. The pump then had to draw the water from the small tank through the tuyere by means of a partial vacuum without which an explosion might as readily occur with the new process as it now occasionally did with the old.

In the addendum there were one or two points which he could not quite follow. For instance, what did the figures relating to coal mean? The Table was not clear on the point.

Mr. ARTHUR RIGG said, that like Mr. Bernays, it was years since he had anything to do with blast furnaces. There was one remark, however, which he should like to make about American practice. In the year 1884 he was in America and he went over a number of works. He observed one fact not mentioned in the paper, and that was that many of the men who really made the great improvements in America were either Englishmen or Scotchmen. If a man made an invention or an improvement in America, it was taken up, but in this country any one who produced an invention or made an improvement upon existing methods was looked upon as an intolerable nuisance.

Then the commercial heads of so many concerns—boards of directors and such like—never seemed able to understand the simple arithmetic that would teach them that a return of even 10 per cent. on new machinery would delight their shareholders. He had seen machinery in London that had now been working for forty years, and the owners would not go to the cost of new machinery, but would keep to the old things as long as they would hold together by the continued efforts of a gang of fitters, although it would be better and very profitable to throw the old machinery away or sell it for scrap iron and have machinery which would do the work at far less current expenses. It was much more the current expenses that made heavy charges upon works than interest on judiciously expended capital.

Mr. J. W. WILSON said that he well remembered when his uncle, Mr. E. A. Cowper, introduced the stove bearing his name. The author had very rightly said that one of the difficulties which arose in its employment was the deposit of dust; and he had mentioned the gun that was employed for clearing the dust from the checkering, thus producing a better effect for a time, the gun having to be fired again at intervals. The author had shown a sketch of the improved form of checker brick which Mr. Cowper introduced afterwards, and which certainly filled up the interior of the stove very well. He (Mr. Wilson) failed to perceive in what respect Stevenson and Evans' checker was better than Mr. Cowper's. They could quite see why the Ford and Moncur checkering was open to objection, for there were in it places upon which the dust might settle; but, as far as he could see, the Cowper stove checker afforded no more opportunity for such settling than the Stevenson and Evans. Perhaps the author would explain in what respect the one was better than the other. The shape of the opening was rather different. That might be an improvement, but it seemed to him that the oportunity for dust settling was much the same in both.

Mr. HEALEY, in reply upon the discussion, said that Mr. Allen had referred to the question of the site. They all knew very well that the position of a works of any kind had to do more or less with its success. In some works the material dealt with was very heavy. The position of a factory making light goods like cabinet work did not matter very much, for in the case of such goods the carriage amounted to very little.

Mr. Thwaite had corroborated the remarks about the necessity of cleaning the blast furnace gases in order to obtain a satisfactory and sustained efficiency of the plant.

Mr. Strong had spoken of the quality of the iron and the increased pressure. The quality of the iron did not depend upon the pressure; it depended upon the ore that was used. If there was good ore and good fuel free from phosphorus and sulphur, good iron could be produced.

With regard to the self-charging gear that had been mentioned, it was not a new idea. The main novelty was in the peculiar way in which the charging gear was operated. If they had six boxes and they admitted that five boxes made a complete round of material—that was, so much ore, so much fuel, and so much flux in the proper proportion—the sixth box must contain part of another round; it formed a spiral arrangement of charge within the furnace. There would be fuel going forward, then limestone, and after that ore. They could always reverse that spiral arrangement. The men could reverse the catches so that the spiral went half the way round and back again. All the while the spiral reversed at certain intervals, so that there was no chance of the charge sticking in the furnace.

Dr. Lee had made some references to the work done in the laboratory. He (Mr. Healey) quite agreed with Dr. Lee in his remarks. The ironmasters in this country did not pay sufficient attention to laboratory work. It would certainly be very much more to their interest and to the interest of the country generally, if they made a more serious study of the scientific part of their business.

As to the remarks which had been made with regard to the granulation of slag, he knew that the process he had referred to was not new. There were many works in the country in which the granulation of the slag was effected, and the systems varied at different works. The reason was that each works tried to satisfy the special requirements of their customers. In some cases a railway company would take broken slag at so much a ton to ballast the railway. Instead of using stones as they formerly did, they now used blast furnace slag in very large quantities.

Mr. Philbrow's remarks about America were very *apropos* indeed. They knew that it had always been the custom, and always would be the custom, for foreigners to come to England and get the benefit of our experience, and they went back to their own countries with the newest English ideas, and with them they commenced structural operations. English manufacturers had had to commence hundreds of years before.

With regard to Mr. Bernays' remarks, as to the tanks for the vacuum tuyere system, it had been found in practice an advantage to admit the water to the tuyeres at some slight elevation above them. That was not exactly what the title of the system would lead them to expect, but so long as the water pressure did not equal the blast pressure within the hearth of the furnace, the principle of the new system was preserved. If the tank was put 5 feet above the tuyeres there would be fully twice that equivalent of pressure within the furnace. There was not an absolute vacuum, but he supposed that they would continue to call it a vacuum. The figures in the addendum to which Mr. Bernays had referred, merely meant that they w working on the same stuff at the same time on both days.

With regard to Mr. Wilson's remarks, he was pleased to meet a nephew of Mr. Cowper. He had a very vivid recollection of having had something to do with designing the Cowper stove. He was then associated with the late Sir William Siemens. As to the ledges, there was no difference whatever. There was no ledge in them except at the top, and that they need not trouble about. There was a difference in the kinds of bricks. The hexagonal brick was nominally hexagonal on the outer lines, but it was not hexagonal internally, and the bricks became of varying thickness. In the Stevenson brick there was more uniformity of thickness, and no very thick corners. That was the only difference between that brick and the others.

PRACTICE. BY B. D. HEALEY.—PLATE 1.

FIG. 15.

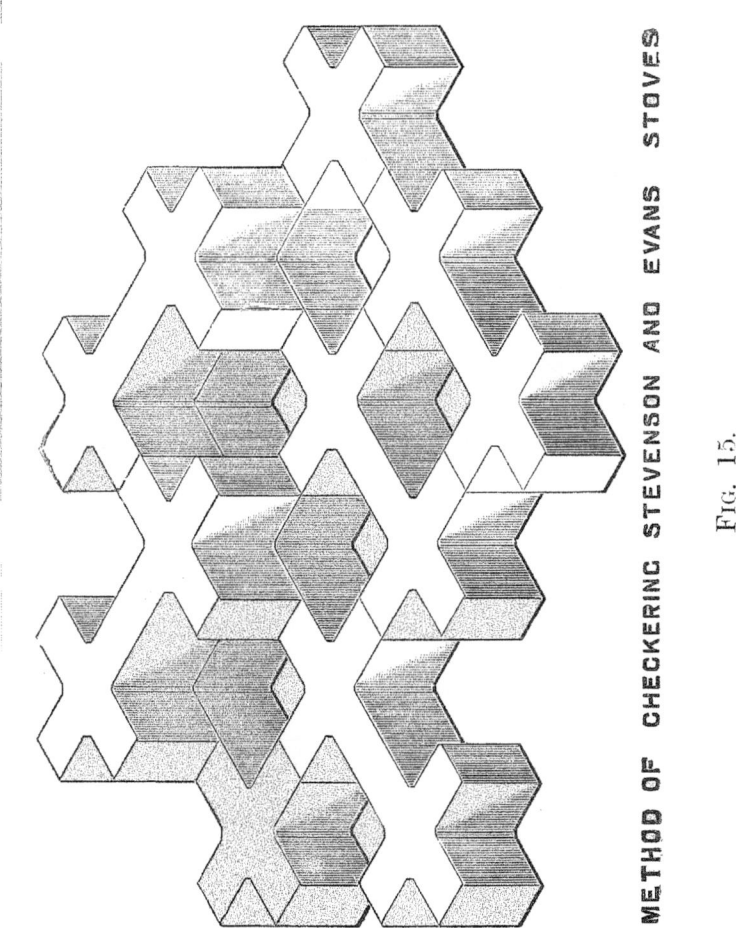

METHOD OF CHECKERING STEVENSON AND EVANS STOVES

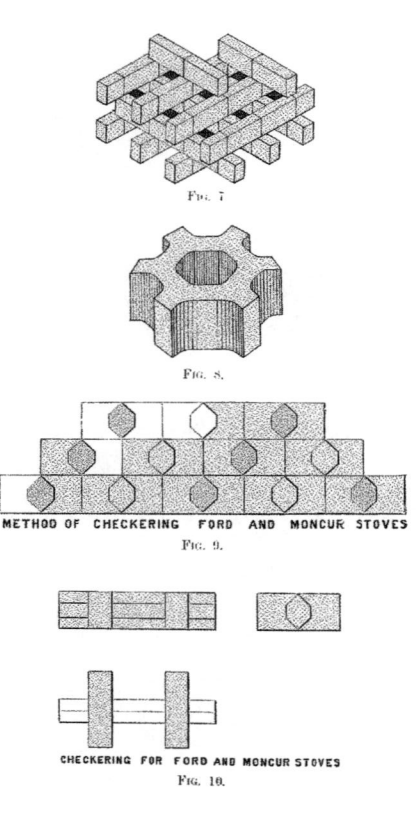

RECENT BLAST FURNACE PRACTICE. BY B. D. HEALEY.—P.

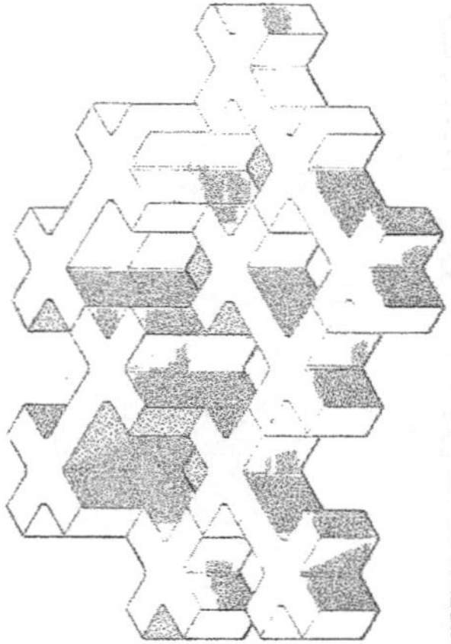

METHOD OF CHECKERING STEVENSON AND EVANS STOVES

Fig. 13.

Fig. 12.

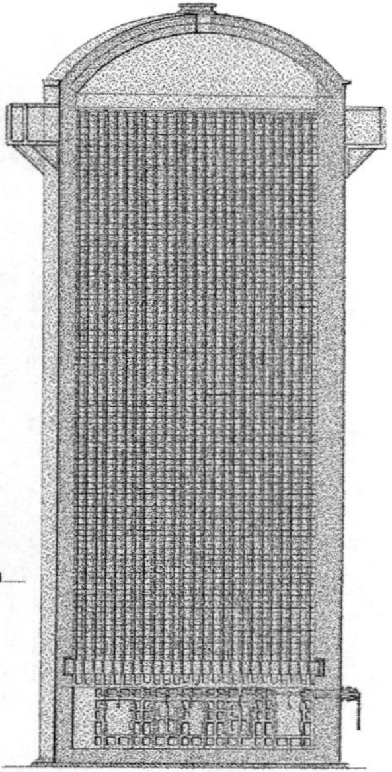

SECTION THRO LINE C.D
SECTION THRO LINE E.F.
DETAIL OF REGENERATOR BRICK
Fig. 14.

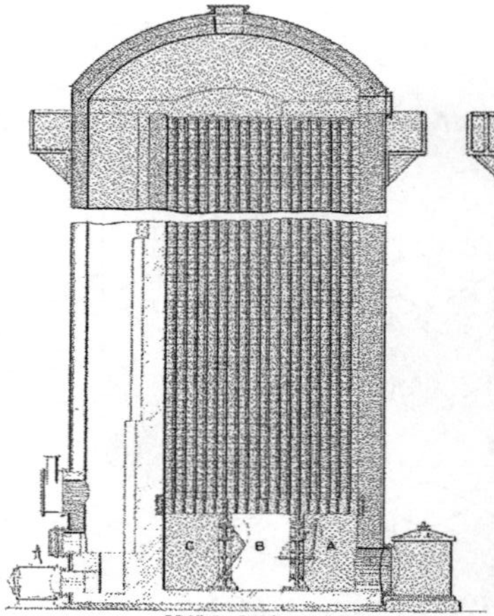

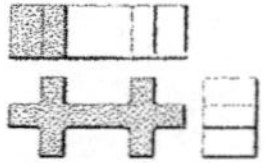

SECTION THRO LINE A.B
SECTIONAL PLAN.
Fig. 11.

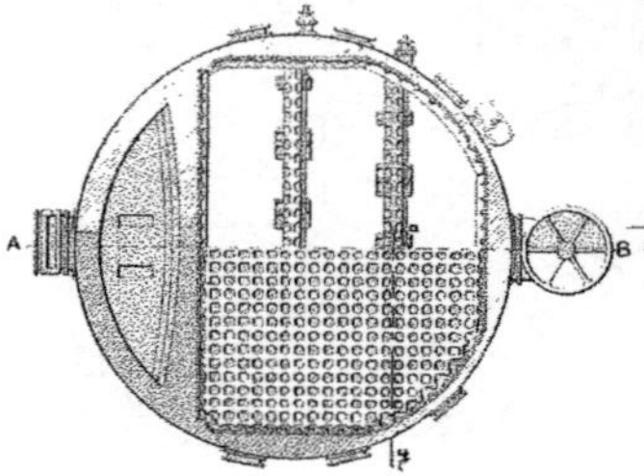

Fig. 7.
Fig. 8.
METHOD OF CHECKERING FORD AND MONCUR STOVES
Fig. 9.
CHECKERING FOR FORD AND MONCUR STOVES
Fig. 10.

RACTICE.   BY B. D. HEALEY.—PLATE 11.

lection of having had something to do with designing the Cowper stove. He was then associated with the late Sir William Siemens. As to the ledges, there was no difference whatever. There was no ledge in them except at the top, and that they need not trouble about. There was a difference in the kinds of bricks. The hexagonal brick was nominally hexagonal on the outer lines, but it was not hexagonal internally, and the bricks became of varying thickness. In the Stevenson brick there was more uniformity of thickness, and no very thick corners. That was the only difference between that brick and the others.

E PRACTICE.   BY B. D. HEALEY.—PLATE I.

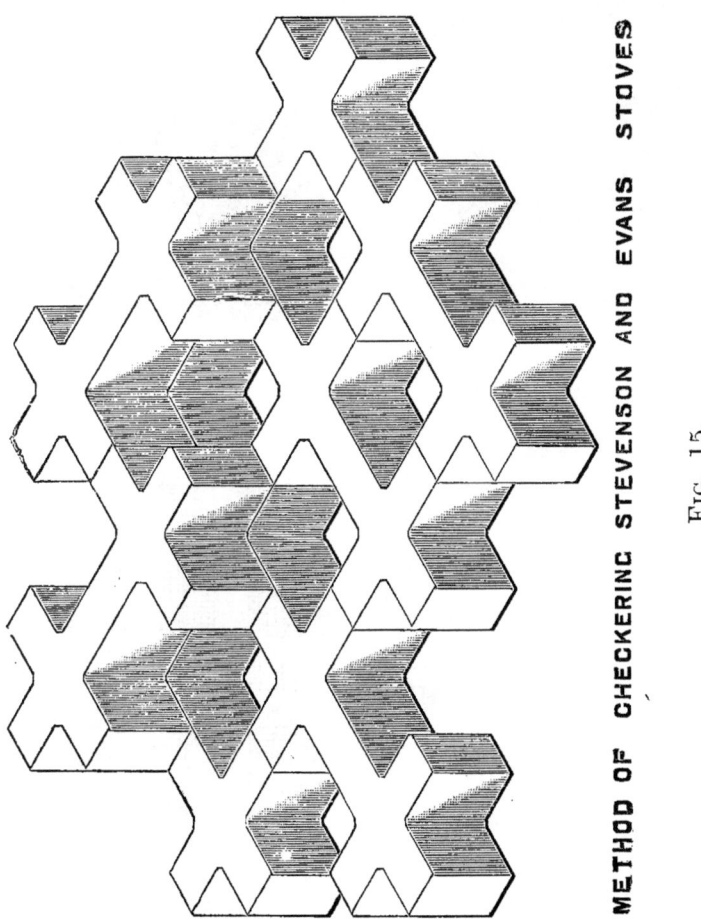

FIG. 15.

METHOD OF CHECKERING STEVENSON AND EVANS STOVES

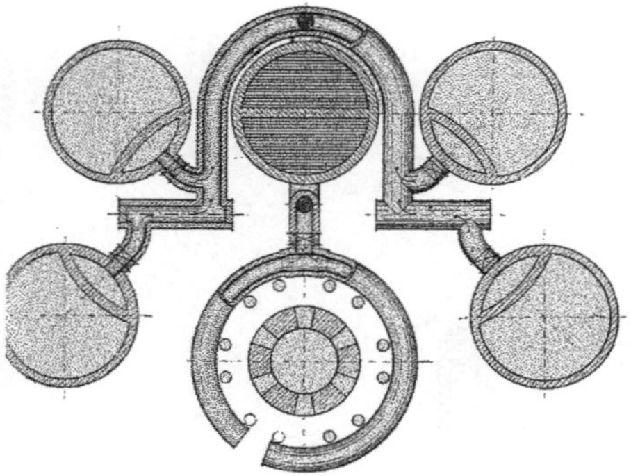

RECENT BLAST FURN

Fig. 16.

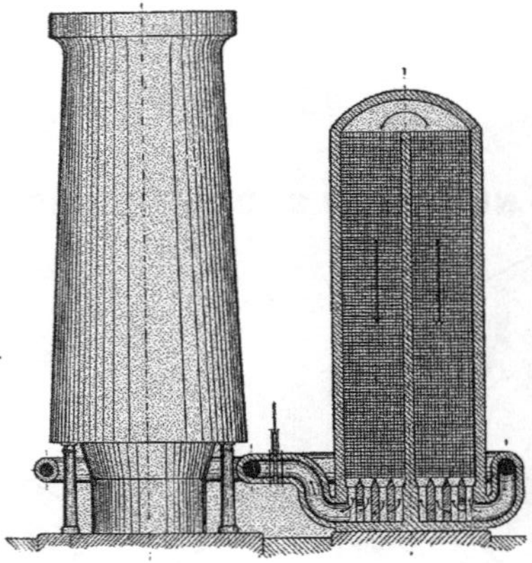

Fig. 17.

Fig. 18.

Fig. 19.

Fig. 20.

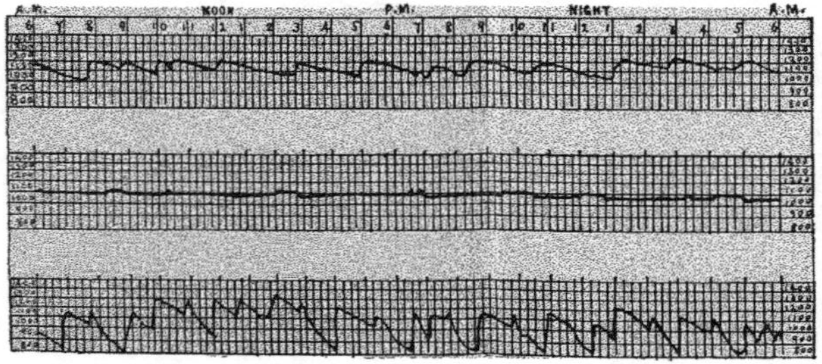

Fig. 21.

PRACTICE. BY B. D. HEALEY.—PLATE III.

FIG. 24.

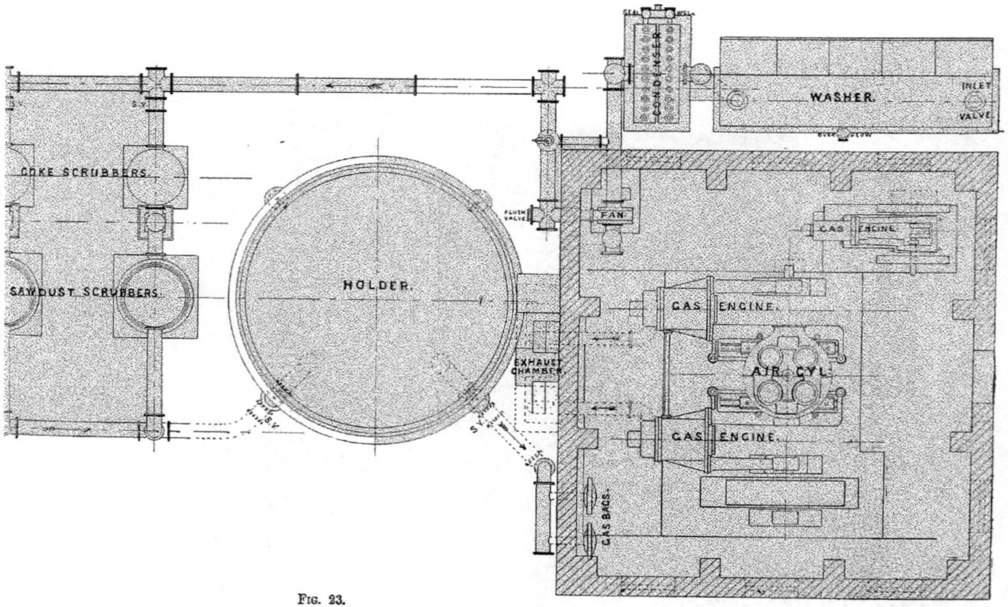

Fig. 24.

Fig. 23.

RACTICE. BY B. D. HEALEY.—PLATE IV.

Fig. 31.

RECENT BLAST FUR

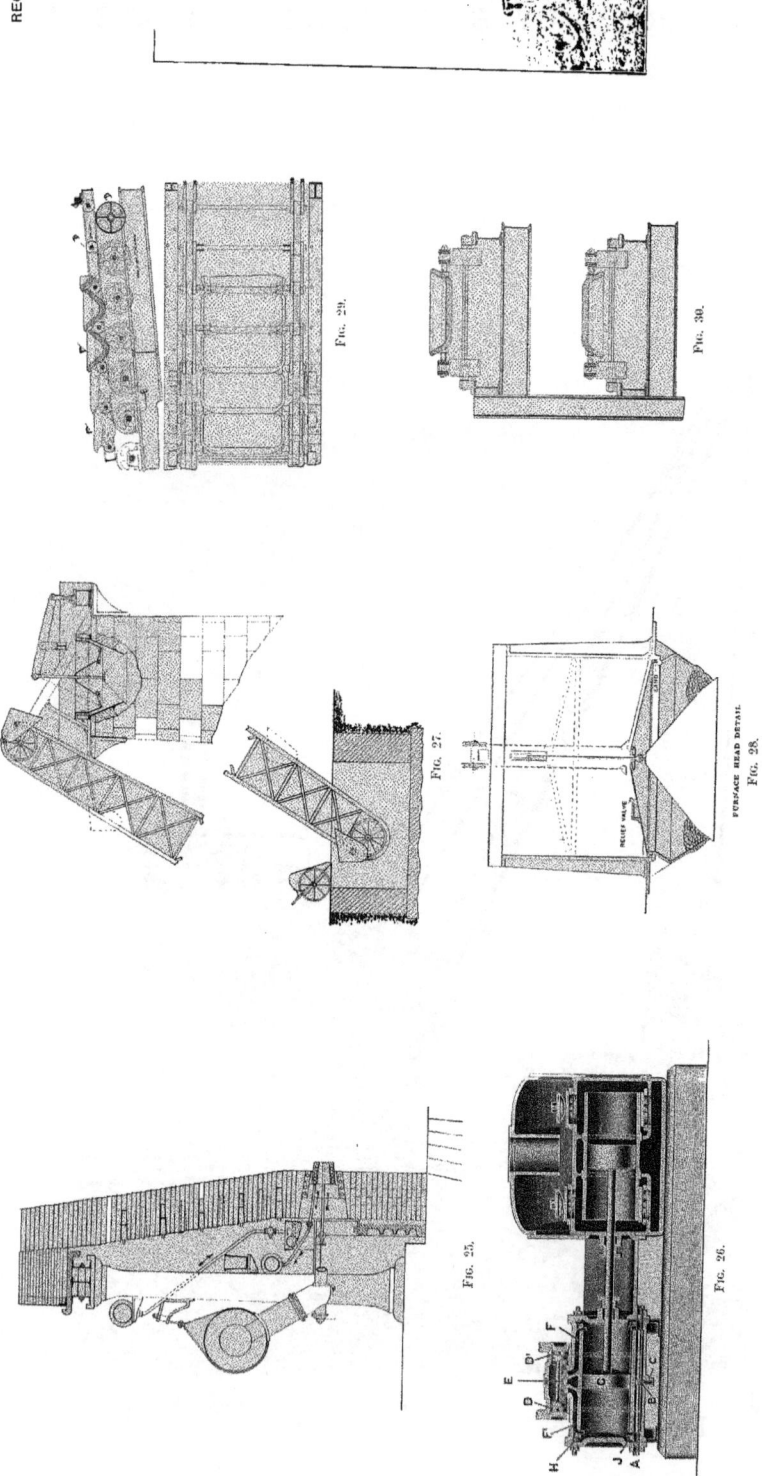

Fig. 25.
Fig. 26.
Fig. 27.
FURNACE HEAD DETAIL.
Fig. 28.
Fig. 29.
Fig. 30.

FIG. 35.

Fig. 32.

Fig. 33.

Fig. 34.

Fig. 36.

June 2nd, 1902.

PERCY GRIFFITH, PRESIDENT, IN THE CHAIR.

NOTES ON
SOME TWENTIETH CENTURY LOCOMOTIVES.

By CHARLES ROUS-MARTEN.

IN order to avoid any misconception as to the scope and purport of this paper, it seems to the author advisable that he should state at the outset what it is and what it is not. It does not profess or pretend to be, or aim at being, a full and exhaustive description of every locomotive at work in the current century or even designed since the year 1900. That would demand many bulky volumes, not a mere paper. It is the author's intention merely to indicate what, in his opinion, appear to be some of the most prominent, interesting and important features in the locomotive practice of the Twentieth Century, so far as it has yet gone.

Even this relatively restricted scope of operations involves so large a field that further restriction is compulsory within the reduced limits indicated. Therefore no attempt will be made to perform the impossible—to give, within the time and space-limits that are imperative in existing circumstances, an essay for which tenfold the time and space would be utterly inadequate. The author will therefore endeavour simply to treat his subject as fully as may prove feasible within the limitations indicated.

The subject divides itself naturally into two main sections. (1) Locomotives built prior to January 1, 1901, but still in regular work. (2) Locomotives built since January 1, 1901. Each of these sections similarly divides itself into two branches. It will be understood that while the date mentioned is taken as forming a convenient dividing line between two arbitrary periods, it will not be insisted upon with such "Chinese exactness" as to invite or warrant any verbal quibbles based upon it. The two periods with which the present paper is concerned are, first, the Twentieth Century, and second, the previous period from which there have been left over many locomotive types of interest to hold their place yet as standard engines in current use. Moreover, the author only professes to offer notes on some—not all—of the Twentieth Century's locomotives; on a few specially interesting types and their salient characteristics.

The surviving locomotives of the whole period preceding January 1, 1901, may be roughly classified into two divisions: (a) those built many years ago but still at work; and (b) those built within a comparatively short time before the close of the Nineteenth Century.

It is always a source of wonder to visiting engineers from America or from the British Colonies to find so many locomotives still at work in Britain bearing building-dates going back 30, 40 or even 50 years. In the United States an engine is deemed old at 10 years of age; ancient at 15; and utterly antediluvian at 20 years—archaic and virtually obsolete. In Britain, however, locomotives only 20 years old are looked on as quite modern, and vast numbers of them are in use. Indeed, one well-known writer on locomotive engines has deliberately adopted "the past 20 years" as the construction period which should entitle an engine to be styled "a modern locomotive." When it is mentioned that—merely to give four instances—the late Mr. William Stroudley's "Gladstone" type on the London, Brighton and South Coast Railway, Mr. F. W. Webb's latest 6 ft. 6 in. coupled of the "Precedent" class on the London and North Western Railway, Mr. S. W. Johnson's "1562" class, with 6 ft. 9 in. coupled wheels on the Midland, and Mr. D. Drummond's "Carlisle" or "476" bogie class on the North British, with 6 ft. 6 in. coupled wheels, date from this period, it will be recognised that locomotives of *circa* 1882 do not yet rank as venerable antiquities, but are deemed capable of doing some of the best work called for on those respective lines, as indeed is practically the case.

But if it be surprising to the American and the Colonial visitor to find 20-year-old engines regarded as modern and almost new-fangled, it is simply paralysing to this typical guest to see locomotives still doing regular duty whose ages range up to 30, 40 and 50 years. Yet this spectacle greets him on most of the British railways. In a single case, indeed, he may behold an express locomotive bearing the date 1847— 55 years ago—taking a fast train in the year 1902; for the old London and North Western 8 ft. 6 in. single-wheeler "Cornwall," which was built by Trevithick in 1847, still works, or did until very lately, expresses between Liverpool and Manchester, and once last year brought up a boat special from Liverpool to Euston, to the great edification of the American passengers. That of course is an exceptional, if not unique, case, but it is also the fact that on the same railway, the premier British line in point of importance, sixty other express engines are still running whose ages range between 40 and 43 years, the earliest having been built in 1859, while a large number more only just miss being 40 years old. Indeed, out of the total number of express engines working on the London and North Western at the present day—not quite 500—no fewer than 327, just about 65 per cent., are from 20 to 43 years old. One, as already mentioned, is 55; but that solitary case, being treated more as a curiosity than on normal lines, may be ignored. The strangeness is to find 65 per cent. of the express locomotive stock of the richest and most important British railway to be from 20 to 43 years old, leaving only 35 per cent. as aged less than 20 years. On the railways ranking next in importance, such as the Great Western, North Eastern, Midland and Great Northern, large numbers of engines are still to be found aged from 20 to 30 years, and a few whose construction dates back even to 40 years. Yet they are still running as Twentieth Century locomotives. This forms a curiously striking contrast to the American method.

It may reasonably be objected here that those ostensibly venerable machines have undergone so many and such extensive overhauls and repairs and rebuildings that in most instances they are virtually new engines. That is quite true up to a certain point. Of the old Great Western single-wheelers, Nos. 157–166, which are still apparently at work as the "Cobham" class, some with brand-new Belpaire-firebox boilers, the author is informed that only the wheel-centres and perhaps also the frames, are those of the original engines. Of the small 6-ft. coupled London and North Western engines of the "Samson" class which first came out in 1863, and still profess to be working express trains, little more than the wheel-centres, name-plates and number-plates, and possibly the framing are to be found in the present engines as rebuilt by Mr. Webb, which do such excellent service on the heavier parts of the line. Other cases of a similar character might be cited. In each instance a weak, old engine, which had become obsolete, has been brought so far up to date as to be about equal in power and efficiency to the normal average of 20 years ago. Engines which had aged to virtual uselessness were thus rejuvenated into distinctly useful machines. This has been practicable wholly through their original excellence alike of material and of construction, which, with the aid of skilful doctoring, has thus at least doubled their lifetime. It would be only natural to deduce the conclusion that a method which produces engines of such remarkable longevity must indeed be a most admirable one, and greatly preferable to any which provides a locomotive destined to last only 10 or 15 or at the most 20 years. But is this so?

Probably no one will have the hardihood to deny the enormous superiority of the British engine—in point of quality of material and finish of construction—over the American. The author believes that any locomotive designed and built by British makers according to British methods, and of materials up to the customary British standard as to nature and quality, would, if a perfectly accurate analytical examination could be made, be found superior at all points per unit of nominal power, to an engine designed and built according to recognised American standards as to materials and mode of construction. That is to say, the purely British locomotive would be made of better (and more costly) materials than the purely American one, and would be put together in an infinitely neater and more finished (and more costly) manner. If the two machines were identical in the number of power-units each could exercise, the British should beat the American when tried under equal conditions.

That is the author's opinion, founded upon long and careful observation of both these classes of engines in Britain and the British Colonies. An inspection of some of the American goods engines imported by the Midland Railway showed him that in the parts not exposed to view there was not even an attempt or pretence at finish. One valve-chest, when the outer cover was taken off, looked as if it had just been blasted out of a solid ironstone rock. Even important working parts in full view, in the case of some of the American engines exhibited at Paris in 1900, were put together with a roughness that was quite startling to a British engineer. So long as the materials and workmanship were good enough to enable the engine to perform the duty for which it was designed, that seemed to be deemed all-sufficient. The external treatment was almost cynical in its audacious roughness. If that were the case with special locomotives made for exhibition, it may fairly be inferred that the normal construction-method would not be more elaborate. No stronger contrast could be imagined than that presented in the Vincennes Annexe by the superbly finished machines shown by Mr. Holden for the Great Eastern, by Mr. Worsdell for the North Eastern, by Mr. Johnson for the Midland, and by Mr. Webb for the London and North Western, all of which manifestly combined the highest excellence in quality of material with the most exquisite finish in construction. A professional visitor hardly needed to be told that the

American engine would be old at 15 years while the British would still be young at 25. That was "writ large" on the machines themselves, and was obviously intended to be so understood.

Yet all those alike were Twentieth Century Locomotives. All were freshly out of the shops and none were put to work until the new century's first year. It was a diametrical opposition of views and aims; two directly opposite modes of promoting the same end—efficient and economic working of the railway to which each belonged. One of the most important questions in locomotive engineering which will have to be settled in this new century is—Which of these two plans is the preferable one for general adoption?

It must be borne in mind that as to several points at issue, British and American practice tends to coalesce. It is no longer an absolute point of difference to use copper instead of steel for fireboxes, steel instead of brass for tubes. Some British engineers are trying steel tubes and fireboxes. Some American engineers are using brass for the one and copper for the other. Outside cylinders which had become obsolete in Britain so far as all new engines were concerned, although universal in America, have in certain cases come in again with the new century, not with the goodwill of British mechanical engineers, but strongly against their will, and simply because with particular new locomotive types necessitated by traffic growth these have become unavoidable. The question whether bar-framing or plate-framing be preferable exercises very few minds at all seriously. But the question whether British methods generally, or American shall prevail, does excite the deepest interest and justly so, because Britain finds America able to compete with her, and on certain grounds to beat her, even in the trade with Britain's own Colonies and dependencies. And so the grave question arises—Will the locomotive practice of the Twentieth Century be British or American?

It must be clearly understood here—although many otherwise intelligent persons fail to grasp the point—that there is no question as to the superiority of British to American engines per unit of nominal power. The sole question at issue is whether the American practice or the British be preferable in all the circumstances of the present day. In the particular phase of practice now under consideration it may be claimed on the part of British practice that it produces engines made of the best materials procurable, admirable as to workmanship, and so durable that they will run for 30 or 40 years without being worn out. For the American method it may be said that it turns out locomotives with remarkable celerity, which are efficient workers, and which by reason of their quickness of construction can be delivered at very short notice, but which are not expected to last more than half as long as the British. Thus, in a recent case, American builders were prepared to deliver locomotives to a British dependency in twelve weeks, whereas the British builders required ninety weeks. In another case which came under the author's personal and professional notice, a British firm took 18 months to complete an order for locomotives given by an English colony, while an American firm delivered the engines in 5 months from the date when the order was telegraphed from the colony, the American locomotives, moreover, costing 400*l.* less apiece than the British. It is obvious that when this sort of thing can be done the interests of British trade imperatively demand that the merits of the case should be clearly discerned in order that, if the British methods be really out-of-date and unsuitable to modern needs, no time may be lost in amending them in the respects shown to call for reform. Hitherto there has been much vague generalisation, but a notable lack of precise thought on the subject.

This particular question—whether the extreme longevity of the British engine be really advantageous or otherwise—has never been thoroughly thrashed out. It may be a worthy matter of national pride that so many of our Twentieth Century locomotives are from 20 to 40 years old, because it shows how splendidly they must originally have been constructed. But is it good business? One result is that two engines are frequently employed on a train which one modern engine could haul with ease. This is a wasteful and objectionable practice. Undoubtedly there are cases when the use of an assistant or pilot engine is defensible, as on excessively steep gradients, or in exceptionally stormy weather. But habitual piloting, or as the Americans term it "double-heading," is extravagant, and is undesirable in other respects. In France it is prohibited in the case of express trains save in wholly exceptional circumstances, and when two engines are used the speed is ordered to be reduced. Taking a pilot in order to keep time by increasing the speed of a train is a purely British practice. Piloting is wasteful in several ways. It involves not only the wear and tear of two engines instead of one, but also the working of the train-engine at a marked disadvantage owing to its being the recipient of smoke and dust from the pilot. It involves also the employment of four men instead of two.

But beside this, it is virtually certain that in such a case twice one does not make two—in respect of power. Two engines on one train will not do twice the work of one. And a further drawback is that they never have the chance. It is not practicable to run trains that would constitute a full load for two engines. Station platforms and siding accommodation will not permit this. It is very seldom when two locomotives are used that there is a load greater by 50 per cent. than one of them could deal with. Much more usually the excess is 25 or 30 per cent. Thus, either each engine must be considered as hauling only from 60 to 75 per cent. of its load, or else if the train-engine be credited with its full load the pilot is taking only from a quarter to half a load. This is manifestly not economical. That there are cases in which piloting or "banking" is admissible the author has already recognised, but he maintains that they are quite exceptional. Here, then, appears one of the disadvantages attaching to the longevity of out-of-date locomotives. "Superfluous lags the veteran on the stage" —and he requires somebody to hold him up. It has to be considered to what extent this dependence on assistance discounts the value of locomotive long-livedness.

But there is another point in this same connection. Where a large number of old locomotives are still at work, it becomes necessary to limit their loads. Take, for example, the Midland line, on which a vast number of veteran locomotives are still to be found, engines originally designed so skilfully by Mr. Matthew Kirtley, constructed of such excellent materials and kept up in such admirable order by his successor, Mr. S. W. Johnson, that they seem likely to last for ever—or thereabout. They do their work perfectly within their limits—but they have their limits; and so the working timetables contain instructions that when these engines are used their load shall be one-sixth less than that given to the modern engines, which therefore take 20 per cent. more than their elders. Now what does this mean? Surely that six engines have to be used instead of five; that 20 per cent. more shed accommodation, engine-sidings, drivers and firemen, cleaners, repairing hands, stores, etc. are needed than would be wanted if engines equal to modern requirements were employed. Nor is even this all. For it follows that in these cases six trains have to be run instead of five; thus 20 per cent. more demand is made upon roads already overcrowded, and so traffic is delayed and extra lines are often necessitated. This consideration also arises in connection with another point which will be treated later. Meanwhile it may be taken as clear that the retention of a large proportion of old engines does most seriously discount the carrying capacity of any railway.

On behalf of the American practice of building cheaply and speedily without aiming at extreme longevity, it has to be said that it possesses the merit of enabling full advantage to be taken of all latest improvements in design or construction, when the more quickly worn-out engines are replaced by new locomotives. The case is analogous to that which is familiar to every married man. When his wife or daughter buys a new dress she is careful to point out that it is far cheaper to pay a higher price, because good things are never really dear, and will last so much longer than inferior materials. That is quite true and just. Only next year that good dress may still be as good as new, but—thought of horror!—it may be out of fashion! Therefore, though as to materials the dress is as good as new, it no longer fulfils an up-to-date duty, and is accordingly condemned. That is exactly the idea which pervades American locomotive practice, and it unquestionably is entitled to consideration.

Yet, of course, as in all mundane matters, there is another side. Obviously the thought arises: Are not these quickly-built locomotives, constructed of cheap materials and roughly finished, likely to prove unduly costly in their upkeep, extravagant in fuel consumption, and in repairs? If so, may not this counterbalance their advantages? That is a perfectly fair question to raise, and, in the author's opinion, it must be answered unfavourably to the American engines as regarded from the view-point of their suitableness to British work. That American locomotives as a rule—there may be many exceptions—are somewhat wasteful of steam and of coal, and need more repairs than a British engine of corresponding dimensions, the author has never doubted. It might reasonably be assumed, on *à priori* grounds, that this would be so. In the United States fuel economy is placed in a wholly subordinate rank as compared with efficiency in the performance of the duty allotted, whether it be the haulage of vast freight trains or the running of expresses at booked speeds never dreamed of in Britain. The conditions of work in America differ in many respects from those which obtain in Britain, and fuel economy holds far lower rank as an item of locomotive merit than it does in Britain. But the other item, repairs, stands on a different footing. It is extremely difficult, in the absence of a common basis, to institute any comparison under this head that would be entirely fair. But so far as can be judged from experience with American locomotives in Britain and in several British dominions, they are more costly in upkeep and repairs than British-built engines.

This drawback, if it be really inherent in American practice, and not merely contingent on British working of American engines—a point which the author does not take upon himself to determine—is undoubtedly a formidable one. It has one of the disadvantages already predicated of the use of old engines. For if an engine spend an undue proportion of its life in the repairing-shop, it thereby loses so much of its effective life, and consequently necessitates a proportional increase in the locomotive stock. It might fairly be assumed *à priori*, as has already been observed, that engines built in the cheap and roughly-finished way mentioned, while they might do their duty most perfectly when in good order, would nevertheless be far oftener in need of repairs than locomotives constructed of superior materials and more highly finished. That they are somewhat wasteful in steam and coal is frankly declared by one of the most eminent American engineering authorities, Mr. Angus Sinclair. It may therefore be assumed that Mr. Johnson's experience with the American locomotives imported for the Midland Railway does not stand alone, although, as the author has had occasion to point out before, no real comparative test of British and American locomotive practice was involved, inasmuch as the American engines were not of a type, as to dimensions, that would ordinarily be employed in the United States on the same class of work.

This leads directly to another branch of the subject—that phase of American locomotive practice which consists in providing an ample margin of power, particularly in the department of steam-generation. For some years past America has been far ahead of Britain in respect of boiler-power. Heating-surface, it is true, constitutes only one element in the conditions which make for that power. But while it is indisputable that surprisingly good results have been got with very small heating-surface—as in the case of the late Mr. Patrick Stirling's famous single-wheelers, the last built of which have only 1031 square feet of heating surface to cylinders having the enormous dimensions of $19\frac{1}{2}$ inches by 28 inches—2290 square feet of heating-surface in the case of the old London and North Western engine "Liverpool," and 2083 square feet in some of Sir D. Gooch's 8 ft. single-wheelers on the Great Western 7 ft. gauge did not afford anything like proportionate boiler-power, the tubes being too much crowded. By means of a more judicious and carefully thought-out disposal of the tubes, and by enlarging the fire-box, Mr. H. A. Ivatt, with his 6 ft. 6 in. coupled "1321" class, on the Great Northern, and Mr. Johnson, with his 7 ft. coupled "2600" class on the Midland, have obtained extremely fine results, although those engines possess only 1250 and 1193 square feet of heating surface respectively. But in the majority of cases British locomotive engineers have found it advisable to follow the lead so long given by America, and to adopt greatly enlarged boiler dimensions, including a vast expansion of the heating surface.

The progress made in this respect during the past five years has been very remarkable. Glancing at the valuable list of dimensions compiled by Mr. J. A. F. Aspinall, in his able paper written for the International Railway Congress of 1895, one notices that the respective heating surfaces then employed by the leading British railways were, in square feet, as follow, given in the order quoted:—Lancashire and Yorkshire 7 ft. 3 in. coupled, cylinders 18 in. by 26 in., 1216; Midland 7 ft. coupled, cylinders 18 in. by 26 in., 1261; M. S. & L. (now Great Central) 6 ft. 9 in. coupled, cylinders 18 in. by 26 in., 1278; London and South Western 7 ft. coupled, cylinders 19 in. by 26 in., 1367; South Eastern 7 ft. coupled, cylinders 19 in. by 26 in., 984; Great Eastern 7 ft. coupled, cylinders 18 in. by 24 in., 1217; London, Brighton and South Coast, 6 ft. 6 in. coupled ("Gladstone" class), cylinders $18\frac{1}{4}$ in. by 26 in., 1492; North Eastern 7 ft. coupled ("1620" class), cylinders 19 in. by 26 in., 1341; London, Chatham and Dover 6 ft. 6 in. coupled, 1110; Caledonian 6 ft. 6 in. coupled, 1184; North British 6 ft. 6 in. coupled, 1262; Highland 6 ft. 3 in. coupled, 1242; London and North Western three-cylinder compound ("Greater Britain" class), 1505 sq. ft.

At that period the American engines mentioned in the same publication, had the following heating surface in square feet:—New York Central 7 ft. coupled, cylinders 19 in. by 24 in., 1930; Baltimore and Ohio 6 ft. 6. in. coupled, cylinders 20 in. by 24 in., 1687; Pennsylvania 6 ft. 6 in. coupled, cylinders 19 in. by 24 in., 1816. Of the French railways the Paris-Lyon-Méditerranée 6 ft. 6 in. coupled, cylinders $19\frac{3}{4}$ in. by $24\frac{1}{2}$ in. had 1522 sq. ft.; Paris-Orléans 7 ft. coupled, cylinders $17\frac{1}{4}$ in. by $27\frac{1}{2}$ in., 1659; Est 6 ft. 10 in. coupled, cylinders $18\frac{1}{2}$ in. by 26 in., 1810; Nord and Midi 7 ft. coupled, compound; cylinders (2) $13\frac{1}{2}$ in. and (2) $20\frac{3}{4}$ in. by $25\frac{1}{4}$ in.; 1670 sq. ft. It was only the two Italian railways, both very slow, that used such small heating surfaces as the British, viz. Adriatic, 6 ft. 3 in. coupled, cylinders 18 in. by $23\frac{1}{2}$ in., 1079 sq. ft.; Mediterranean, 6 ft. 10 in. coupled, cylinders 17 in. by $24\frac{1}{2}$ in., 1090 sq. ft. On the other hand the Australian Colony of New South Wales used as much as 1916 sq. ft. of heating surface for its express engines.

All those locomotive types are still at work as Twentieth Century engines. But shortly before the close of the Nineteenth Century, when it became fully realised in Britain what vast advances had been made by France and America in respect of boiler power, with the results, *inter alia*, that both of those countries had so completely distanced Britain in respect of booked railway speed that whereas in 1896 Britain was easily first and the rest of the world nowhere, in 1899 Britain had become a bad third, not only being beaten by both France and America, but also having fallen back enormously from her own high level of 1896, then the movement in the direction of augmenting the boiler power, which had virtually been initiated by Mr. J. F. M'Intosh in 1896 with his Caledonian "Dunalastairs," steadily spread over the whole length and breadth of the Kingdom until we find that the new engines actually built in or for the Twentieth Century, have such heating surfaces as those to be quoted directly, with proportionately enlarged dimensions of the boiler barrels and fireboxes. The prominent mistake of early days, that of getting increased heating surface by crowding tubes together until insufficient room was left for water and steam, has been carefully avoided, and a barrel of large diameter is provided. This is usually accompanied by an enlarged firebox, which often has the advantage either of the Belpaire extension or of the water-tubes traversing it, according to the methods respectively of Mr. D. Drummond (London and South Western) and of Mr. W. Worsdell (North Eastern), while the steam pressure, which in 1895 seldom exceeded 160 lb., is in 1902 seldom below 175 lb. to the square inch, and often higher.

Taking some of the most noteworthy locomotives which have either been actually built in the Twentieth Century or have come into work since that period began, the following are noteworthy for their increased boiler dimensions which may be instructively compared with those of 1895 already given:—
Mr. J. F. M'Intosh's new eight-coupled goods type on the Caledonian has 2500 sq. ft. of heating surface, with cylinders 21 in. by 26 in.; Mr. Dean's new six-coupled Great Western express has 2400 sq. ft. with cylinders 18 in. by 30 in.; Mr. Aspinall's 7 ft. 3 in. coupled "Atlantic" type has 2052 sq. ft. with cylinders 19 in. by 26 in.; Mr. P. Drummond's six-coupled Highland express has over 2000 sq. ft. with cylinders $19\frac{1}{2}$ in. by 26 in.; Mr. Worsdell's six-coupled express with 20 in. by 26 in. cylinders has 1750; Mr. Holden's 7 ft. coupled, with cylinders 19 in. by 26 in. (Great Eastern); Mr. Dean's 6 ft. 8 in. four-coupled, with cylinders 18 in. by 26 in. (Great Western); Mr. Billinton's 6 ft. 9 in. coupled, cylinders 19 in. by 26 in. (London, Brighton and South Coast); Mr. M'Intosh's "900" class with 6 ft. 6 in. coupled wheels and cylinders 19 in. by 26 in. (Caledonian); Mr. Johnson's Belpaire class, with 6 ft. 9 in. coupled wheels and cylinders $19\frac{1}{2}$ in. by 26 in. (Midland); and Mr. D. Drummond's new 6 ft. 6 in. coupled, with cylinders $8\frac{1}{2}$ in. by 26 in. (London and South Western), all have

1600 sq. ft. or upward, or its equivalent with the aid of fire-box—water-tubes, Belpaire box, etc., while other new Twentieth Century types, such as Mr. Ivatt's "990" (or "Atlantic") type (Great Northern), Mr. H. S. Wainwright's 6 ft. 8 in. coupled (South Eastern and Chatham), Mr. J. G. Robinson's 6 ft. 9 in. coupled (Great Central), and Mr. F. W. Webb's four-cylinder compounds, have from 1400 to 1500 sq. ft. Here is indeed a noteworthy and gratifying advance.

Further, the steam pressure has similarly increased. In 1895, out of eighteen British non-compound engine types mentioned, only two had more than 160 lb. steam pressure, one being Mr. W. Adams' last express engines of the "680" set on the London and South Western, and the other being the latest class of Great Northern 8 ft. single-wheelers, which were built by Mr. P. Stirling just before his lamented death. These two had 175 lb.; eight had 160 lb.; six had 150 lb., and two only 140 lb. But among the newest British locomotives the Great Western, North Eastern, and London and North Western have 200 lb., and nearly all the others 170 to 180 lb.

In both these respects, however, Britain is still outstripped by France and America. The newest French engines on the Nord and Est lines have 2275 and 2100 sq. ft. respectively of heating surface, and 228 lb. steam pressure, while in America the boilers surpass those of Britain as much as they did seven years ago, such heating surfaces as 3000 to 3500 sq. ft. being relatively common among the newest locomotives, while some run up as high as 4000 ft. It is true that abroad engineers are not "cabined, cribbed, confined" within such severe restrictions of loading gauge as afflict our British designers. With an extra foot of height and width available in Europe, and 2 ft. extra height in America, a vast enlargement of scope is afforded. Still we have not even yet built up to our full capability in Britain. It may be quite true that Mr. Aspinall, Mr. Worsdell, and Mr. Dean have gone up as high as they can with safety, so that if they maintain the present large diameter of their coupled wheels, larger boilers will be impracticable. But as equally good results are obtained in all respects with wheels from 3 in. to 12 in. smaller in diameter, it is manifest that at any rate from 1½ to 6 in. can still be added to the feasible diameter of boiler-barrels even in the case of Mr. Aspinall's huge "1400" type. Thus we have not yet reached an absolute limitation, even irrespective of the question of compounding, which is not touched on in the present paper, having been treated by the author in a previous paper.*

* "Notes on English and French Compound Locomotives," Transactions, 1900, p. 173.

Another feature of American practice which is steadily and rapidly becoming acclimatised in this country consists in the employment of six coupled driving wheels for passenger express train service. Although this method only made its appearance in Britain at the extreme end of the Nineteenth Century, it had long been in use in the United States, and also in such British dominions as Canada, Australia and New Zealand, and it had been tentatively employed on the European Continent. In its essence this practice represents the modern efforts to obtain as much adhesion weight as possible, distributed as widely as possible upon the metals. The old-fashioned idea that the single-wheeler was the only locomotive type admissible for fast running was perfectly just at the time it was conceived, when express trains rarely weighed more than 50 or 60 tons, exclusive of engine and tender. A single-driver engine had abundant adhesion weight for the haulage of such a train at high speed, and that type, when drawing a train proportioned to the tractive force it could exercise, was and is the ideal one, having superior ease and steadiness of motion, and being notably economical in the consumption of fuel and lubricants. As train-loads increased, so, *pari passu*, did the difficulty of providing a single-wheeler with sufficient adhesion weight to utilise the tractive force it derived from the firebox, boiler and cylinders, estimated by the accepted method of squaring the cylinder diameter, multiplying the result by the length of piston-stroke, and dividing the product by the driving-wheel diameter, all in inches, which formula gives the approximate tractive force in pounds for every pound of effective steam pressure on the pistons.

But of course only so much of that is actually available as can be taken up by the driving wheels by means of their adhesion to the rails through the weight placed upon them. Thus the driving wheels began to be coupled to a second pair even for express work, as had long been done for goods and slow passenger traffic, and in this way an adhesion weight which, placed on a single pair of wheels, would have been highly dangerous on the light rails of those days, was so distributed that it could be safely and profitably employed. For while 20 tons would have been an imminently unsafe weight on the single-driving wheels of an express engine running at high speed upon iron rails weighing 72 lb. or less per lineal yard, that weight, and indeed five or six tons more, became entirely unobjectionable when shared by two pairs of wheels. So far back as 45 years ago some expresses were run by coupled engines on the Great Western, Great Northern, South Eastern, and London, Brighton and South Coast lines. Still, locomotive engineers, as a rule, did not like the practice, and most railways used single-wheelers as long as possible. It was just 40 years ago that the Midland Railway started building coupled express engines, and thenceforward the practice persistently gained, until, early in the eighties, the single-wheeler appeared on the verge of final extinction. But then came the Gresham sand-blast, which so valuably enhanced the adhesion, especially on a slippery rail—the weight of the steel rails being also increased to 85, 90, and even 100 lb. per yard—that the single-wheeler obtained a new lease of life, and this Twentieth Century sees a large number—more than 450—still doing good service.

But the increased difficulty of the work has gone heavily against them. It is not that the maximum weight of trains has increased. The traffic has enormously expanded, but it is conveyed by a far larger number of express trains, and as third-class passengers are carried by all expresses on every British railway excepting only the two slowest and least progressive lines—those of the South of England—a vastly greater number of heavy fast trains must be run now. Consequently, while in the sixties only an occasional heavy express had to be provided for, this being generally done by applying an assistant engine, in the current century the great majority of main line trains are expresses, and the great majority of these are heavy. Even a quarter of a century ago the standard weight of an express was reckoned as 150 tons behind the tender. Now the average is more like 250 tons. Seven years ago the West Coast afternoon corridor dining-car train between Euston and Scotland used often to load up to 320 tons, and yet was taken between Euston and Crewe in excellent time by one of Mr. Webb's three-cylinder compounds. But now most of the Anglo-Scottish expresses on both East and West Coast routes, and some of the London-Manchester and London-Liverpool trains quite commonly weigh as much as that. Sometimes this weight may be exceeded, but as a rule the accommodation available at station platforms, etc., precludes any material excess above 320 to 340 tons on British railways.

Even a four-coupled engine cannot always keep fast booked time with such a load on moderately steep gradients, such as 1 in 150 to 1 in 300, without pilot assistance. When the grades stiffen to 1 in 100, as at Settle, 1 in 90 as at Peak Forest—both on the Midland—or 1 in 75 as approaching Shap on the London and North Western, and Beattock Summit on the Caledonian, a four-coupled engine cannot take up a fast-timed 300-ton train, and is almost invariably helped by a pilot or bank engine. But where grades of 1 in 150 to 1 in 300 extending for several miles together are freely scattered about a line, as is so often the case in Britain, it is wasteful to use a second locomotive, and yet the ordinary four-coupled engine which can keep time in fine weather on a dry rail and in a calm atmosphere, will lose time badly when these favourable conditions are absent. When, for instance, a drizzling rain or fine snow is falling, when the rails are greasy or when a strong side wind is blowing, it is undeniable that a four-coupled engine does often lose time through slipping, just in the same way as a single-wheeler would, although of course in a far less degree. These conditions often occur in Britain, particularly in the winter season, which not infrequently encroaches in both directions alike on the preceding period popularly characterised as autumn, and on the subsequent season which is politely flattered by being termed spring. That is to say, provision has to be made for wintry conditions which may extend from September to June.

In the meantime certain French railways, following the example of America, began about the year 1898 to employ on passenger express duty some powerful six-coupled engines with 5 ft. 9 in. driving wheels, and leading four-wheeled bogie which had been designed by M. du Bousquet on the de Glehn four-cylinder compound system to work fast goods trains or heavy passenger excursions to the seaside or at race times. The result proved entirely satisfactory, as it had in the United States, in Canada and in various British Colonies, the six-coupled engines running with good steadiness at ample speed and hauling extremely heavy loads. So then after several attempts to provide a four-coupled engine which would fulfil the requirements indicated, Mr. Worsdell designed and built for the North Eastern Railway of England ten very fine six-coupled express engines with leading bogies—after the American "Ten-Wheeler" type—and followed them up with five more differing only in having driving wheels of somewhat larger diameter, 6 ft. 8 in. instead of 6 ft. 1 in., and in weighing 5 tons more, viz. 67 tons instead of 62 tons, the cylinders in each case being 20 in. by 26 in., and the boiler having 1750 square feet of heating surface.

This new departure elicited, as might have been expected, the customary conservative criticism which almost invariably assails novel methods. The experiment, however, appears to the author to have been emphatically successful in its outcome. In numerous tests he always found the six-coupled engines to haul huge loads with comparative ease at speeds fully equalling those not only of four-coupled engines, but even of single-wheelers. Practically the 6 ft. 8 in. six-coupled class can run, and have run, quite as fast as the best speed ever made by a Great Northern single-wheeler with 8 ft. drivers. As to the possible qualifying conditions of wear and tear, and of coal and oil consumption, it is too soon to speak. The engines have not yet been long enough in use to enable a really trustworthy comparison to be made with the four-coupled type.

But the example set by Mr. Worsdell has already been followed on other lines. Mr. P. Drummond has brought out for the Highland Railway an excellent express engine of very similar type but having somewhat larger boiler and slightly smaller wheels and cylinders than the North Eastern engines. These are doing valuable service on the long grades of 1 in 70 and even 1 in 60 with which the Highland line is infested, and should save a great deal of piloting, especially on such gradients as the Struan bank going north from Blair Atholl which previously required no fewer than three engines—two in front and one behind—to be used with expresses of at all substantial weight. Mr. Dean on the Great Western and Mr. Robinson on the Great Central have been the next to fall into line with the new departure. The former has recently brought out a new express locomotive which enjoys the distinction of being the heaviest ever yet placed upon British metals, weighing no less than 69 tons in working order. It has six coupled wheels 6 ft. 8 in. in diameter, outside cylinders—a cataclysmic novelty on the Great Western—18 in. in diameter, with the extraordinarily long piston stroke of 30 inches, and a large boiler with enormous Belpaire firebox and a total heating surface of 2400 sq. ft., pressed to 200 lb. per sq. in. This notable engine is at present in its experimental stage, and the author has no information as yet as to its work. Mr. Robinson's new Great Central engine is similarly in its period of infancy, but also seems full of promise. It is not described specifically as an "express" engine, but as it has 6 ft. drivers, six-coupled, cylinders $19\frac{1}{2}$ in. by 26 in., a larger boiler and a leading bogie, it is manifestly adapted for heavy express duty on grades so continuously trying as those of the Great Central between Sheffield and Manchester. Doubtless it will be utilised on that length so soon as the Great Central main line express trains shall be heavy enough to call for more power than the standard four-coupled type can exercise.

Thus then, it will be seen, no fewer than four leading British main lines have adopted the plan—formerly deemed a rank Americanism—of building and using six-coupled locomotives for fast passenger service. This and the increase in boiler power and steam pressure, have been distinctly the salient features of British locomotive practice at the outset of this new century in respect of express passenger work, and it is noteworthy that the employment of six-coupled wheels is also largely extending in the case of the tank engines used for suburban and branch traffic. At one time the Brighton railway with Mr. Stroudley's small tanks, having six-coupled 3 ft. 10 in. wheels, and the Lancashire and Yorkshire with Mr. Barton Wright's six-coupled tanks having a trailing pair of carrying wheels, stood virtually alone in this respect. Then Mr. Holden followed on the Great Eastern with his very efficient little 4 ft. wheeled six-coupled class of tanks, and more recently the London and North Western has followed the Lancashire and Yorkshire, and Mr. Billinton has done likewise on the London, Brighton and South Coast, the new tank engines in each of the two latter cases having six-coupled 5 ft. wheels and a carrying pair of trailers. This plan is certain to come into wider use as suburban traffic continues to increase with giant strides. Mr. Holden is even building a "decapod" or ten-coupled tank engine, with three high-pressure cylinders for the enormous suburban work of the Great Eastern Railway.

On the same principle there has come in with the Twentieth Century a British tendency to employ eight-coupled wheels in goods engines. This again follows a European and American and Colonial precedent. For many years past the heaviest goods traffic on the European and American Continents, and in the British Colonies, has been worked by eight-coupled locomotives. These were introduced in New Zealand so long ago as the year 1880, and they have since been largely multiplied, being found useful even in passenger train service. In Britain Mr. Webb, on the London and North Western, was the pioneer of this new departure. He brought out a number of large eight-coupled engines for mineral and heavy goods traffic. The earlier ones were compounds on his three-cylinder system, but the latest which came out this year are on his four-cylinder compound plan. Next came Mr. Aspinall on the Lancashire and Yorkshire Railway. A year or two back he brought out a batch of very large non-compound goods engines with eight-coupled wheels, inside cylinders, and very large boilers, having 175 lb. steam pressure. More recently Mr. H. A. Ivatt, on the Great Northern, and Mr. Worsdell, on the North Eastern, have built eight-coupled locomotives—the latter with outside cylinders—and now Mr. Robinson, of the Great Central, is doing the same; while for the Caledonian Mr. J. F. M'Intosh has designed and built some new engines which are by far the largest and most powerful goods engines ever yet seen in this country. They have eight-coupled wheels, inside cylinders 21 in. by 26 in., and a huge boiler with 2500 sq. ft. of heating surface, the largest in the kingdom. Thus, no fewer than six British railways have now adopted this type for goods and mineral service, and so far all appear to have given most excellent results.

Of splendid new locomotives belonging to the more ordinary types, *e.g.* four-coupled for passenger service and six-couple for goods, the name is legion, but space will not permit a detailed notice of each. Prominent among the former are a fresh batch of Mr. Dean's very efficient "Atbara" type on the Great Western, which have 6 ft. 8 in. wheels, cylinders 18 in. by 26 in., and 1663 sq. ft. of heating surface; Mr. D. Drummond's "300" class on the London and South Western with 6 ft. 6 in. wheels, cylinders 18½ in. by 26 in., and 1500 sq. ft. of heating surface supplemented by his very valuable system of water-tubes passing through the fire-boxes; Mr. R. J. Billinton's "Emperor" class on the London, Brighton and South Coast, with 6 ft. 9 in. wheels, cylinders 19 in. by 26 in., 1600 sq. ft. of heating surface and 175 lb. steam pressure; Mr. H. S. Wainwright's "730" class on the South Eastern and Chatham, with 6 ft. 8 in. wheels, cylinders 19 in. by 26 in. and 1505 sq. ft. of heating surface; Mr. Holden's "Claud Hamiltons" on the Great Eastern, with 7 ft. wheels, cylinders 19 in. by 26 in. and 1600 sq. ft. of heating surface; Mr. Ivatt's "1321" and "990" classes on the Great Northern, with 6 ft. 6 in. wheels, the former having inside cylinders 17½ in. by 26 in., the latter of the "Atlantic" type, outside cylinders 19 in. by 24, and 1444 sq. ft. of heating surface; Mr. Webb's "Alfred the Great" type of four-cylinder compounds with 7 ft. wheels, and 1557 sq. ft. of heating surface and water-bottoms to the fire-boxes; Mr. Johnson's 7 ft. and 6 ft. 9 in. classes on the Midland, the former having cylinders 19 in. by 26 in. and 1193 sq. ft. of heating surface, the latter 19½ in. by 26 in. cylinders, and 1600 sq. ft. of heating surface; Mr. Worsdell's "2011" class on the North Eastern with 6 ft. 9 in. wheels and cylinders 19 in. by 26 in.; Mr. Robinson's "1015" class on the Great Central, with 6 ft. 9 in. wheels and cylinders 18½ in. by 26 in.; and Mr. M'Intosh's "900" class on the Caledonian, with 6 ft. 6 in. wheels, cylinders 19 in. by 26 in. and 1600 sq. ft. of heating surface. All are fine engines, and have done satisfactory and sometimes brilliant work under the author's personal observation.

It may seem an anachronism and an inconsistency that the Twentieth Century should see a new batch of single-wheelers appearing on the Great Northern, but it must be confessed that the ten brought out last year by Mr. Ivatt, which have 7 ft. 6 in. wheels and cylinders 19 in. by 26 in., have justified their existence by doing much fine express work. It is doubtful, however, whether any more single-wheelers will be built, in view of the drawbacks already referred to which tend to become aggravated as the traffic increases in weight.

Of the more ordinary six-coupled type of goods engine, some excellent representatives possessing many improvements, have been turned out by Mr. Drummond for the London and South Western, by Mr. Holden for the Great Eastern, by Mr. Robinson for the Great Central, and by Mr. Wainwright for the South Eastern and Chatham. For the Great Western Mr. Dean has built two curious variants of this type, each having inside cylinders, outside bearings and cranks, and large Belpaire boilers. One set has a four-wheeled bogie under the leading end; the other a two-wheeled pony truck. Both appear to be very efficient.

Summing up the general character of Twentieth Century locomotive practice in Great Britain, it must be said to be following largely in American and European Continental footsteps so far as concerns the general plans and schemes of the engines, that is to say in respect of increased boiler-power and steam pressure, and of multiplication of coupled wheels, both being clearly useful and beneficial movements. As regards the question which was dealt with in the earlier part of this paper, as to whether the American or the British practice in respect of building for longevity or for " a short life and a merry one," with the idea of quickly using up the machine and then obtaining a new one with all the latest improvements, the author has endeavoured to set out the salient recommendations of each method. Either plan has its advantages and its drawbacks. To what extent these may be held to balance one another will probably depend upon the circumstances of each individual case, and perhaps in some degree on the personal idiosyncrasy of each locomotive superintendent. The author prefers to express no final judgment on that score, although he unhesitatingly declares his advocacy of the other phase of American practice which has come so largely into British use during the last few years, namely that of augmenting the boiler-power of engines and enlarging their reserve of force available in case of emergency. He is aware that in adversely criticising the prevalence of piloting and other methods which seem to him inconvenient and archaic, he lays himself open to the gently courteous reproof so often administered by railway officials to their critics, even professional ones, that the criticism shows how impossible it is for outsiders to realise all the difficulties under which British railway traffic is conducted. This hint is invariably accepted by the author with becoming meekness. In that spirit he concludes his present remarks.

DISCUSSION.

The PRESIDENT said that the session had been very prolific of papers bringing before the Society the subject of the competition between America and this country, with regard to the engineering industries, the present paper being the third which had dealt with the subject. The author had brought forward a problem which would have to be faced not only in that branch of engineering to which it related, but in others. The point was mainly whether English engineers should sacrifice the very high class work for which they were now distinguished, in order to reduce the cost and time of manufacture, and secure the more frequent replacement of old designs by new ones, and thus keep themselves up to date. The author had summed that question up in a way which it would be well for engineers to imitate whenever any problem of this nature was brought before them. He had asked, "Is it good business?" He (the President) did not think that any engineer would contend that good engineering was not compatible with good business. Surely in every commercial undertaking engineers must consider themselves in some respect the servants of the financiers or, in the case of railways, of the shareholders. It surely must be the engineer's object to make the concern with which he was connected pay. He thought that English engineers had some lessons to learn in that respect from their American friends. The author had carefully refrained from pronouncing any definite judgment upon that difficult problem as regarded locomotives; but he had, perhaps, sufficiently indicated the trend of his own opinion, which might be taken as running very largely on the lines of the advice which they had received from other authors of papers in connection with other subjects. It was that which was brought prominently before the British public by His Royal Highness the Prince of Wales when he had returned from a close inspection of Colonial practice in connection with various manufactures. His Royal Highness's warning to England was, "wake up." In other words they must move forward in order to keep themselves up to date.

In the present case the same point had been brought before the Society by a gentleman who was eminently qualified to deal with the subject before them. Probably, there were few men in this country so expert in the technicalities of general locomotive design and performance as the author of the paper, and they were extremely grateful to him for giving the Society the advantage of his experience, and the opinions he had arrived at arising from that experience.

The paper contained a vast amount of information, which would be useful to those who were not technically associated with railway work, if only for the purpose of reference, but it especially presented points which would he was sure be most highly appreciated by those who were themselves connected with locomotive engineering. He therefore had the greatest pleasure in proposing a vote of thanks to the author for his paper.

The motion was carried by acclamation and acknowledged by the author.

Mr. J. E. DARBISHIRE said that he had been amongst locomotives all his engineering life, and that he took very great interest in them. But, at the same time, he felt that it was almost impossible to criticise Mr. Rous-Marten, for his own knowledge of the subject was very small compared to that of Mr. Rous-Marten. There were, however, one or two small points in which he might, perhaps, add to what Mr. Marten had put before them.

With reference to the use of iron tubes and steel tubes in this country, he had recently got a considerable amount of information, and one piece of news which he had got from an English railway was so remarkable, that it might be of interest. Mr. Pettigrew had recently told him, that on the Furness Railway he had used iron tubes which had been put in by his predecessor; when taken out of the engine in which they were first used, they were cut shorter, and put into an engine which had a shorter boiler, and then again had their ends cut off and put into a third engine. He suggested to Mr. Pettigrew in jest, that possibly the water used contained so much iron that the tubes would get thicker. It was a most extraordinary use of iron tubes. He believed that on the South Eastern Railway the water was so chalky in many parts, that they could not use iron or steel tubes; so the matter was one of local practice, and not one of national practice.

The author had observed that the British builders took much longer than American builders to turn out locomotives. He thought that he knew the particular instance referred to, in which the American firm delivered engines in five months, and the British firm took eighteen months. First of all, the British firm had been very busy for the last few years, and the Colonial orders had to take their turn; and, secondly, the American engine was not the same engine as the British, but was the American builder's type, whereas, the British engine was built to the purchaser's design and specification, so that it was rather hard to blame the British builder.

With regard to piloting, he need hardly say that every person connected with locomotive running must be of Mr. Rous-Marten's opinion. Last January, while he was standing at the station at Wolverton, he noticed a great many expresses passed through, and he thought that half those expresses had a big four-cylinder compound engine with a pilot in front. To his mind it would have been better to have two four-coupled of the "Precedent" class, than one of the enormous four-cylinder compounds, with a pilot in front. He was, however, not a railway man, but only a builder, and therefore he was looking at the matter as a layman from the railway point of view. He had frequently had the question brought before him, as a locomotive builder, as to which was the right practice—the American practice of making engines to run for ten or fifteen years, and then throwing them on the scrap-heap, or the English practice of making better engines and keeping them going. His view was that for financial reasons every case must practically stand by itself. There were cases in which he would have to recommend the American system, and say, "buy the cheapest thing you can, and throw it away when worn out." On the other hand, there were cases in which, coal being very expensive (as in Brazil and some parts of the Argentine Republic), if the line could buy an English engine which would burn about 20 per cent. less coal, and run for ten or fifteen years, even if it was then to go on the scrap-heap, it would have saved the increased cost by the decreased fuel consumption and less cost for repairs, so that even in ten or fifteen years an expensive engine in first cost, might be thrown away on the scrap heap and still have been less expensive in the end, than an American one. But the reason of that would be, because it would be running in a country where the coal which was suitable for it was dear. It might not be so if the coal was purchasable at 8s. or 10s. a ton.

He had seen a good deal of English practice and of American practice, and there were advantages in both. In some cases the right thing was to take an American engine, and use it up, and throw it away. There were some cases in which new railways could not pay for English engines at all, being unable to spare the capital, and yet they could get the line to start with an inferior article at a lower price.

Then there was the same question as it affected workshop practice. From many years' experience of works, his own opinion was that they could not do two classes of work in one shop. He could not ask everybody to agree with him in that view. He thought that if they were going to make British locomotives, as British firms did, of the very best materials combined with the very best workmanship and the best finish, even down to the varnish of the cab panels, they must confine their shop to work of that class. It was no use to put up a workshop and say, "We are going to make all first-class engines on the right-hand side of the works, and we will make second-class engines of the American style on the left-hand side." Try as they liked they would not be able to do that. They would find that the whole of the work in the shop would gradually work down to the second-class quality, and they would have no first-class work produced in the place. Therefore engine builders must make up their minds whether they would stick to first class work, or whether they would do second-class work; but they would not be able to do both. But he was not by any means certain that his views upon the matter would be backed up by people who had had different experience.

The only possible point that he thought he could criticise in Mr. Rous-Marten's paper was where he spoke about getting boiler barrels larger. They could get the boiler barrels larger, but he was afraid that they had got to the end of enlarging the fire-box. If the fire-box were made larger they would have to get bigger men to put the coal in. The limit of the size of the fire-box seemed to be the limit of modern locomotive design. The size of fire-boxes and consequently of grates could only be increased in the direction of length, except by spreading the fire-box laterally above the frames and wheels, as in some American engines; and one of the reasons why the American locomotive engine was so much more wasteful of coal than the English was the small distance below the tubes to the fire-bars in an arrangement of that sort.

Mr. W. ORRINSMITH said that his experience in locomotives had been very moderate up to the present, but he should question that it was impossible to increase the length of the fire-box. Fire-boxes were in use up to 11 feet long in America. As to the question of coupling the wheels of express engines, the six-coupled engine seemed to do wonderfully well on the North-Eastern and Highland Railways. The larger boiler and longer fire-box could be certainly got in in that type of engine if the fire-box was made less deep. He believed that on the North-Western Railway a new rule had been issued by the Locomotive Department to use pilots on all trains exceeding $17\frac{1}{2}$. That length did not represent the maximum capacity that the engines could haul, as even the earlier 3-cylinder compounds regularly ran to time with loads up to $20\frac{1}{2}$ equal to about 320 tons.

Mr. W. B. THOMPSON said that with regard to the question of piloting, he could not help thinking that the old engines which still survived after thirty or forty years were not working at the greatest advantage yet. The modern engines, as one speaker had already pointed out, which the North-Western sent out, were ridiculously underloaded. There was far too much piloting, and he considered that the engines, whether old or new, were not generally driven so as to get the utmost out of them. His recent experience was an illustration of that. On Saturday last, he left Rugby at 8 P.M. with a four cylinder compound engine, No. 1930, in the train due into Euston at 9·45, bringing portions from Manchester and from Birmingham, and weighing, as far as he could guess, about 250 tons. Owing to a hot box in the tender, the engine, which was a compound, had to be detached at Bletchley. A delay of eight minutes occurred there, and an old-fashioned 7-feet 6-inch single-wheel engine, built in 1863, was produced, and attached to the train and sent up to London to do the best it could. It ran up to Tring in 24 minutes, and from Tring to Willesden in 26½ minutes, doing the whole distance from Bletchley to Willesden in 50½ minutes.

Mr. ARTHUR RIGG said, that when he was in Philadelphia in 1884, the Baldwin Locomotive Works were making a locomotive with the fire grate 8 feet wide, which was an enormous width, and as far as he could recollect, it was about 8 feet long. It was an exceedingly ugly engine, but if he said more about it he should be running away from the scope of the paper and incurring Mr. Rous-Marten's displeasure. There was one remarkable thing about English in comparison with foreign locomotives which the author had not stated. The English locomotive was a good-looking thing, but the Continental locomotive was the ugliest collection of parts that could well be put together, and the American locomotives were just as bad. He could not see why good taste should not prevail in regard to locomotives as it did with other things. Some of the more modern locomotives which Mr. Rous-Marten spoke of were very far from being monuments of beauty, he must say, but the old-fashioned English locomotives were among the best looking designs that had ever been put together. He believed that it was Brunel who said that any piece of engineering that did not look well could not work well, and he thought that there was a great deal of truth in that remark.

Mr. W. C. ROWCLIFFE asked whether Mr. Rous-Marten could tell them what was the approximate weight of the 17 or 17½ coach limit in excess of which the North-Western made it a rule to pilot; and could he give the meeting some account of the Great Eastern Railway "Decapod" locomotive which he mentioned in his paper?

Mr. PERRY F. NURSEY said that a circumstance which had a very important bearing on twentieth century locomotives was brought before the public through the *Times* in the beginning of the present year, respecting which, he would ask the author of the paper if he could give the meeting any information. That circumstance was the public announcement by Mr. H. S. Cautley, M.P., that we were "on the eve of a mechanical revolution such as had never been seen since the introduction of steam." It was further announced that that revolution was to be brought about by means of a new valve gear, invented by Mr. J. T. Marshall, of Leeds. That valve gear, it was stated, had been fitted on one of the Great Northern mineral locomotives, built in 1882 by Dübs and Co., and known as No. 743. A number of trials were made with that engine in March 1901. The load hauled was a train of 280 tons, which was taken up an incline of 1 in $71\frac{1}{2}$ with 140 lb. steam. The load was a 50 per cent. increase on the ordinary load drawn by that engine. She then took a load of 307 tons up the same gradient, and then 280 tons up 1 in 40. Then the coal consumption was tested against a non-converted sister engine built by the same makers and which had exactly the same power and was of precisely similar construction. The result of the trials was that the coal consumption on the non-converted engine was 72 lb. per mile, while that of the converted engine was 39 lb. per mile—a very considerable reduction. Further comparative trials were made with those engines, and the average of the trials was 57·7 lb. for the non-converted engine, and 46·2 lb. for the converted engine, the latter running with a steam pressure 30 lb. lower than that of modern practice. A number of other details were published at the time, and great expectations were indulged in respecting the application of Mr. Marshall's valve to some of the Great Northern locomotives. Unfortunately, Mr. Ivatt, the locomotive superintendent of the Great Northern Railway was not present, although he hoped to have been. He (Mr. Nursey) would therefore ask Mr. Rous-Marten if he had investigated the matter, and if, with his wide experience, he could enlighten the meeting on the subject.

Mr. ROUS-MARTEN said that he would answer the speakers in the reverse order to that in which they had spoken, as he did on the occasion of his previous paper.

Mr. Nursey had put a question about a new valve gear.

He (the speaker) was in some difficulty as to answering that, because, in the first place, he had been professionally consulted in connection with it, and, in the next place, his personal experience of it was still to come. He did not know that he ought to say more than that the statement from which Mr. Nursey quoted, and which appeared in the *Times*, about the performance of the valve gear, he had reason to believe was in the main accurate. He had been told by one in whose trustworthiness and capacity he had full confidence, that the valve gear was substantially doing what was claimed for it. It was, however, quite possible that the results might not be so markedly gorgeous as those which Mr. Nursey had quoted. He was afraid that what had appeared in the *Times* was perhaps a little "too previous," as the Americans said. Since that report came out, the valve gear had been fitted to one of the newest class of express engines on the Great Northern, and he was hoping to have an opportunity before long of seeing what it could do.

A question had been asked by Mr. Rowcliffe as to the weight of the "$17\frac{1}{2}$ coaches" which, on the London and North Western Railway, constituted the prescribed limit of load to be taken without pilot. He did not know that there was any actual standard weight given. The train was supposed to be computed on the system by which the North Western reckoned; that is to say, a six-wheeled coach was estimated as one "coach," an eight-wheeled coach as a coach and a half, and a twelve-wheeled coach as two coaches. It had always struck him that that was a rather clumsy way of calculating, but some railway people are particularly strong on "half-coaches," and they never seem to be satisfied with the train load unless a half-coach came in. So far as he could calculate roughly, he should say that the North Western train of 17 coaches would probably weigh from 260 to 280 tons behind the tender. He did not think railway officials were so particular here as they were abroad in reckoning passengers and luggage, when estimating the full weight hauled by the engine.

As to the Great Eastern Decapod, he could not say more than he had said in his paper. He was informed that there would be three $18\frac{1}{2}$ inch high-pressure cylinders with 26 inch stroke, and an enormous boiler with about 2800 square feet of heating surface. He understood that the engine would be quite a new departure, and he should think that it would be a very valuable one.

Mr. Rigg appeared to be sorry that he (the author) had not laid more stress upon the æsthetic aspect of locomotives. He reminded Mr. Rigg of the proverb—"Handsome is that handsome does." And in railway matters they must have the "handsome does." Certainly some of the old single-wheelers were among the most beautiful engines ever built. He quite agreed with Mr. Rigg that it was desirable to make the engines as æsthetically perfect as they could be, but he thought that the question of efficiency must take a place ahead of mere beauty. Still, there was no reason why the engines should not be handsome as well as efficient. The new six-wheel coupled express on the North Eastern Railway had struck him as being exceedingly handsome, and the one of that class which he saw at the Paris Exhibition appeared to him certainly among the handsomest of the 68 locomotives there present. Some of the most recent engines built on the Continent were not by any means ugly machines. In France, for example, and on some of the Swiss and German railways, the engines used were largely of one standard design, namely of the de Glehn four-cylinder compound type. England might claim some share in them, because Mr. de Glehn was English by birth, and was educated in England. It fortunately happened that the great manufacturing company at Belfort recognised what a remarkable man he was, and since that time he had been the leading spirit there. More than one thousand engines compounded on his system were now at work on European railways. It could not be said that the de Glehn engines were ugly, but he (Mr. Rous-Marten) quite admitted that some of the old engines on the Orleans line were perfect marvels of ugliness. They were all sheeted with polished brass, everything except the chimney being covered with that material, and they had two huge domes connected by a large pipe which passed through a big cylindrical sand-box, all brass-sheeted and looking like a sort of brewery or distillery on the top of the locomotive's boiler. They were most extraordinary things, and some tank engines used to have a sort of iron coffin on each side. But Continental officials were improving in that respect, and were making their engines much handsomer.

Mr. Darbishire referred to the question of boiler extension and the size of the fire-box. As to that, he (Mr. Rous-Marten) could only say that, whether it was impracticable—as had been alleged—or not to extend the fire-box, a good many engineers were doing it up to a certain point. Of course, he quite admitted that a fire-box which required a man 10 feet high to stoke, would not be a useful one. But he thought it would be found that some engineers were making the fire-boxes 8 or 9 feet long, and yet normal-sized firemen managed to stoke them. Then there was mechanical stoking as used abroad, which he understood promised very well, but he had not yet observed it personally. It had been said that some British boilers were as big as they could be now. Of course, if engineers insisted on having such a big wheel as 7 feet, or 7 feet 3 inches in diameter, they no doubt were; but then he could not see why they wanted a 7-feet 3-inch wheel, as at least equally good results were obtained on the Caledonian Railway with 6-feet 6-inch wheels.

As to the size of the fire-box, it must be remembered that a good deal had been done with the Belpaire and other systems in developing its steam-generating capacity, and that Mr. Worsdell and Mr. Drummond with their water-tubes got a great deal more value out of the fire-box. Of course, the greater the amount of steam generation they could effect by direct contact with fire, the better it was; and the further they got away from the fire-box the less valuable the tubes were. If they compared the practice of engineers year by year, they would see that there was a steady increase, not only in the tube surface of the boilers, but also in the capacity of the fire-box. No doubt they might not be able to go on extending indefinitely, but Mr. Johnson got remarkable results with such an exceedingly small boiler because he had a large fire-box and a very well placed tube-surface.

As to the question of the piloting of North Western compounds, it was done, as had been pointed out, by direction of the Traffic Department. He could not go into that matter so fully as he should have liked without committing a breach of confidence. He could only say most emphatically that it certainly was not the wish of the Locomotive Department that all this piloting should take place. As to the necessity, it was quite clear that if an engine once had drawn a certain load in a certain time she should always be able to do it on other occasions, when the conditions were identical. An engine could not do more than she was able to do. If she was able to take a certain load in a certain time once, she ought to be able to do it again. If the engine did not do it again, it seemed to him that the defect was not in the engine; but, it was not for him to say where it was. He was quite sure that any train that ran on the North Western in the present day with a pilot engine, could be run by one of the new big engines without a pilot over the greater portion of the North Western main line.

The President had mentioned the fact that be (Mr. Rous-

Marten) appeared to have indicated a trend of opinion in the direction of that expressed by the Prince of Wales in his famous Guildhall speech, when he called upon England to "wake up!" It seemed to him that the great desideratum was that these questions should be considered and talked over. He did not think that we should jump to the conclusion that our own way must be right because it was our own, and that other people must be wrong. That was unbusiness-like. But, on the other hand, he did not at all believe that in everything "they managed things better in France" or that all the American methods must necessarily be right. He did not think that either mental attitude was a proper or a scientific or a business-like attitude. What he contended was that they should take all points into consideration fairly, and think them out, but not jump to conclusions. The great mischief of the present day was that people jumped to conclusions in accordance with their own prejudices. He thought that many English engineers needed to "wake up" in more senses than one, and to bring up to date not only their machinery, but also their tone of mind and habit of thought. Engineers needed to get more into the way of receiving novelties with respect, even if they did not care about them at first, and they should consider whether, after all, other people might not be right and themselves wrong. With regard to the particular question of building engines to last for a short or a long time, it might be that liability to get out of repair counterbalanced the advantages of building the cheaper sort of engines. Nobody could decide offhand how that might be. It was perfectly idle to theorise upon such a point, which could only be determined by experiment and experience.

He thought that some English engineers needed to "wake up" also in respect of forming a clearer view of what they wanted done by their engines. For then there would be little difficulty in designing or arranging the dimensions and parts so as to compass what they wanted. In spite of all that had been said on the subject, trains had not really increased in maximum weight during the past few years, but heavy trains were more numerous. If the trains were bigger they could not be dealt with at the platforms of the present stations. There were certain maximum train dimensions which they could not go beyond; and these were reached in former years. Therefore the arguments used now and acted upon now in favour of having bigger boilers and greater power were equally applicable before, although that was not admitted then. The President had quite rightly judged as to his trend of opinion. There was room in British engineering practice to improve a good deal. He did not say for a moment that we should abandon our practice of making engines of good material and well finished; but the chief thing was to ensure that the engines should be entirely suitable for the work which they were required to do.

VACATION VISITS.

Three visits to works of professional interest were made during the vacation of 1902. The first of these, on June 11, was to the pumping stations and works connected with the water system of the Crystal Palace, and the new roof over the centre transept. The second visit, on July 16, was to the works of the new graving dock and the widening of the Old Extension Quay at Southampton, and to the Southampton Water-works at Otterbourne. The third visit, on September 24, was to the works of Messrs. Joseph Baker and Sons, manufacturers of industrial machinery and appliances, and to the machine works of the Wicks Rotary Type-casting Company, both of which establishments are situate in Hythe Road, Willesden Junction. The latter works were visited in 1899, since which date further interesting developments have taken place in the details of the Wicks Rotary Type-Casting Machine.

The Crystal Palace Water System.

There are few, if any, of the members of the Society who have not, at one time or other, paid a visit to the Crystal Palace, and it may be that there are equally few who, although fully appreciating the varied attractions of the building and grounds, have ever bestowed a thought upon the multifarious engineering details which are necessary to the production and maintenance of those attractions. It was considerations such as these that suggested the present visit of inspection, for which permission was most readily accorded by the directors of the Crystal Palace Company. The inspection related mainly to the pumping stations and works connected with the water system of the Crystal Palace, the following description of which is compiled from an article on the subject which appeared in the *Engineer* of February 1, 1901.

It will doubtless be a matter of surprise to all to learn that the ordinary amount of water requisite for the needs of the Crystal Palace varies between 250,000 gallons and 300,000 gallons per day throughout the year. This is enough—at 30 gallons per head—for the supply of a town of from 8000 to

10,000 inhabitants. This amount leaves out of account altogether any which may be used for dietetic purposes. Such water is obtained from another source than that from which the general supply is drawn. For the most part the Crystal Palace depends for its water supply upon the amount of rainfall collected from its grounds—nearly 200 acres, including the roofs, are available as catchment area—and supposing a rainfall of 25 inches a year, which is the amount calculated by the authorities, the total rainfall is over 117 million gallons. Of course, only a proportion of this is caught, and the actual amount used is about 90 million gallons. Some of this is used twice over, so that the actual amount of water in circulation is less than this. In dry years the supply is not equal to the demand, and the mains of the Lambeth Water Company are drawn upon. The average quantity taken from this source is about 10 million gallons. In 1900 it was between 13 and 14 million gallons—an expensive substitute for rain water. Naturally enough, seeing the elevation of the Crystal Palace and the huge tanks at its southern end, the water company charges high rates for water supplied. In fact the company does not care much about supplying at all even at the price it charges—1s. per 1000 gallons. Taken at this price the average expenditure of the Crystal Palace on water is some 500l. a year, and the amount purchased forms but little more than one-tenth of the total consumption.

A reference to the annexed plan of the Palace grounds will serve to make clear the following explanation. The drainings from nearly all the land within the Palace boundaries fall by gravitation to the boating lake—No. 14 on the plan—which forms, as a fact, the main storage reservoir. Its area is some $5\frac{3}{4}$ acres, and its depth varies from 6 feet to 9 feet. Its surface is, at an average, 155 feet below the mean level on which the Palace itself is built. The tanks in the towers come some 280 feet above this again, a total difference of, say, 435 feet. The fall of 155 feet takes place in a distance of 800 yards. The slope of the hill is therefore very fairly steep. The ground, however, is of a retentive nature, holding the water to a large extent; moreover, the loss by reason of evaporation from the numerous surfaces of water must be considerable. Such surfaces, in addition to the boating lake, are distinguished on the plan by Numbers 1 to 7, 10, 11, and 15 to 19. The total exposed surface amounts to nearly 15 acres. The various basins, for the most part, drain into one another in succession, but all the overflow water eventually finds its way into either the boating lake or the cooling pond—No. 17 on the plan. In order that it can be utilised, however, it is necessary that the water should be raised to a height from which it can command every part of the grounds and buildings. Placed on the top of each of the towers is a large circular storage tank. These are 47 feet in diameter, and they are provided with inverted funnel-

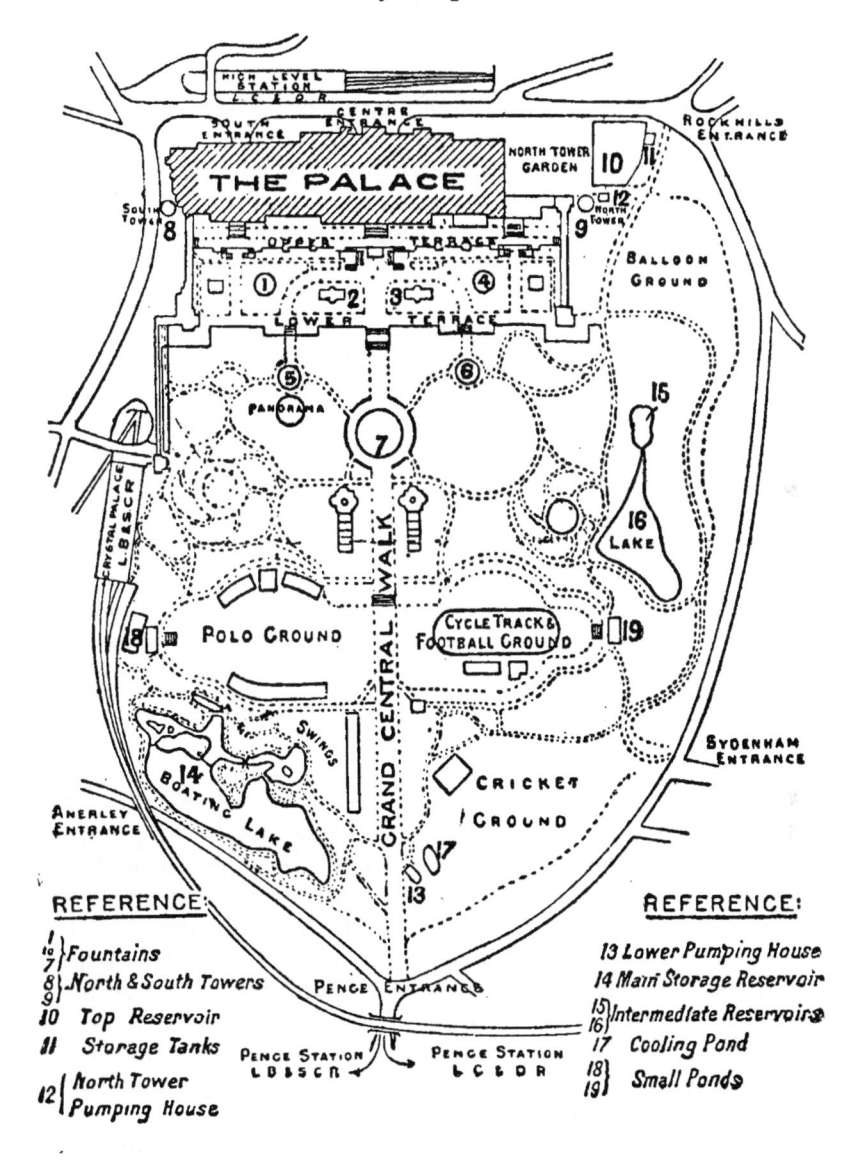

REFERENCE:

$\left.\begin{array}{l}1\\\text{to}\\7\end{array}\right\}$ Fountains
$\left.\begin{array}{l}8\\9\end{array}\right\}$ North & South Towers
10 Top Reservoir
11 Storage Tanks
$\left.\begin{array}{l}12\end{array}\right\{$ North Tower Pumping House

REFERENCE:

13 Lower Pumping House
14 Main Storage Reservoir
$\left.\begin{array}{l}15\\16\end{array}\right\}$ Intermediate Reservoirs
17 Cooling Pond
$\left.\begin{array}{l}18\\19\end{array}\right\}$ Small Ponds

shaped bottoms. The capacity of each is, with a depth of 35 feet, some 290,000 gallons. Some idea of the strength of these towers can be obtained by considering what this volume represents in weight—it is nearly 1300 tons.

The water in the tanks at the top of the two towers is

devoted almost entirely to reserve in case of fire, though some of it is used for blowing the bellows of the great organ, and some for actuating hydraulic lifts. The tanks are joined by two separate 8-inch cast-iron mains, which have fixed to them some fifty hydrants. The water in the two tanks, therefore, always stands at the same level, since there is open communication between the two. In these two tanks there is therefore a possible storage of nearly 600,000 gallons. The water is always at a pressure of about 120 lb. at the floor level of the Palace. At a height of some 80 feet above the level of the ground near the bottom of the north tower are two storage tanks, marked 11 on the plan. These tanks are each of them 48 feet square by 16 feet deep, and together they can hold some 460,800 gallons. There is, hence, a possible storage of water under pressure of over one million gallons, taking the towers also into consideration. As a matter of course, the pressure of the water from the tanks—which to distinguish them from those in the towers may be called the lower-level storage tanks—is less than that from the towers, being only some 35 lb. per square inch. A 15-inch main from these tanks is led through the building, where the water is used for general purposes. At the foot of the north tower is an irregular-shaped storage reservoir—marked 10 on the plan—which is capable of holding $5\frac{1}{2}$ million gallons. Its area is about $2\frac{1}{4}$ acres, and it can be filled to a depth of 9 feet. The water from this reservoir can be pumped into the low-level tanks, or into those in the tower, and it is also available for the fountains, but not, of course, for the buildings, since its level is too low. It is, however, 50 feet above the highest fountain, and 124 feet above the lowest.

There is yet another source of water supply at 13, at the bottom of the plan, which number marks the lower pumping station. Here there is a well, which is sunk through the superincumbent strata till it meets with water-bearing greensand at a depth of some 300 feet, and chalk at some 360 feet below the surface. Its total depth is about 250 feet, but three bore holes, 10 inches in diameter, are sunk from the bottom of the well to a further depth of 250 feet, making the total depth to the bottom of the bore holes some 500 feet. The bore holes are cast-iron lined down to the greensand. The well is brick-lined for its full depth of 250 feet, and is 8 feet 6 inches in diameter. At a depth of 150 feet there are four headings at right angles to each other. These vary in length, the longest being 30 feet, and are 8 feet 6 inches in diameter. Down in the well, at a depth of about 150 feet, is a set of three-throw pumps made by Hunter and English. The delivery from these is 4 inches in diameter, and is taken up to the reservoir

—No. 10 on the plan—and a branch is also taken to the lower-level tanks. The ordinary water level in the well is at a depth of some 95 feet from the surface, and the totals of the lifts to the reservoir and the tanks are respectively 268 feet and 350 feet. The yield of the well is about 40,000 gallons a day, and when it was originally sunk it was intended to use the water from it for drinking purposes. It was found to be somewhat hard, however, as well as being rather saline—though otherwise of excellent quality. It is now only employed for general purposes. The drinking water used is obtained from the Lambeth Water Company's mains, and there is a tank near the south tower which is capable of containing 12,000 gallons of this drinking water. A 4-inch main takes the supply throughout the building. The amount used during the year is some 4,000,000 gallons.

The pumping machinery employed is of an antiquated pattern, and most of it dates from the earliest days of the Crystal Palace at Sydenham. In the north tower pumping-house, which is situated close beside the foot of the north tower, there is a pair of engines made by William—now James—Simpson and Co. These are of the beam type, and each has one steam and two hydraulic cylinders. The steam cylinder is 25 inches diameter, with a stroke of 42 inches. The pumps are of the bucket-and-plunger type. The larger pump cylinders —which are used for lifting the water from the reservoir to the low-level tanks, a height of, say, 83 feet—have 19-inch buckets and 16-inch plungers, with 51-inch stroke; and the smaller cylinders—which are used for pumping the water from the reservoir to the tank at the top of the tower, a height of, say, 245 feet—have 14-inch buckets, 11-inch plungers, and a 28-inch stroke. Both these hydraulic cylinders are, in the case of each engine, worked from one end of the beam, but only one of the cylinders is in use at one time. Thus, it is not customary to pump into the tanks and up to the top of the tower at the same time. These engines bear the date 1853. They are in a wonderful condition, and are still at work. The steam to drive them is obtained by two Galloway boilers, 20 feet by 6 feet, which were put in in 1875, and work at 30 lb. pressure. The engines work condensing, the condensers being of the jet form. A vacuum of 26 inches is obtained. The horse-power developed at 22 revolutions is about 70 from both engines.

Down at the lower pumping station there is a cross-coupled double horizontal pumping engine. This is used for pumping water from either the boating lake or the cooling pond into the reservoir or into the lower tanks. The engine was originally made by Cox and Wilson—a firm now extinct—in 1853-4.

Subsequently it was altered—the pump rams being reduced in size—by Thomas Middleton and Co., in 1877. The two steam cylinders are 35½ inches in diameter, and have a 36-inch stroke. They drive direct the piston-rods of two double-acting pumps —the cylinders of which are 15 inches in diameter. By means of curiously-shaped fork connecting-rods, the two piston-rods are also connected to the two cranks of a shaft carrying a fly-wheel, placed at the opposite end of the engine to the steam cylinders, and a jet condenser is worked from a crank connected to the piston-rod. The level of the boating lake is some 16 feet above the level of the pumps, and therefore, ordinarily, the pumps have the water supplied to them under this head of water. If, however, the pumps are drawing from the cooling pond, as sometimes occurs, this is not the case, and there is, on the contrary, a slight lift. The engine makes 14 revolutions a minute, and the pumps, as already stated, are double-acting. In a trial carried out not very long ago, this engine pumped 198,000 gallons of water into the low-level tanks—a lift of 253 feet—in 3 hours 17 minutes. A 21-inch main runs from this engine-house to the reservoir, No. 10—a distance of some 1150 yards—and there is also a connection to the lower-level tanks. Jet condensers worked off the piston-rods are fixed under the engines, and the condensing water is obtained from the cooling pond—No. 17 on the plan.

The overflow, which in times of heavy rain takes place from the boating lake and cooling pond, is led away by a culvert which passes out of the Palace grounds near the Penge entrance, and eventually finds its way into the Ravensbourne. It would be a great saving if the storage capacity in the grounds were greater, for then much, if not all, of this waste of water could be prevented. The boating lake, however, is artificial and not of natural formation, and it would be impossible to raise safely its banks without very considerable expense. Things are now, as a fact, rather worse in this respect than formerly, since the cycle track, polo ground, etc., have been constructed on the sites of what were once large reservoirs. These in their time did much by acting as subsidence reservoirs to clear the water by allowing time for sedimentation before it reached the boating lake. Now, however, in addition to a smaller storage capacity, the water in times of rain reaches the boating lake in a very turbid condition. In this condition, of course, it is sometimes necessary to use it.

The engine for working the well pumps was made and installed by Easton and Amos—now Easton, Anderson and Goolden—in 1853. It is a compound vertical engine, working on to a beam. The cylinders are 13 inches and 22 inches, and the stroke of the low-pressure cylinder is 50 inches. The length of the crank is 25 inches, and the engine makes 21 revolutions a minute. The pinion on the crank shaft has 50 teeth, the spur wheel on the pump shaft has 92 teeth, and the speed of the pump shaft is about 11 revolutions a minute. The pumps are three-throw, the three barrels being each 7 inches in diameter, and the stroke is 2 feet. Both this and the larger pumping engine are supplied with steam from two Lancashire boilers, 30 feet by 7 feet, made by Jones Brothers, of Millwall. These work at 40 lb. pressure, and it is curious to note that while the well engine is compound and condensing—the jet condenser being used—the cylinders are not lagged, and the pressure is only 40 lb. at the boiler.

A certain amount of the water, when it has been used for general purposes in the grounds, eventually finds its way back to the boating lake, and the cycle of operations is repeated. The greater part, however, is lost by evaporation, absorption into the ground, or by being delivered into the sewers or the main culvert. In spite of the antiquated nature of the pumping machinery it runs smoothly and well. The engines certainly have not the appearance of having been at work for nearly fifty years, and of having done a fairly heavy duty all the time. Of course, as compared with a water-works, the Crystal Palace water arrangements are not on a very extended scale, although there are at least 25 miles of piping of one sort or another connected with the water supply. Some of these pipes are as much as 3 feet in diameter. What is remarkable, however, is that there should be all this machinery in connection with one single establishment, a circumstance which appears to be unique.

The Centre Transept Roof.

Forty-eight years ago, namely in 1854, the Great Exhibition building of 1851 was reconstructed on the Norwood Hills, at Sydenham, as the well-known Crystal Palace. The glass roofing, of which there is about 14 acres, is on the ridge and furrow system of Sir Joseph Paxton, the designer of the original Exhibition building. The glass was placed in position upon the main framework of the roof by the alternate fixing of glass and sash bar, the latter resting upon the ridge timber at the upper end and upon the Paxton gutter at the lower. The glass was pressed home into grooves formed on either side of the sash bars, which were coated with paint immediately before the glass was placed in position, the paint successfully taking the place of putty.

The maintenance of this roof in efficient repair proving a source of considerable expense to the Crystal Palace Company, it was resolved, about three years since, to re-glaze the whole of the roofs, commencing with the great arch over the centre transept of the Palace, upon the "Eclipse" principle of Messrs. Mellowes and Co., of London and Sheffield. To this end the iron main framing of the roof was left unaltered, but the old Paxton gutters and ridges were removed and replaced by new ones of wood covered with lead. The sash bars consist of light T steel, entirely encased in a drawn coating of lead with an admixture of tin, as shown in the annexed engraving, which represents the bar full size.

These bars spring from gutter to ridge, in place of the old timber sash bars, the lead being drawn with projecting strips on either side for embracing the edges of the glass. The sheets of glass used measure 51 inches by 18 inches, instead of 49 inches by 10 inches as formerly, and the weight is 26 oz. per square foot, with 31 oz. glass on the crown instead of 16 oz. and 21 oz. In fixing the new glass neither putty nor paint has been used, and none will be required in maintaining the roof, the glazed area of which is about two acres. The lantern which surmounted the old roof was removed, and the new roof is completely semicircular. There is now practically no external surface which will require painting, and should any glass need replacing it can be done with a minimum of trouble and expense. The cost of maintenance is thus greatly reduced especially when the comparative inaccessibility of the roof is taken into consideration.

The New Graving Dock at Southampton.

The graving dock now being constructed for the London and South-Western Railway Company at Southampton will be the largest of six dry docks belonging to the company, all of which are in constant use. About eight years ago a somewhat smaller dock (but at that time the largest in the world) was opened by His Majesty the King (then Prince of Wales). Since that time the size and number of vessels frequenting the port of Southampton has been steadily increasing, and it has become necessary to provide more dry dock accommodation for vessels of the largest class. Accordingly, towards the end of 1899 the Directors of the Railway Company instructed their Consulting Engineer, Mr. W. R. Galbraith, M. Inst. C.E., to prepare designs for the present work, the contract for which was let to Messrs. John Aird and Co. early in 1901.

The dock will be approached by vessels directly from the estuary of the River Test, which forms one boundary of the dock estate. The river channel is being deepened by dredging to form an ample approach to it. The dock itself will be 860 feet long (clear inside gates), with provision for extension if necessary; 90 feet wide at entrance, with the same bottom width inside; and 125 feet wide at cope level. The whole depth from cope to floor will be 43 feet, giving a depth of water over the keel blocks of from 29 feet 6 inches neap tide to 33 feet spring tides at high water. It is built almost entirely of Portland cement concrete, the floor, altars, and walls being faced with this material. The bulk of it is made in the proportion of 8 to 1, the facing to walls and floor, together with the culvert linings, being somewhat stronger, generally 4 to 1. A skin of 4 to 1 concrete is also laid on the underside of the floor, and carried up the back of the walls, so as to prevent any water which may accumulate there from soaking through the more porous concrete into the dock. This 4 to 1 concrete is made with special care, the large stones in the gravel being reduced in a crusher. A mass of 5 to 1 concrete is built at the skew-back, where the walls and floor meet, to resist the heavy crushing stress at that point.

The altars, seven in number, are all grouped near the top. They are 2 feet 6 inches wide and only 2 feet 9 inches high, and thus very safe and convenient to work on. Access to the floor is given by eight flights of steps, and four slides for timbers are provided. The entrance gates, which will have a span of 90 feet and a rise of 16 feet 9 inches, will be of steel, and will be worked by direct-acting hydraulic rams. The hollow quoins and sill quoins will be of granite from the Shap Quarries in Westmorland. The steps and timber slides will also be faced with this stone.

For emptying the dock the water will fall into large pits near the entrance. From these three large culverts will lead it to the pump wells, placed at some little distance behind the eastern wall, and at a depth of 10 feet below the dock floor. Over these the pump house will be built, and will contain two 48-inch centrifugal pumps, capable of emptying the 85,000 tons of water in the dock in a little over two hours. Space will be provided for a third pump in case the dock should be lengthened. The boiler house will be built alongside the pump house.

The site of the dock originally formed part of the extensive mudlands on the River Test shore, which were covered by every high tide. To reclaim the site a bank was tipped round it with chalk taken from the Railway Company's cutting at Micheldever. While tipping this bank a dredger was at work within removing the top mud, which was very soft and foul. The sea-face of the reclaiming bank was pitched with stone to preserve it from storms, and covered with clay to render it water-tight. The latter process was a very troublesome one, owing to the difficulty of finding suitable clay. A supporting toe of 3-inch sheet piling was provided in places where necessary. As soon as the bank was sufficiently sealed to keep out the tidal water the enclosure was dried out with a 9-inch centrifugal pump and engine, erected just inside the bank.

Meanwhile a 1 in 12 incline road, with a winding engine, had been built at the bow end of the dock, and as soon as the site was dried, excavation by hand gangs, by steam crane, and by steam navvy was commenced, the earth wagons being hauled up the incline and emptied on to other mudlands on the dock estate. The surface was thus lowered to a depth of from 26 feet to 30 feet below cope, at which level it was decided that timbered trenches should be commenced, in which the dock walls should be built.

At the entrance end the chalk enclosing bank approaches very near the dock, and protection was given to it by driving a complete belt or dam of 12 inch sheet piling across the entrance, which in its turn is supported by raking struts. Inside this again a cross trench was sunk, so as to build the first 20 feet of dock floor, which now forms a massive toe for the timbering. The arrangement of this timbering is worthy of notice. The wall trenches are now being sunk and timbered in a substantial manner, and directly a length is sunk to its full depth the cenerete wall is started inside it. When these walls are completed the dumpling between them will be excavated in a similar fashion and the concrete floor built in, length by length. The fexcavation for the pump wells, etc. is being similarly dealt with, the trench being heavily piled to avoid all risk of dislocation. Any water which finds its way into the enclosed area is led by pipes and grips to one or other of two pumps, which are provided to keep the works dry.

The Old Extension Quay Widening.

The Old Extension Quay was built in 1875, and has been used principally by the large vessels of the Union Castle Line. It is a Tidal Quay, well equipped with hydraulic cranes and good cargo sheds. and having a minimum depth of water of 20 feet at L.W.O.S.T. This depth, though at the time of its construction considered very ample, has of late years become quite insufficient to accommodate the large mail boats of this fine fleet. As the quay is close to the repairing factory and stores of the Union Castle Co. and so very convenient to them, it was thought better to deepen it rather than to construct another quay elsewhere. To effect this it was resolved to dredge to the required depth (30 feet below low water), at a distance of 50 feet from the existing wall, and to cover the intervening space with a platform, which would serve to widen the existing quay to deep water, so the vessels could still lie close alongside.

The design of this platform was the next problem. It is found that timber is very soon destroyed in Southampton Water by the ravages of the wood shrimp, which in a few years eats a large timber through. Accordingly it was decided to build the platform of ferro-concrete on the Hennebique system. Piles of this material are built up in vertical moulds, in which are placed the long steel rods, which really give the required strength. These are laced together with wire stirrups, and Portland cement concrete of excellent quality is carefully put into the moulds and rammed round the steel. After a month the pile is taken out of its mould and conveyed to the quay, where it is driven in position much like a timber pile would be. The ram is exceptionally heavy, however (30 cwt. generally). The head of the pile, too, is protected from being bruised by covering it with a helmet or iron case filled with sawdust. Moreover, a timber dolly is always used. The whole process is novel and interesting.

The work is being carried out in three lengths, so as to

interrupt traffic as little as possible. Over the first length a temporary timber stage has been erected, on which two large Lacour pile-driving engines for the permanent work have been placed. Each engine is provided with an 8-ton steam winch with which to pitch the piles and to lift the ram and other gear into position.

In the present work it was found impossible to drive the piles down to the required depth, especially in the front row, where they had to go 22 feet into the ground through gravel and hard sand. To overcome this difficulty the water-jet system of sinking was introduced. As beds of gravel and clay occurred it was impossible to sink with the water-jet alone. Accordingly arrangements were made to drive and pump simultaneously. The water is fed down a $\frac{3}{4}$-inch pipe buried in the centre of the pile and ending in a $\frac{3}{8}$-inch nozzle at the point of the shoe. It is supplied from the hydraulic pressure mains, containing water at 750 lb. per square inch. This pressure is reduced by throttling to about 300 lb. per square inch for this work. The result has been entirely successful, and the piles are driven to their depth in less than an hour. Progress is retarded, however, by the fact that the piles and their guide timbers can only be fixed at low tide.

When the piles have been driven the concrete round the head is to be stripped off, and the steel rods for the various beams and struts laid in position. Each set of rods is laced with wire or hoop steel, and surrounded with a timber casing. Into this casing concrete is poured and rammed, and each beam or strut thus built up *in situ*. A flooring of rolled joists and timber decking is laid on top of all. This will contain rails for the trucks and for the hydraulic cranes.

The engineer for the new dock works is Mr. W. R. Galbraith, M. Inst. C.E., the resident engineer being Mr. F. E. Wentworth-Sheilds, Assoc. M. Inst. C.E. The contractors for the works are Messrs. John Aird and Co., who were represented by Mr. J. W. Landrey.

SOUTHAMPTON CORPORATION WATER-WORKS, 1290–1902.

The water-works of Southampton have a history going back to a more remote date than can probably be ascribed to any other such undertaking in this country, it being recorded that on June 16, 1290 (Edward I.), one Nicholas de Shirlee granted to the Friars Minor the right to take water from a spring at Colwell (now called Spring Hill, Hill Lane) to Achard's Bridge, and thence by the King's highway to their church in the town of Southampton. It is further recorded that upon the Feast of the Purification, 1310 (Edward II.), the Friars granted the use of the water to the town. On October 3, 1420 (Henry V.), they conveyed to the mayor and community of Southampton all their rights and title in the springs, conduit and pipes, and the water-works of the town have ever since, for a period of 482 years, remained in their possession.

The original vaulted chambers covering the Colwell spring and one of the old water-houses (adjoining St. Peter's Church), to which the water flowed, may yet be seen. On June 1, 1515 (Henry VIII.), another spring at Lobery Mead (now Grosvenor Square) was presented to the town by John Flemynge. The water was led to a water-house (which until recently could be seen in Waterhouse Lane), and thence to the still existing house, which was quite close to it. From this water-house lead pipes conveyed the water to the town, and, together with sundry wells of a purely local character, including the Houndwell Well, 1490 (Henry VII.), constituted the water supply until 1803. The first Act of Parliament was obtained in 1747, followed by others in 1803 and 1810.

About 1804 the No. 1 reservoir was constructed. It collected surface water from the Common by means of earthenware pipes, and a line of elm pipes conveyed the water to the town. This reservoir has been abandoned, and the banks levelled down.

About 1811 the No. 2 reservoir, and about 1832 the No. 3 reservoir, were made on the Common. They also collected the surface water, which was conveyed to the town by a line of 10-inch cast-iron pipes. They continued in regular use until 1852, after which, and until recently, the water was used only for road watering. They have now been converted into ornamental waters.

In 1838 the deep well on the Common was commenced, and the work was carried on intermittently until 1883, when it was finally abandoned at a depth of 1317 feet (842 feet in chalk), having involved an expenditure of 20,000*l*. for a yield of only 130,000 gallons per day.

In 1851 a supply of water was obtained from the River Itchen, at Mansbridge, the works comprising a brick and masonry-lined subsiding reservoir ($3\frac{1}{2}$ million gallons), a pair of Cornish engines and three boilers; at the same time the two upper reservoirs, Nos. 4 and 5, on the Common were constructed. The old supplies were then discontinued, and in 1865 the Mansbridge works were increased by the addition of a second engine and boiler-house, containing a pair of coupled rotative engines and five Lancashire boilers. An additional boiler had meanwhile been added to the older plant.

In 1884, the water in the River Itchen had become so liable to pollution that it was determined to abandon that source and obtain a supply from wells sunk in the chalk at Otterbourne, to accomplish which an Act of Parliament was obtained in 1885. The works were put in hand at once, and the original portion was completed by June 1888. In 1896 large additions were made, and yet further extensions are now in progress.

The Otterbourne works, when the extensions are completed, will comprise two wells and a shaft 100 feet deep, about 1500 feet of adits, a pumping station at a level of 90 feet above Ordnance datum, containing four compound beam pumping engines and three boilers; a workshop; softening plant, consisting of lime mills, lime cylinders, lime storage tank, mixer, softening tank, and filters; four lime kilns, a quarry with hydraulic hoist, waste deposit tanks, a railway siding, roadway, and seven cottages for the working staff. The area of land acquired is 13 acres, to which 35 acres and four houses are about to be added for protective purposes.

The quantity of water pumped during the year 1901 averaged 3¾ million gallons, and often exceeded 4 million gallons per day, the population supplied being about 77,000. The trade supplies equal over one million gallons per day. The capital outlay upon these works has been about 85,000*l.*, to which must be added about 12,000*l.* for extensions in progress.

The water is pumped from these works through a 24-inch main—about to be duplicated—to a covered reservoir on Otterbourne Hill, whence it gravitates to Southampton through 24-inch and 16-inch trunk mains, the surplus water going to the reservoirs Nos. 4 and 5, which were covered in in 1897.

The total capital outstanding upon water-works account is only about 170,000*l.*, and in connection with the very moderate working expenses incurred, enables water to be supplied for domestic purposes at the exceptionally low rate of 10*d.* in the £ on assessable value (without extra charge except for hose pipes), and for trade purposes at 8*d.* per 1000 gallons, there then being still a surplus sufficient to reduce the general district rate to the extent of 4*d.* or 5*d.* in the £.

At the time of the visit, there was being erected an additional independent compound rotative beam engine, of the Woolf receiver type, driving a high and a low lift pump. The cylinders are respectively 28½ inches and 38¼ inches diameter, with strokes of 4 feet 9 inches and 7 feet. The distribution of steam is effected, in the high-pressure cylinder, by means of Meyer's variable expansion slides, driven off the crank shaft and the radius rod of the parallel motion; and in the low-pressure cylinder by means of Cornish valves, driven, by means of tappets and cams, off the crank shaft. The normal speed of the engine will be 18 revolutions per minute, and the indicated horse-power about 125, when working against a head of 235 feet. The steam pressure is to be 60 lb. per square inch, and the coal consumption is guaranteed (under the pain of severe penalties) not to exceed 2·15 lb. of the best Welsh steam coal per actual horse-power of work done by the pumps themselves. The condenser is of the multitubular surface type, the whole of the water raised by the engine passing through it. The air pump is 22 inches diameter, with a stroke of 2 feet $6\frac{3}{4}$ inches.

The low-lift pump, which draws water from the bottom of a well and delivers it to the softening works, is of the bucket and plunger type, the bucket being 22 inches diameter with a stroke of 5 feet. The high-lift pump, which receives the water from the softening works and delivers it into the Otterbourne reservoir, is of the by-pass bucket and plunger type (having suction and delivery valves of exactly the same size and pattern), the pump piston being $19\frac{1}{2}$ inches diameter with a stroke of 7 feet. This pump delivers 10 per cent. more water than the low-lift pump, this quantity of water being pumped for use in the softening process. The pumps are driven from the outer end of the engine beam, the connecting rod centre being between them. This engine will be identical with the three already at work. The water-works engineer is Mr. William Matthews, M. Inst. C.E.

Joseph Baker and Sons' Works.

These works are situated in Hythe Road, Willesden Junction, upon a branch of the Grand Junction Canal, which connects them directly with the Thames. They are also within a few minutes of the London and North Western and Great Western Railway Companies' lines. The company have been in their present premises for about twelve years, and their shops are well equipped with modern tools and machinery by English and American makers. The manufactures carried on are chiefly machines for the bread, biscuit, chocolate and confectionery trades, also garbage destructors, and several other kinds of industrial machinery. The first, or old, building is 270 feet in length, and is built along the banks of the canal. At one end of this building are situated the forge and the coppersmiths' and boilermakers' shops. The central portion of the building is the machine and fitting shop. The company have adopted the principle of keeping their machine tools and fitters' benches as near together as possible, in order to avoid the necessity of removing the parts of machines from one portion of the shop to another. This is a one-story building, and at the further end is the packing and despatching department, which adjoins the wharf. On the further side of the wharf is the timber store and experimental shed. The cast iron and tool stores are arranged along one side of the main building.

The new shop is 230 feet long and has a gallery about 20 feet wide running round three sides of the building. The centre of this building is used as a fitting and erecting shop, the machines being ranged in double row against the sides of the building. In this shop there are several large vertical lathes, and a large assortment of modern tools by the best makers. In the gallery are constructed the lighter machines, biscuit cutters, engraved wafer plates, speed-reducing gears, and the like. The woodworking building is situated on the west side of the company's premises, and is 82 feet long by 52 feet wide; it is fitted with circular saws, planing machines, and various other woodworking machinery. The upper story is used as the patternmakers' shop, beyond which is the drawing office and photographic room. In the yard are the iron and timber stores, the centre of the yard being used as an erecting ground for travelling ovens and garbage destructor furnaces.

Baker's Speed-Reducing Gear.

At the date of the visit the company had recently brought out Baker's patent speed-reducing gear, of which the following is a general description, accompanied by a perspective view.

On the fast speed coupled to the motor shaft there is keyed a pinion, which gears with three loose or floating pinions, which in turn gear into an internally toothed ring. This ring, when not transmitting power, is free and revolves, but when power is to be transmitted the ring is held by a brake band and the intermediate pinions revolve inside the ring. In one piece with these intermediate pinions are three other pinions which gear in turn into a second internally toothed ring, which is held in a disc, the slow-motion shaft being keyed on to the boss of this disc. By this arrangement of gearing a very high ratio of reduction may be easily attained. The special centripetal oil circulating disc with which the apparatus is furnished ensures complete lubrication of every part of the gear, and assists in producing the high efficiency of 85 to 95 per cent. shown in repeated tests. The pressure on the teeth of the wheels is small, as they all run at a high speed, and this conduces to the good working of the apparatus. The Baker gear can be driven at the highest possible motor speed. Owing to the balanced drive, all side strain on the driven shaft is avoided, and there is no tendency to bind. The lubricating device ensures a constant flow of lubricant to all the working surfaces, and the gearing being cased in, is protected from dirt and dust. The motor can be started independently of the gear, and the load gradually put on by the hand-wheel or lever. In the same way the gear can be stopped, and the motor kept always running if

desired. A speed-reducing gear was fitted up to a motor, and testing gear applied in order that the members of the Society might see the gear at work and its efficiency demonstrated.

The Wicks Machine Works.

These works are remarkable on account of the conditions under which the resultant machine operates with an almost unprecedented accuracy. The idea of constructing an apparatus in a temperature 60° Fahr., that should work in conjunction with molten metal at a temperature of over 700° Fahr. and produce casts with a limit of error of only two ten-thousanths of an inch, is a problem that few engineers would attempt to solve. In these works, however, the problem is not only solved, but the resultant machine produces automatically millions of these casts with an invariable accuracy, and without any appreciable wear of the machine itself. As an instance it may be stated that a single machine has been running continuously for two years and has produced 200 million types, most of which have been used in the production of the *Times* newspaper, and the machine is still running.

The engineering works of the Wicks Rotary Type-casting Company occupy a considerable area, and include a long two-story building, standing back from the road and running parallel with and closely adjacent to the railway, with office and various smaller buildings in a large front yard. These detached structures are specially arranged for certain sections of work demanding a special light, freedom from vibration, and other conditions of an unusual character. The ground floor of the main building forms the general machine shop. It is lighted on both sides, and fitted with benches under the windows, whilst a large number of machine tools are arranged in rows on the floor, which is paved with wood blocks. In the section containing the heavier machinery, the work connected with the bases and driving gear of the casting machine and the construction of the pumps is carried out. Here are some high-class milling machines, with shaper's lathes, planing machines and a radial drill. In the centre of the building, and separating the two main machine shops, are the engine room and the electric plant room. The former contains a couple of powerful gas engines. The electric plant consists of a 110-volt continuous current dynamo with switches and batteries, the arrangement being such that the external load of lighting and supply to motors driving light machines is performed by the dynamo whilst charging a battery of sixty-one E.P.S. cells situated in a separate room outside the works. The batteries enable light machinery to be run independently of the main engines. A smaller dynamo working up to 4 volts supplies current for the electric deposition of the formers from which the punches are cut. A compressed air plant driven by the main gas engine supplies air for working blowpipes and for other purposes. Adjacent to the electric plant are the store-rooms. The grinding section of the shop includes a Brown and Sharpe Universal grinder, a Universal cutter-grinder, and kindred appliances. In the mould wheel shop, the wheels are finally turned up and fitted with the segments which form the moulds. Many of the appliances used for this work have been specially designed. One of the most noteworthy parts of the Wicks machine is the delivery chain which is fitted with sliding leaves for receiving the type as it leaves the wheel and delivering it later. This chain is composed of intricately shaped links, the details of which at first sight appear almost impossible to produce by machine, but in consequence of the adoption of special appliances there is practically no handwork on these parts, except the removal of the sharp edges left after machining.

On the upper floor of the building are the punch-cutting, justifying, and the matrix departments. The hand punch-cutters have a room to themselves, possessing the best possible light, and hard by are other skilled workers engaged in the operations of justifying matrices and testing. The making of a matrix involves an extended series of operations requiring many machines and tools and a number of workmen. Close by is the drawing office, in which a pantograph and apparatus for coating plates with wax are noticeable, for here are made the drawings which serve as the basis for originating founts. Drawings to scale are made from the artist's sketch many times larger than the largest size of the destined type; these are copied in wax by means of the pantograph and then electrotyped. In some instances the formers are cut from solid metal by a revolving cutter. A large room is devoted to the Wicks type-setting machine, now in process of development. Beyond this is the pattern-making and joinery section. The most important shop in the yard is the machine punch-cutting department, where the punch-cutters are at work. The machines at which they are occupied are vertical cutters, each being separately driven by an electric motor. An adjoining building forms the testing shop, where the machines are finally assembled before being put to work. Its equipment includes a gas-engine, with base plates, metal-pot and pump, so that all the regular working conditions of the type-casting machine are ensured. The whole of the works is lighted by electricity produced by the company's own plant. The heating is effected by a hot-water system, much importance being attached to keeping the shops at a proper temperature, this being essential to accuracy of work.

THE WICKS TYPE-CASTER.

The type-casting machine, which is the invention of Mr. Frederick Wicks, is mounted on a heavy bed-plate, to which is secured a cast-iron fire-box carrying a metal-pot, in which the pump is fitted. The pump body is immersed in type-metal, which is kept in a molten state by gas burners beneath the pot. The pump is driven by a belt from underground shafting from which the machine is driven. The metal-pot, which holds 12 cwt. of metal, is fitted with a pyrometer. The pump has four plungers which are driven by eccentrics; the molten metal is sucked in through inlet-valves and delivered into a pipe fitted with a device for regulating the fluid pressure at the nozzle.

The machine proper consists essentially of a mould wheel, cam head, shield, the matrices and the delivery chain, and galley. The mould wheel is a disc of cast iron on which the moulds are built out of separate pieces of steel in such a manner that the slots formed by these are accurate for size of type to be produced within one-thirty-thousandth of an inch. The wheel and segments together form three sides of the mould. In the moulds the matrices slide and form the end of the mould nearest the centre of the wheel. As the wheel revolves on its vertical axis the moulds are in turn presented to an orifice in the shield against which the stream of molten metal from the pump is playing. For some 3 inches each side of this casting point the moulds pass under a top cover, which forms the fifth side of the mould, the sixth being formed by the shield itself. As each mould is filled, and the wheel rotates, the matrix is acted on by cams which cause it to advance towards the periphery of the wheel and so eject the type from the mould.

In the latest form of machine, which is illustrated herewith, the types are cast with one or more deep nicks, an improvement upon the earlier practice. When the type has been ejected about one-sixteenth of an inch, it passes in front of a nicking saw, which cuts other smaller nicks in the heel; after rotating some further distance, the ejection continuing meanwhile, other nicks are cut by saws revolving on a horizontal axis; following the type further round the wheel, the ejection continues, and when half a revolution has been completed it has been received on the delivery chain, from which it passes under a retaining cam which prevents its return into the mould with the matrix, and having left the wheel entirely it proceeds with the chain into the delivery galley. On each wheel 100 moulds are formed of different sizes corresponding to the "set" or width of the letters, and the varying demand for letters of the respective "sets." The delivery chain consists of 100 links in which sliding leaves are fitted, which rise up to receive the type and to support it on both sides until it arrives in the galley. Each of the chain links comes opposite the same mould in turn. When the type has been delivered on the chain, it proceeds between the leaves until it enters the delivery galley, when the leaves fall and the type is pushed along the galley. One man and one boy attend to each machine, the latter taking off the type and transferring direct to galley. The machine casts 100 types of all sorts and sizes at a single revolution, delivering them finished and arranged in line ready for use. No heating-up is required, finished types being produced at once. The average output of each machine is 60,000

finished types per hour, this being a rate of delivery twenty times greater than that of the ordinary machines, and, as stated, at one-twentieth the cost of production.

THE WICKS TYPE FOUNDRY.

Besides the machine factory at Willesden, the Wicks Company have an extensive type foundry at 154 Blackfriars Road, London, where type is produced for the market. The front portion of the building comprises the general offices, whilst to the rear is the type foundry, which covers an area of 8000 square feet. It consists of a ground floor and two spacious galleries running around the entire building. The machines are arranged on the ground floor, and are driven by underground shafting. The products of the machines are conveyed by a lift at each end of the foundry to the galleries to be there dealt with. As the types are produced they are arranged in blocks of about 15,000 types, and sent to the sorting department, where they are sorted and packed for the market. By the courtesy of the Wicks Rotary Type-casting Company, the members of the Society were invited to inspect the type foundry, of which invitation a number of members and associates subsequently availed themselves.

October 6th, 1902.

PERCY GRIFFITH, PRESIDENT, IN THE CHAIR.

THE HENNEBIQUE SYSTEM OF FERRO-CONCRETE CONSTRUCTION.*

BY AUGUSTUS DE ROHAN GALBRAITH.

INTRODUCTORY.

INASMUCH as ferro-concrete construction does not appear to be so generally known as its merits justify, the author ventures to hope that the present paper may prove of service to those who are, or may become, engaged on works in one or other of the many directions in which this class of special construction is applicable.

History.—Ferro-concrete, as the name implies, is the combination of iron, or more strictly speaking, steel and concrete. The use of iron, such as hoop iron bonding in conjunction with concrete is of ancient date. Napoleon the Great utilised these materials in the erection of the fortifications near Strasburg in Alsace Lorraine about the end of the eighteenth century, this being it is believed the first authenticated instance of this kind of construction. The more scientific application, however, of this combination to the many uses to which it is now applied is a comparatively modern development. The use of concrete in buildings was in vogue with the Romans, 509 B.C., and the Moors of Granada, in the year 1248, built the walls of the Alhambra of lime concrete run into wooden moulds.

Initial Experiments.—More than 17 years ago experiments on an exhaustive scale were made in Great Britain, the United States and Germany with the object of ascertaining the best medium to protect iron and steel work against the action of fire, the oxidising effects of the atmosphere, and other deteriorating and destructive agents, with the result that it has been found that the most reliable and suitable material for the purpose was Portland cement concrete. At the same time the amount of concrete necessary to completely envelop the metal in a

* The Bessemer Premium was awarded to the Author for this paper.

structure had serious drawbacks : it increased the weight and bulk and consequently the cost; therefore engineers had to solve the problem of the proper disposition and economical correlation of the materials, so as to best withstand the stresses and strains they were called upon to sustain. It is the various solutions of this problem which are responsible for the many different systems that have been evolved and brought before the public. Ferro-concrete was first applied in the construction of slabs, pipes, etc., in a tentative manner, but as the results of the experiments were so successful and the enduring nature of the materials so evident, Continental engineers and architects speedily felt themselves justified in giving its application a far wider scope, with the result that its use was quickly extended to the building of floors, bridges, arches, and the like.

Different Systems.—The original invention, as a system, is generally ascribed to Joseph Monier, in 1868, and is called, after him, the Monier system. It comprises a network of round iron rods of small section placed near the underside of the concrete. Besides this system there are several others, amongst which are the following, which are in use on the Continent and in America :—the Cottancin, with its wavy cross wires and straight longitudinals, the Hennebique, the Expanded Metal, the Bonna, the Mattray, the Mueller, the Klett, the Armoured, and Columbian systems. The difference in these or any systems lies only in the details of their construction, the principle being the same : the concrete being used where it will display its characteristic resistance to compression and the iron and steel being only utilised in tension, the desideratum being to produce a homogeneous structure with two heterogeneous materials. Owing to its importance, and in view of its adoption in this country during the past five years, the author proposes to confine himself to the system invented by Monsieur François Hennebique, an eminent French engineer.

The Hennebique System.—This system is distinguished for its compactness and simplicity. Entire buildings, with piles for foundations, foundation blocks, piers, columns, beams, walls, floors and roof have been constructed on this system, and it would be difficult to mention a form of construction, either architectural or engineering, to which it has not been applied.

Such has been its success in France and Switzerland that last year it was applied to over a thousand structures, and in 1900 it was adopted in the erection of several important buildings at the Paris Exhibition. It was introduced into this country about five years ago, when a retaining wall was constructed of sheet piles driven 17 feet through hard sand and shingle at the northern end of the London and South Western Railway Company's coal wharf on the River Itchen, Southampton. Since then its progress in England has been steady and continuous.

APPLICATION OF SYSTEM—MATERIALS.

Constructions in ferro-concrete derive their value not only from the proper distribution of their compound elements, but also from the care which should be taken in the selection of the materials employed, and in the execution of the works. All materials must be of the best quality, and the closest attention must be given to their admixture and application. The sand, gravel and ballast must be selected with the greatest care on account of the important part they play in the compressional resistance of the work.

Sand.—Coarse sand, very gritty and of the greatest cleanness, must be used in preference to fine, but whatever the composition of the sand, it is absolutely necessary that it shall be entirely free from all traces of any argillaceous or earthy matter, and if no other quality is obtainable, it must be most carefully washed.

Gravel and Ballast.—These remarks as to cleanness also apply to the gravel and ballast. Should the gravel or ballast contain sand, it is advisable that they should be separated through a screen to be mixed again in their proper proportions.

Broken Stones.—If it should not be possible to obtain either gravel or ballast, broken stone of a very hard and gritty nature, such as flint, granite, porphyry, or crystallised silica, etc., may be used, but limestone, even of the hardest quality, should be avoided and absolutely prohibited in all buildings intended to resist fire. All gravel, ballast and broken stone must be screened before use through a ¾-inch mesh sieve, and stones of a larger size must not be permitted to pass into the work.

Cement.—The cement must be Portland of the highest class. It should not leave a residue superior to 10 per cent. in a sieve of 5800 meshes to the square inch, and it should give a tensile resistance of 400 lb. per square inch after seven days, and 500 lb. per square inch after twenty-eight days immersion in water. No fresh cement, or cement containing free lime, should be allowed, and the tests should be made almost daily. Every new consignment of cement received should be tested before use, and rejected if it fails to reach the required standard. It is also important to ascertain its age, and fresh or new cement should not be accepted for any work. Such cement may very well come up to the foregoing tests, but give very poor results in practice.

Concrete.—Experience has repeatedly demonstrated that in order to obtain uniformly good concrete or béton with sand and cement, it is necessary, firstly, to regulate in a systematic manner the amount of water employed in the manufacture thereof; secondly, to obtain with a minimum quantity of water the cementing material or matrix in a state of plastic or viscous paste; thirdly, to cause every grain to be lubricated with a thin film or coating of this paste; and fourthly, to bring every grain into close and intimate contact with those surrounding it.

Proportions.—The usual proportions adopted for ordinary work in the construction of buildings with their floors, pillars, etc., are as follows: gravel, clean and free from sand, of all sizes, mixed between $\frac{1}{8}$-inch and $\frac{3}{4}$-inch (gravel of a uniform size should be avoided) 27 cubic feet; sand, gritty and very clean, $13\frac{1}{2}$ cubic feet; cement of the best quality, 6 cwt. occupying a volume of $\frac{672}{81}$ lb. to $8\frac{1}{4}$ cubic feet. This mixture, when in place and properly rammed, yields a volume of concrete of 31 cubic feet, corresponding to a proportion of five to one. In piling, and works of a similar nature where impermeability is essential, the proportion of cement is increased to three to one, and as an extra precaution when piles are used for sea work, it is desirable to give them, when set, a good coating of cement grout.

Mixing.—It is better that the mixing be done mechanically, preceded by a thorough admixture by hand, first in the dry state and then in the wet, before going through the pug mill. When entirely mixed by hand it is well to make an increase in the proportion of cement by $\frac{1}{2}$ cwt. to $6\frac{1}{2}$ cwt.

The concrete should be in a plastic and not in a liquid or dry state. It should be used fresh, in small quantities at a time, and well rammed until the moisture rises to the surface. Care should be taken that the concrete be not made too liquid, as if so, its compressional resistance is considerably reduced, besides the invariable accompaniment of cracks in the setting. On the other hand, should there be insufficient water, its proper setting will be correspondingly affected.

General Precautions.—When the construction of a beam or floor, etc., has to be abandoned before completion the edge or end of the work must be roughed with a cutting tool, and thoroughly cleansed from all foreign matter before work is resumed on it. Pure fresh cement grout must be poured on the surface of the edge before commencing the concreting operation, and care must be taken to ram the fresh concrete well on to the old concrete. All concrete must be used up before work is suspended, and any concrete left over must not be again used.

Fresh concrete work should be freely watered for several

days, and when this cannot be done it should be kept in a moist state, especially where it is exposed to heat. To prevent concrete adhering to the moulds a coating of soft or yellow soap boiled and applied while hot will answer this purpose, besides giving a fairly good face to the concrete. Where an especially smooth face is desired, sheets of metal may be placed against the inner sides of the moulds.

In working to a smooth surface sand should be sprinkled over the concrete, as it greatly lessens any likelihood of peeling that may be occasioned by changes in temperature. Where a large surface of concrete facing is exposed to considerable heat or great variations of temperature, it should be cut up with $\frac{1}{2}$-inch wooden strips about every 10 feet, to prevent cracks that may occur from expansion and contraction.

Rendering on concrete is to be deprecated, as it is liable to flake off in patches, producing an objectionable appearance. For rendering the bottoms and sides of conduits, reservoirs and tanks thoroughly impervious to water, the application of concrete with a sheeting of bitumen to the wetted surfaces is economical and effectual.

The greatest care must be taken in removing centres and moulds, and the props and struts supporting beams, floors or walls should on no account be removed until the concrete has sufficiently set. No load should be placed on a green floor. As a rule the moulds or shuttering can be removed after two or three days, but it is usual to allow at least a month to elapse before the entire removal of the temporary supports, when the structure will then be ready for the tests. In turning concrete arches it is a good plan to do so in layers: as each layer is put in place it will assist the centre, which can therefore be of a lighter construction than would be necessary under other conditions. Arches generally have a rise of about one-tenth to one-twelfth of the span.

Iron and Steel.—The iron should not be too soft, for a high coefficient of elasticity is of value, and for this reason a hard steel is to be preferred. Owing to its purer and more uniformly good quality, Siemens-Martin steel is preferable to Bessemer. The bars are from 0·39 inch to 1·57 inch diameter, and should sustain a tensile stress of 28 to 32 tons per square inch of section with an elongation of not more than 20 per cent. in 8 inches. Nearly all bars are cleft at their extremities as an extra precaution for gripping the concrete. In work where they have to be forged great care should be exercised, as careless forging may so seriously injure the metal as to render it practically valueless. All rust should be thoroughly removed from the iron and steel work, as although the rust adheres to the concrete, its coherence to the metal is so slight in itself that there is a tendency of the bars to slip through the enveloping concrete through the medium of the rust. Should any work require to be welded, care should be taken in deciding upon the position of the welds. Most joints are formed by the overlapping of the bars, and all iron and steel both round and flat should receive a good coating of cement grout as it arrives from the makers. Care should be taken to see that the inside curve of the stirrups in all cases is in actual contact with the underside of the bars, and the prescribed position of the stirrups must be always strictly adhered to.

General Construction and Principles.

Stirrups.—The most prominent feature of the Hennebique system are the elongated U-shaped wrought hoop iron straps called stirrups (see Fig. 1). They are 1·18 to 2·36 inches wide and ·0945 inch thick. These are designed to take the shearing stresses, as it is well known that concrete is ill adapted to withstand any strains of this nature. They are distributed at proper intervals along the beam, being passed underneath and in contact with the tension bars, and are continued vertically through the height of the beam, being turned slightly over and outwardly at their extremities, so as to obtain a firmer hold on the concrete. The upper ends of these stirrups project a little above the surface of the beams into the lower part of the floor to act as a tie and convey the strains well up into the concrete.

Beams.—In a ferro-concrete beam the axis of compression is at the upper side, and the axis of tension on the under side, similar to the action of these forces on an ordinary steel joist. (See Fig. 2, which is a section through A B Fig. 3.) The tension bars are of two kinds, viz. straight bars parallel to the under side of the beam embedded about one inch in the concrete, and cranked bars placed above them but in the same vertical plane. These cranked bars are in the form of a truss (Figs. 3 and 6), the bent portion being not quite one-third of the span, the upper bends being close to the points of support and carried generally into the floor slab. (See Fig. 6, which shows the junction of two beams and a pillar.)

The cranked bars acting in conjunction with the straight bars and the stirrups, which stirrups are placed closer together at the ends of the beam near the uprights, where the shearing stress reaches its maximum (Fig. 4), constitute a statical triangle, and somewhat resemble a timber beam trussed with iron rods and brackets. Fig. 5 shows a section through A B (Fig. 3) resisting the maximum bending moment. When head room has to be taken into consideration and the depth of the beam correspondingly reduced, compression bars can be introduced into the upper portion of the concrete for the purpose. Fig. 6 shows the respective position of the bars and stirrups, and the way in which any bending strains of the lower bars become transmitted to the upper part of the beam and transformed and distributed in the way of compressive strains into the mass of concrete. Fig. 7 is a cross section through this beam, showing longitudinally above it the floor constructed in a precisely similar manner to the beam. The depth of these beams is usually one-twelfth to one-fifteenth the span. Beams constructed on the foregoing principles will support the heaviest loads over large spans, 43 feet and 47 feet being not unusual; a beam at a cattle wharf at Liverpool has a span in the clear of 64 feet. It should be borne in mind that in all calculations for the strength of concrete beams with wires embedded therein, a portion of the concrete floor itself of a width equal to about one-third the distance between the adjacent beams should be taken along with the beams.

Tests of Beams.—In order to demonstrate the value of the stirrups in the Hennebique beam, two trial beams were made with exactly similar quantities of materials, but the one with and the other without the stirrups, and were subjected to similar loads until they showed signs of cracking. The beam without the stirrups was the first to show this weakness. In this beam the concrete broke away in lumps from around the tension bars, leaving them exposed, whereas under the same load the beam with the stirrups showed only small vertical fissures which were not sufficient to practically endanger its safety.

Foundations.—Foundation rafts and mattresses are formed in a similar manner to the floors, having due regard in calculating the proportions of the metal and concrete for the loads they may have to carry. Where it is not convenient to float the building, and the nature of the ground will permit it with advantage, piles are used with highly satisfactory results. Fig. 8 shows a foundation block, which is constructed of concrete with a network of steel bars crossing at right angles, the lower bars being supported at the intersections by the stirrups carried up to a point just underneath the surface of the concrete. At the plane of intersection of the block and the pier or column, a steel plate is inserted, the better to distribute the vertical pressure.

Columns, etc.—Columns, piers, pillars, etc., unlike piles, are generally moulded *in situ*, the rods being first placed in position and kept there by pieces of wire or flat iron distance pieces (Figs. 9 and 12) placed about fifteen inches apart. These rods terminate just over the plate in the foundation block previously referred to, and at their upper extremities are either connected to the column or superstructure by overlapping, or are cut off a little distance below the upper tension bars of the beam (Fig. 6). When connecting the columns to foundation piles the concrete surrounding the tops of the piles is broken away for the required length after the piles have been driven to the desired position. The rods of the columns are then placed as before described, the moulds fixed, and concreting commenced. These pillars and columns can be moulded to almost any shape, and they offer great resistance to oblique strains.

Floors.—In a floor supported by ferro-concrete beams main and secondary beams are employed. Fig. 10 shows a section of the main beam constructed to receive the heaviest loads, whilst Fig. 11 is the secondary beam which is connected up to the main beam, which latter receives a flat slab constituting the floor, and which is formed similarly to the beams. The bars are crossed at right angles in the lower portion of the slab, and covered with about half an inch of concrete. These bars are supported by stirrups, as in Fig. 7, for a short distance from the beams, but the stirrups are discontinued as the centre of the floor is approached, as they are not then required. These bars are sometimes crossed diagonally where lateral stresses are at all likely to be set up. The arrangement of main, and subsidiary beams and floor is shown very clearly in Fig. 12. The beams and floor form a monolithic mass, the whole series being joined together so as to form a continuous structure without scarf or joint. This form of floor presents a good ceiling for decorative treatment, and it can be either panelled and moulded in plaster or left plain, or simply whitened or coloured. For ordinary purposes the simple flat beam makes an excellent floor in itself, but for floors of warehouses, factories, or other large buildings, which have to sustain heavy loads, beams become necessary, especially when, in order to economise space, it is desirable to minimise the number of supports. These floors are well adapted for loads of 1·83 to 3·66 cwts. per square foot. A floor at the Co-operative Society's warehouse at Newcastle-on-Tyne, designed to carry a load of 6 cwts. per square foot, was satisfactorily tested to 9 cwts. per square foot.

Arched Floors.—Arched floors are constructed on this system, of which there are some very good examples in the Palais des Beaux Arts, Paris, one having a span of 18 feet with a 9-inch rise, 18 inches deep at the haunches, 3 inches thick at the centre; and another with the large span of 60 feet.

Walls.—Partition walls, and occasionally external walls, are constructed with vertical and longitudinal rods; the longitudinal rods being placed in the centre of the wall with the vertical ones disposed alternately on either side equidistantly apart, and tied to the further surface of the wall by stirrups (see Fig. 13). These walls are light in construction and of but slight thickness compared with their height, the walls of the highest buildings being seldom more than 9 or 10 inches thick. In the case of retaining walls the vertical rods are embedded only in that face of the wall which directly sustains the thrust.

Roofs.—Roofs on this system are generally constructed flat, similar to the floors, with sufficient cross falls to drain off the rain-water. At the Southampton Cold Storage Building the roof is to be utilised as a cooling tank, and consequently it will have a depth of two feet of water over the greater part of its surface equal to a load of about 123 lb. per foot super. The roof of the engine house connected with these buildings will carry the condensers belonging to the engines, and, further, the roof of the boiler house adjoining the engine house will also be used as a water reservoir. Domes, cupolas and similar forms of roof structure have been constructed on this system, notably one of a lanterned cupola over the main hall of Shepheard's Hotel, Cairo.

Lintels.—Lintels, sills, mullions and other small framings are generally moulded prior to their being fixed in position, the rods being tied together with wire distance pieces in the same manner as the columns. These are moulded and ornamented as required, the arrangement of the rods differing accordingly.

Piles.—Piles are constructed in vertical timber moulds supported by frames, the inner section of the mould corresponding to the size and shape of each pile. The working face of the mould is left open, care being first taken to see that everything is perfectly plumb. The steel shoe is then inserted in the bottom of the mould, with its upper ends turned over inwardly to form a key to the concrete. The vertical rods are then placed in position, about an inch below the surface of the concrete, and connected together with distance pieces dropped from the top as required. Concreting is then commenced, and the working face of the moulds is gradually closed with shuttering fixed about every six inches in height by the workman as he proceeds with the punning. After about thirty-eight hours the concrete is sufficiently set for the moulds to be stripped, and the piles are allowed to remain from twenty-eight to forty days to dry preparatory to driving. It is sometimes more economical and convenient to make the piles in horizontal moulds, but in that case the greatest care must be observed in obtaining the right consistency of concrete, so that in the punning operation the cement be not worked out too much to the upper surface.

Fig. 14 is a typical example of a sheet pile in elevation, transverse and longitudinal sections, showing the disposition of the steel work and drifted shoe. Fig. 15 is a plan of the same pile, showing the arrangement of the distance pieces, stirrups, etc. These piles are fitted with a semicircular groove on either side, which extends from the upper end of the shoe to the top of the pile; and at the lower end of the pile, on its longer side, is fixed a metal spur which fits into the groove of the pile preceding it, and acts as a guide in driving. After driving, these grooves are carefully cleaned out by a water-jet and filled with cement grout, forming a solid watertight joint between the piling. These piles are made in lengths of 46 feet and 48 feet, and have all the resiliency and elasticity of timber piles. As an instance of this, a 14-inch by 14-inch pile, 43 feet long, suspended in the middle, will bear a deflection of from $3\frac{1}{2}$ inches to 4 inches, and, unlike timber piles, they can be easily lengthened and joined to the adjacent work.

According to the nature of the ground the piles can be driven hydraulically, or by the ordinary pile-driver, or a combination of both, but the Lacour pile-driver, with its quick succession of blows combined with a heavy ram and short drops, gives the best results. Piles 14 inches by 14 inches have stood being driven by a 2-ton ram with a 6-feet drop, and at Southampton Old Extension Quay, they are being driven 22 feet through a stratum of gravel and hard sand on the combined system, as owing to the presence of clay and fine gravel it was impossible to sink with the water jet alone. The water is conducted from the hydraulic mains down a $\frac{3}{4}$-inch pipe in the centre of the pile, terminating in a $\frac{3}{8}$-inch nozzle at the point of the shoe, the working pressure being throttled down to about 300 lb. per square inch, the piles being driven to their required depth in less than an hour.

In driving, the heads of the piles are protected with a steel helmet (Fig. 16), the space between the helmet and the head of the pile being filled with sawdust. A tight joint of hemp and clay is made between the lower part of the helmet and the pile to prevent the escape of the sawdust, which becomes so compressed after the driving as to be almost solid. Occasionally fine sand is substituted for sawdust, and a timber dolly is interposed between the ram and the pile head in the usual way. These piles are especially adaptable for foundations of piers, bridges, and locks of docks, saving the heavy cost of cylinders, and in the case of masonry piers the necessary but expensive cofferdams and pumping. They are sometimes moulded hollow and applied to foundations in rocky ground.

Cantilevers.—Cantilevers are built on the Hennebique system with very successful results, especially in the quay constructed at Chantenay Nantes on the river Loire. These cantilevers have an overhang of 24 feet 7 inches, and support a platform 27 feet wide, being anchored to the warehouse erected at the rear of the wharf, the warehouse thus practically supporting the quay itself. They are calculated to carry two cranes of 20 tons, a goods train and locomotive, besides a load of 1 ton per square yard. The quay was tested by rolling a truck loaded with 42 tons along the platform. Cantilevers have also been constructed with an overhang of 11 feet 5 inches at the new grain warehouse, West Wharf, Plymouth, and were successfully tested to carry a load of 4½ tons per lineal yard.

Cylinders.—Ferro-concrete cylinders are constructed in various lengths similarly to the piles, but their chief advantage over the ordinary cylinders lies in the fact of the hearting being dispensed with, the whole weight being taken by the cylinders themselves. They are sunk in the usual way, but owing to their extra weight they do not require so much pressure to sink them as ordinary cylinders. Diaphragms are built in and fastened to the inside of the cylinder at intervals, according to the size and supporting power required, and the usual methods of excavating in ordinary work of this nature can be adopted.

Adherence Tests.—According to experiments carried out by Professors Baushinger and Ritter, the adherence between the steel and the concrete is estimated at about 570 lb. per square inch. It has been thought that the adhesion between the iron and the concrete would be destroyed by variations of temperature, but the researches of M. Durand Claye, Chief Engineer of the French Bridges and Highways, show that the coefficient of expansion and contraction for both materials is identical to the fifth decimal.

Proportions of Materials.—The proportioning of the iron and concrete is dependent upon the condition that the former must have sufficient sectional area to sustain all the tensile stresses in the beam or slab, and the disposition of the iron bars is found from the fact that the value is in direct proportion to their distance from their neutral axis. All the tensile stresses are calculated to be taken up by the iron or steel bars, but it should be observed that these stresses should not exceed the coefficient of elasticity of the concrete, or else cracks will inevitably occur. The safe tension load on the iron bars is 6·5 to 7 tons per square inch.

Breaking Strain of Concrete.—The breaking strain of concrete, mixed in the proportions generally used in ferro-concrete, is taken between 3000 and 4000 lb. per square inch, and the coefficient of safety is usually fixed from 400 to 500 lb. per square inch.

ADVANTAGES.

Fires.—Amongst the many advantages of ferro-concrete construction is its fire-resisting nature. Iron or steel used in the ordinary manner in any building and not encased in an incombustible matrix is, when subjected to a fierce heat, so liable to distortion that the walls and floors are torn asunder, almost invariably necessitating the entire rebuilding of the structure.

Fire Tests.—Many experiments have been made to demonstrate the protection ferro-concrete affords against fire, the most important being those carried out at the Ghent Exhibition by the Belgian authorities with the primary object of testing the efficacy of various fire-extinguishing appliances. From the Official Report of Commander Welsch, Chief Officer of the Ghent Fire Brigade, it appears that on September 28, 1899, a concrete building on the Hennebique system was, for the second time, tested to ascertain its resistance to fire. The first floor supported a load of 47 cwt. per square yard with a deflection of barely 0·02 inch, with a fierce fire maintained for two hours on the ground and first floors. Cracks were produced here and there, but upon the structure cooling they closed up. The radiation of heat was very small. A spinning mill at Court St. Etienne, built on the Hennebique system in 1898, was burned out in 1901, but the structure itself was uninjured. Tests applied to the floors after the fire showed that, notwithstanding the expansion and contraction by fire and water, they retained their normal elasticity and strength.

Durability.—As regards durability, there has not yet been sufficient time to test this, but, so far, everything points to a favourable conclusion in this respect. Some ferro-concrete drain pipes laid in 1886, in Grenoble, were taken up in 1901, and were found to stand their normal pressure of 80 feet head of water, and in every respect the pipes were found to be constructively sound.

Freedom from Vibration.—Owing to the monolithic nature of the work, vibration is reduced to a minimum. This is a consideration of the utmost importance with regard to mills, factories, workshops, subways, and the like. This immobility is also characteristic of ferro-concrete bridges. Some tests were carried out at the electric works at Austerlitz station, by the engineers of the Paris and Orleans Railway Company, to ascertain the effect of shocks. Two floors were constructed, one with iron girders and jack arches, and the other of ferro-concrete, each of the same bearing and calculated for a similar load. The dead weight of the iron and brick floor was 100 lb. per square foot, and that of ferro-concrete 62 lb. per square foot. A weight of 1 cwt. dropped from a height of 6 feet 6 inches on the iron and brick floor produced vibrations of $\frac{5}{16}$ inch amplitude, lasting two seconds, while a weight of 220 lb. falling 13 feet on to the ferro-concrete floor only caused vibrations of $\frac{1}{16}$ inch, lasting $\frac{2}{5}$ of a second. These results are highly satisfactory.

Hygienic Advantages.—Hygienically this method of construction has considerable advantages. It is impervious to damp, whilst the hard and gritty consistency of its materials defies the attacks or harbouring of vermin.

Labour.—Owing to the simplicity of the system, only unskilled labour need be employed, with the exception of the carpenters to erect the necessary staging, casing and shuttering, and of the smith for the iron and steel work.

Besides the foregoing important economical advantages there is no loss of available section due to riveting as in ordinary girders, and the materials lend themselves readily to any manner of architectural treatment. Buildings erected on this principle are found to be warm in winter and cool in summer. It is claimed by the inventor for ferro-concrete that in comparison with other building materials it shows a saving of between 15 and 20 per cent., and that a building can be erected in ferro-concrete in about one-third the time required if specially built up steel stanchions and girders are used. Apart from this no outlay is incurred for the annual painting as required by ordinary steel construction, the cost of maintenance of the ferro-concrete structures being practically *nil*.

Cost.

Owing to fluctuations in the cost of materials and labour in different localities, it is impossible to give any information as to prices which would be reliable for general application; the following prices are therefore given, with due reservation.

Bridge.—A bridge of three arches, the centre one being 172 feet span, and the two side arches each 135 feet span, was built on the Hennebique system at Chatellerault, France, the total width being 26 feet 3 inches. The cost of this structure was about 18*l*. 12*s*. per lineal foot. A description of this bridge is given later on.

Grain Warehouse.—A two-story grain warehouse at Plymouth, having an area of 5500 square feet, with a load on the floors of a quarter of a ton per foot super., supported on 12 inch by 12 inch columns 10 feet and 11 feet 6 inches apart, connected to 14 inch by 14 inch foundation ferro-concrete piles 30 feet long, cost, including piling, $4\frac{1}{2}d$. per cube foot of space.

Flour Mill.—The whole of a flour mill at Swansea, constructed in Hennebique's ferro-concrete, cost about 4½*d.* per cube foot. The foundations are on made ground, consisting of sand and rubbish brought as ships' ballast and tipped on soft mud on the estuary of the Swansea River. It was therefore necessary to float the mill on a ferro-concrete sill. The height of the structure from the foundations to the roof is 112 feet, the latter being used as a reservoir containing about 100 tons of water.

Grain Silos.—Some grain silos at Swansea are constructed on similar foundations to the mill, and contain over 7000 tons of grain. The height from the foundations to the roof of the cleaning tower is about 130 feet, the thicknesses of the walls being as follows:—External walls, at foot 12 inches, at top 4 inches; internal partitions, at level of hoppers 5½ inches, at top 3 inches. The capacity of each silo is about 80 tons of grain; the cost per cube foot being about 6¼*d.*

Simplon Canal.—The canal of the Simplon, 9800 feet in length, internal section 6 feet 7 inches square, with sides 4 inches thick, carried on ferro-concrete trestles from 16 feet to 33 feet high, was executed at a cost of about 4*l.* per lineal metre.

Water tank.—A cylindrical water tank at Bournemouth, 45 feet high, containing 15,000 gallons, with a diameter of 20 feet 10 inches, depth 10 feet, thickness of sides of tank 5 inches tapering to 4 inches, supported by six pillars, 5 inches thick, 35 feet high, with 18 inch by 18 inch bases, cost about 9*l.* 6*s.* 8*d.* per vertical foot.

Slipway.—The slipway, Woolston Ferry, some 280 feet long, and capable of lifting vessels of 300 tons displacement, was constructed of three parallel rows of piles, each pile being about 20 feet long and 10 inches by 10 inches in section, supporting the groundways 24 inches deep, fixed to the tops of the piles. The cost of the piles, including driving, was about 8*s.* 6*d.* per cube foot, and the groundways, inclusive of the cradle rails, 4*s.* 6*d.* per cube foot. The total cost of the slipway with the machinery, cradle, etc., was 2350*l.*

Coal Hoppers.—The coal hoppers built at Rudmore Wharf, Portsmouth, containing over 2400 tons of coal, cost, including foundations, about 7¼*d.* per cube foot.

Sheet Piling.—The sheet piling at the Hard, Gosport, cost, inclusive of concrete backing and patching to old work, about 2*l.* 10*s.* per lineal foot.

It is stated that wharves, piers and jetties constructed on the Hennebique system cost a great deal less than if constructed in masonry or steel, and are only about three times as expensive as similar structures in timber, besides being of a much more permanent character.

The Chatellerault Bridge.

The Bridge of Chatellerault, France, is considered to be the chef-d'œuvre of ferro-concrete construction on the Continent. The foundations, piers, abutments, starlings, arches and floor are constructed on the Hennebique ferro-concrete system. The bridge crosses the River Vienne in three spans; two side spans of 135 feet, rising 13 feet, and a central span of 172 feet, with a rise of 15 feet 8 inches; the length over all being 442 feet. The flooring, 25 feet in width, is carried on four ferro-concrete arches 20 inches high, tied together with 8 inch by 8 inch braces, the footpaths being partially supported on cantilevers of the same material. The piers and abutments are composed of four beams corresponding to the arches, and connected by a 5-inch concrete curtain; they are filled with weak hydraulic lime concrete. The calculations were based on the assumption that the bridge would have to support a load equalling the passage of two files of four-wheeled carts weighing 16 tons each, the footpaths carrying at the same time a dead load of 100 lb. per square foot.

The foundations finished, and the centres in position, concreting was begun on August 15, 1899, and completed on November 5 following, the centres being removed a month later. Then followed a series of elaborate tests, under the direction of Engineer Aubin, of the Ponts et Chaussées. In the dead weight tests each bay was first loaded over its entire length, then on each half, and then in the middle. The load consisted of moist sand, laid at the rate of 165 lb. per square foot on the roadway and 123 lb. on the footpaths. The engineer in his official report of the trial states that: "The maxima of depressions were $\frac{1}{4}$ inch for the arch of the left shore, $\frac{7}{32}$ inch for the arch of the right shore, and $\frac{13}{32}$ inch for the central ditto. The mean depression of the lateral arches $\frac{1}{7300}$ of the span, and for the central arch only $\frac{1}{5000}$. When the superloads were removed, and the bridge completely cleared, the arches returned exactly to their original position." The bridge was afterwards tested with a moving load composed of—one steam road roller of 16 tons, two four-wheeled carts of 16 tons each, and six two-wheeled carts of 8 tons each; which, together with the teams, gave a total weight of about 40 tons passing simultaneously on the bridge, the footpaths of which bore an additional load of 80 lb. per square foot.

As a further test, 250 infantry soldiers crossed the bridge, first at a slow march and then in double-quick time; after that a steam road roller was run over the platform, upon which chocks of wood 2 inches thick were strewn in order to produce a series of shocks. Besides these trials the bridge successfully withstood other severe tests with moving loads, and in all the tests the maxima of depression attained did not exceed $\frac{1}{9000}$ of the length of the arches. The deformations of the arch with reference to its median line always remained inferior to those caused by the dead weight tests. There was never any permanent deflection, but it was found during the trials that the load which caused a bending in one arch caused at the same time a corresponding rising in the other, thus establishing in a remarkable manner the fact of the solidarity of the three arches. As evidencing the lightness of the structure, it may be stated that the total thickness at the centre of the middle arch is only 28 inches, the entire dead weight being only 250 lb. per square foot, and its whole inclusive cost being a little under 8000*l*.

Another example of construction is a quay wall built in Hennebique's ferro-concrete. It consists of a front row of sheet piles, 16 inches wide, with a ferro-concrete platform supported by 3 rows of square piles, 7 feet apart, centre to centre. Upon the sheet piling is fixed the superstructure of the quay, stiffened with angle pieces continued right across and intimately connected to the platform and pile superstructure. The bollards are bolted to ferro-concrete slabs fixed between the angle pieces. Elm fenders are attached to the quay side by iron bolts fixed through piles and the flange of the superstructure. The fender commences at the waling, just above low-water mark, and is continued upwards well beyond the quay coping. On the superstructure is laid a granite coping, behind which is a square concrete culvert to drain off the surface water. The working load is practically carried by the platform, the angle pieces bracing the superstructure against the impact of vessels, while the fenders help to distribute the shocks over a large area. In the calculations for these structures the materials are assumed to be homogeneous, and treated as a retaining wall.

The final example which the author proposes to give is one of jetty construction on a rocky foundation. A hole is blasted in the rock to receive the end of a ferro-concrete pile, which is moulded hollow. Steel bars are then lowered in the rock through the central cylinder of the pile, which is afterwards filled up with rich cement concrete. If blasting proved undesirable the rock could be drilled through the central cylinder of the pile. The openings left in the lower end of the pile permit a sufficient quantity of grouting to escape, thus forming about the foot a block resting on the surface of the rock. The steel bars being perfectly embedded in the concrete, the pile becomes properly fixed into the ground.

CONCLUSION.

In conclusion, the author would point out the rapid development of ferro-concrete work on the Continent, owing to which the French Government recently created a special Department at the Ministry of Public Works to take into consideration all matters relating to such construction. It has, however, so far made but slow progress in this country, although it would appear that its advantages are being recognised and that its adoption is increasing.

In compiling this paper the Author has consulted the following works, viz.: Professor W. Ritter's 'Die Bauweise Hennebique' (Schweizerische Bauzeitung, 1899): Herr J. Rösshander's 'Theory and Practice of Steel and Concrete Construction' (Schweizerische Bauzeitung, 1900); Mons. L. G. Mouchel's Pamphlet on 'Hennebique's Ferro-Concrete Construction'; Mr. Walter Beer's Paper on 'The Monier System of Construction,' Proceedings of the Inst. C.E., vol. cxxxiii.; and Mr. C. Fleming Marsh's 'Reinforced Concrete,' Proceedings of the Inst. C.E., vol. cxlix.

He also desires to thank Mons. L. G. Mouchel, M. Inst. C.E. of France, for a great deal of information with reference to the subject. He further thanks Messrs. H. C. Portsmouth, M.S.A., P. Hastie, and C. J. Leather for particulars as to the cost of the various works in ferro-concrete with which they were respectively connected.

DISCUSSION.

The PRESIDENT said he was sure that those present would appreciate the care with which the author had dealt with the subject of his paper. He had confined himself to examples in actual practice, and had indulged in the least possible amount of speculation. In all these new systems, the art of dealing with them in a paper consisted in presenting proved facts and avoiding speculation as much as possible, and he considered that the author had accomplished that task in a most thorough and successful manner. His illustrations were also to the point, and brought before them prominently the salient points of advantage which were claimed for the Hennebique system over others. The author had selected one out of many systems to bring before them, and he (the speaker) thought that it was better to be instructed fully and accurately with regard to one system, than to have a smaller amount and possibly less accurate information about a larger number of systems. He would ask those present to endeavour to supplement what the author had told them by definite and accurate information, with regard to other systems to which the author had referred only by name. He would ask them to accord a very hearty vote of thanks to the author for his excellent paper.

The vote of thanks was carried with acclamation.

Mr. F. E. WENTWORTH SHEILDS said that he was sure that they would all very heartily congratulate Mr. Galbraith on having most lucidly set before them a very interesting system of construction. It was not the only good thing that they had had from France. One had only to turn over the pages of any French journal, say the Annals of the Ponts et Chaussées, to see how freshly and originally the French treated their constructive problems. Not only was the whole idea of ferroconcrete very unique, but the constructive details had some interesting features. He had never seen concrete so well and carefully made as that used under the Hennebique system, where the materials were so carefully selected, and the whole workmanship was of such an excellent quality, as to make something which seemed entirely different from the old-fashioned rough English concrete. He would not go so far as to say that no such good concrete had ever been made in England, but at all events it had been the exception and not the rule.

A point referred to by Mr. Galbraith in his paper was that very great stress was laid on the fact that gravel materials in the concrete should not be of the same size, but that both large and small stones should be used, the largest not to exceed say three-quarters of an inch. He (the speaker) would be inclined to go further than that, and to say that the most excellent concrete could be produced by using sand of different sizes, especially where a watertight concrete was desired. He had noticed that the watertightness could be very greatly increased by using not only coarse sand, but also a proportion of fine sand. He recently made an experiment in Southampton, and had found that a very good sample of gravel from the Portsmouth district contained air spaces to the extent of something like 25 per cent. The sand in that sample was unusually coarse, and by adding a small quantity of finer sand, the air spaces were reduced to 16 per cent., thus showing very markedly how a small proportion of fine sand added to a coarse sand reduced the air spaces, and consequently increased its watertight qualities.

Mr. W. R. Galbraith, of Westminster, was now constructing a large platform or deep-water quay at Southampton, and he (the speaker) had been engaged in looking after it. The author had alluded to that work, and to the fact that they were sinking piles with the combined system of water jet and ramming. He must say that they had been exceedingly pleased with the results. They had driven altogether 130 piles by that method. Those piles had to be driven through about 20 feet of very ugly material including a hard ballast, and underneath that a close fine sand. That was about the most awkward combination of strata that one could wish to drive piles through. It was perfectly hopeless to try to sink them with the water jet alone, and they found that even if they could drive them through the ballast, directly they got on to the close sand the piles absolutely refused to go further. But by means of the water jet and ram combined, they had got the piles down without trouble. He would not say that the water jet would make them go very easily through the ballast, but they would go more easily than without it, and directly they got through the ballast on to the close sand they went down with very little resistance. About 1 foot above the final depth, they shut the water off, the effect of which was very marked; and after a few more blows the piles refused to go further. Some of the members had visited Southampton and had seen the work. He was pleased to be able to say that every one of the piles they had driven on the Hennebique system had gone down successfully.

They were very much handicapped at Southampton by the fact that most of their work had to be done between high and low water level, and a great deal of bracing had to be put in a very little way above low water, so much so that at neap tides they could not get on at all. He had been hoping on that account that they might be able to devise a system, by which the bulk of the low water bracing could be built on the quay, and lowered down into position. Unfortunately their present staging was planned in such a way that it could not be managed very well, but he would commend the idea to any one who had similar work to do.

The author had disclaimed holding a brief for the Hennebique ferro-concrete system, and no doubt that was correct. But in the paper he had been very loud in its praise and perhaps quite rightly. That being so he was sure that all present would wish to discuss any drawbacks that they might think of, or that experience might have brought out. Moreover, the system was so good, that he was sure its promoters would welcome any adverse criticism, in order that any real disadvantages might be done away with. There were one or two drawbacks, by no means fatal, that occurred to him from a very short experience of the system. He noticed the author said that, owing to the simplicity of the system, only unskilled labour need be employed, with the exception of that of the carpenters and smiths; but he (the speaker) must say that the exceptions were rather large. The sight of a piece of ferro-concrete work going on represented to his mind an army of smiths and carpenters, who must be by no means unskilled, and even the concreters themselves needed a certain amount of training to do their work satisfactorily. He would not say that that was a great drawback. He thought that all concrete work ought to be done by men more or less skilled, but he certainly did not think that unskilled labour could predominate in ferro-concrete work.

Mr. Galbraith pointed out in his paper several advantages of ferro-concrete cylinders. He (Mr. Wentworth Sheilds) believed that ferro-concrete cylinders were coming into favour very much, at all events with the promoters of the Hennebique system, and no doubt they had several advantages; but he had doubts with regard to the advantages over cast-iron cylinders which were claimed for them in the paper. The author said that, owing to their extra weight, they did not require so much pressure to sink them as ordinary cylinders did. He (the speaker) had had no experience in sinking ferro-concrete cylinders, but he had had experience in sinking plain concrete cylinders. Now at Southampton they had found that there were many situations where they could not possibly get an ordinary concrete cylinder to go down, but where they could sink a cast-iron cylinder quite easily. That was especially the case when there was water inside the cylinder, and when it was therefore difficult to get the excavating tool near the edge of the cylinder. It had often happened that the broad cutting edge of a concrete cylinder had rested on a bed of gravel, which refused to fall away from it, and the cylinder would not go down whatever weight was applied to it; whereas in a cast-iron cylinder, with its fine cutting edge and comparatively small thickness, the grab or miser could cut close to the edge, and the cylinder sank without any difficulty whatever.

Another point that was most interesting to engineers was the matter of cost. Unfortunately he could not compare the cost of the Hennebique system or of any other system with the cost of the more old-fashioned types of construction; but certainly the statement that piers and jetties constructed on the

Hennebique system cost a great deal less than if constructed in masonry or steel might be open to challenge, although he must admit that in the particular work with which he had been connected at Southampton which Mr. W. R. Galbraith was carrying out, they found that the ferro-concrete platform cost less than a solid concrete wall would have done. That, however, was partly because with a platform design they were able to make use of a wall which was already there, whereas if they had tried to add to or to underpin the existing wall, it would have been almost as costly as building a new one. It might be that the ferro-concrete system would work out cheaper in other cases, but at present he reserved his judgment on that point.

A more important matter than the first cost, and one that after all the engineer looked to most of all in his work, was the question of durability, and here it certainly looked as if the ferro-concrete system had strong points in its favour. The author told them of a very elaborate test that was carried out to show its fire-proof qualities, and that of course was a most important point.

Again, for constructing piers or jetties in sea water, there was no doubt that ferro-concrete work would score considerably over timber, in a situation such as that at Southampton, and such as occurred in most sea waters, because the ferro-concrete would be free from the ravages of sea worms, which so very soon destroyed even creosoted timber. But, on the other hand, he supposed that it would be impossible for some years to say whether sea water would not destroy ferro-concrete in other ways. It yet remained to be seen whether ferro-concrete, built on the Hennebique or any other system, would withstand the action of rust in case by any accident the water got to the steel, and again, whether it would be possible, supposing that the concrete by any accident got knocked off, to efficiently repair it. That was a point worthy of great consideration, and ferro-concrete was hardly old enough for them to pronounce a final judgment upon it. They had had only a few years' experience, but, so far, it seemed that no grave difficulty had occurred. For himself he hoped that no grave difficulty would occur, and that the system would prove to be as lasting as any other form of construction; in fact he was so pleased with the ingenuity and the novelty of the Hennebique designs that he wished, both to the system and to the individual constructions, a long career of usefulness.

Professor HENRY ROBINSON said, that he joined in the

thanks to the author for laying before the meeting such an interesting record of facts and results which distinguished the paper from some where opinions only were given. In this case they had a very good series of practical results, which commended themselves to all who had had to do with that branch of work. He thought that the author said very properly that English engineers were conservative, and very much behind the age with regard to the use of a combination of iron and concrete such as the paper referred to. He had known of the combination for some years, but until last year he had had no occasion to deal with it in his own practice. Then he had to advise about the reconstruction of an old storage water-works reservoir on the top of a hill, which had been disused for many years owing to the subsidence of the ground due to coal workings. The great cost of reconstructing it by any of the ordinary methods led to his being called in to advise with reference to the adoption of what was called "Armoured Concrete," which was a combination of iron or steel and concrete. He went very carefully into the whole of the calculations, and supported the proposal which ultimately was passed by the Local Government Board. They had to construct a reservoir to be self-sustaining, self-supporting, and to carry a large floor over fissures, which required a very considerable amount of care. The method of construction was somewhat similar to that which the author had described. The arrangement of rods was not quite the same as illustrated in the paper, but the principle was the same. The calculation as to where the rods should be placed required a considerable amount of care, but admitted of being very definitely and accurately determined. He had to go through the calculations very fully in order to satisfy those who were responsible for the outlay. That was not the place or the time to show the exact method of the arrangement of the rods which he adopted as compared with the arrangement of rods in the illustrations. He thoroughly believed that, by the combination of iron and steel rods and concrete, engineers would meet with the solution of many problems not so easily or so cheaply solved by other methods.

With regard to the work being carried out by unskilled labour, he disagreed with the author. He should be very reluctant to see any work with what was called "Armoured Concrete" or "Ferro-Concrete" entrusted to those who had not some knowledge of how to calculate stresses very carefully and accurately. Having determined the exact form of construction, men were required to carry out the work who understood the use of materials in every detail right away from start to finish. Work of that kind wanted to be entrusted to very good and very careful men in the various departments.

As a matter of history, he might refer to a catastrophe which occurred abroad a few years ago when some one erected a pile of buildings on a method of utilising iron or steel rods and concrete, but being designed unskilfully the whole thing came down like a house of cards and caused a great loss of life. He mentioned that to the younger members of the profession because he was certain that no work needed to be more carefully designed or carried out than that class of work.

He had a large number of photographs of works which had been carried out all over the world, which showed to what varied purposes the system could be applied. He wished to support the author, and to thank him for bringing before the profession so many interesting facts. He hoped that what he had said with regard to the necessity for very great care and considerable skill on the part of those who had to deal with the system—although he said it very earnestly—would not in the least depreciate the system. He wished his observations to be taken in the direction of supporting the system, hoping that it would be more frequently adopted in this country.

Mr. EASTON DEVONSHIRE said that he had spent many years on the Continent amongst the different systems, and he could give the result of his own experience and the opinions of many engineers who had used the Hennebique and other systems. He endorsed what previous speakers had said as to the thanks due to anyone who brought before engineers any information on that subject which would advance their knowledge of an art in which they were certainly fifteen or sixteen years behind Frenchmen. It had been said in France that the principal merit, or one of the principal merits, of the Hennebique system was to be found in the energy and business-like way in which the system had been brought forward. He thought that there was a great deal in that, and the merit he gave M. Hennebique was no slight one.

As far as the slab floor was concerned, Hennebique applied principles introduced originally by Monier and Coignet, the latter being one of the pioneers not mentioned in the paper. It was considered on the Continent, and he thought that it was the case, that Coignet used those principles according to a formula of his own. It was generally admitted in what he might call generically the Monier system of construction, that the steel rods forming the tensile member of a slab of concrete, should occupy a space in that slab equal to 1 per cent. of the volume of the slab, those rods and those alone which lay in the direction of the tensile strains being taken into consideration. Generally speaking, they ran in one direction and parallel. Therefore, in a floor 10 centimetres thick, there should be a volume of rods equivalent to a sheet of steel 1 millimetre thick, and taking the specific gravity of steel as 8, for the sake of easy calculation, it meant a steel bond of parallel rods weighing 8 kilogrammes per square metre for a floor of 10 centimetres in thickness. As he understood it—although no formula was given in the paper—in the Hennebique system, the whole of the rods, longitudinal and cross, were taken into consideration. If that were so, he thought that the margin of safety in the Hennebique slab was insufficient. In order to properly utilise the full strength of both longitudinal and cross rods, the rods must be absolutely united at the joints, and no ordinary attachment was sufficient so to unite them. Cottancin, whose system was mentioned, saw that and elaborated (that was the right word, because it was very elaborate) a system of looping and knotting together wires in order to make a knot at every joint in the square or rectangular mesh.

The tests made by Messrs. Fowler and Baker of concrete slabs containing expanded metal, afforded the opportunity of comparing the effect of a steel network where the joints were absolutely solid with slabs containing crossed rods wired together by hand. A slab 10 centimetres thick, containing a sheet of expanded metal weighing 4 kilogrammes per square metre, was 25 per cent. stronger than a slab containing 8 kilogrammes of crossed bars fastened together by wires, the explanation being two-fold. First of all, the bars, working independently, were not united in strength; and secondly, there was not the compression of the concrete in each mesh which was produced where there was a network with absolutely solid joints. Therefore, he had come to the conclusion from experiments which had been made, that if Hennebique calculated on the whole weight of the rods he used, the margin of safety was not sufficient. By including the strength of the cross rods, he took into his calculations twice as much steel as he really could usefully employ. Engineers should look at facts, and it was a fact that in many cases on the Continent serious accidents had happened with the Hennebique system. It had been carried out by persons whom he called his agents, and for whom he said he was not responsible, and they had worked with too small a margin of safety, and very frequently in hurrying to remove the centring the structure had collapsed. In a case in Basle two years ago a top floor gave way, and carried away the lower floors with it.

With regard to what were termed in the paper the main and secondary beams, he (Mr. Devonshire) must say that he was not in favour of that form of construction as being either the strongest or the best for its cost. That was, perhaps, saying a great deal against the Hennebique system. The application of the Monier system to the construction of the slab was perfectly sound, but Hennebique had endeavoured to carry it to such a point that he would not use any material except his rods and concrete. He thought that that was wrong. You could use steel joists of known strength and protect them against fire, and use them in conjunction with the Monier slab in such a manner as to give in every way the efficiency required, and to give it more cheaply. He said that because he had had the opportunity of comparing the cost in the two cases. The construction of girders and beams on the Hennebique system required very heavy centring and strutting, which must remain for a month; it further required skilled labour in every respect, as other speakers had said. For the combined steel joist and Monier slab floor skilled labour might be required, but there were many adepts and it was not expensive. By using joists they saved the cost of strutting, and had something solid to go upon. The cross centring and boards to form the floor slabs were supported by the steel joists, and the fire-proofing of the joists was very economical and very readily carried out by surrounding the girders with concrete. Joists lent themselves to the attachment of suspended ceilings, giving an air space under the floor, which was an enormous advantage in case of fire. The Hennebique system presented difficulties in that respect. He (the speaker) advocated the combined use in a proper and scientific way of materials which they had at command, and of a strength which they knew—viz. steel joists, expanded metal and Portland cement. He did not like the construction of a beam, especially of great length, when he was dependent on the flawlessness of the rods that he was using, and in many cases, perhaps, as the author had indicated, upon the strength of a welded joint.

As to the mixture of the concrete, there were many ways of doing it, but he (Mr. Devonshire) thought that the sound principle to follow was not to lay down a formula at all as regards the proportions. That could not be done. They must gauge the aggregate, and fill the spaces it contained. The plan that he had always found best was to commence by gauging the aggregate and then to add the concrete mortar, which could be of a fixed composition—say 1 of cement to 2 of sharp sand, so that it filled the interstices, giving a margin of, say, 10 per cent. for faulty mixing.

Mr. H. SHERLEY-PRICE said that the author had referred to a large building situated at Swansea. Some time ago he had the melancholy pleasure of passing that building twice a day for a considerable time. He was sure that the author would not call it a beautiful building. Everybody in Swansea said that it was the ugliest building there. It was a corn-mill eight or ten stories high. If he read the paper rightly, the wall at the base was 12 inches thick, and the top wall was 4 inches thick, and the cost was $4\tfrac{1}{2}d.$ per cubic foot. You could get a splendidly faced corn-mill of the same size built any day with a solid brick wall 2 feet 6 inches at the base, the top wall being certainly not less than 12 inches, and you could have cast-iron columns and steel girders (the usual roller girders that they had heard of), and the cost would be $4\tfrac{1}{2}d.$ at the outside for a building of the same dimensions as the one he referred to.

Mr. G. W. HUMPHREYS said that he had the opportunity two or three years ago of seeing under construction the west wharf at Plymouth, which was alluded to in the paper. He was much struck with the very practical method of construction which the whole work revealed. He summed it up in as few words as he could by saying that it was ironwork without any smithing. It seemed, from a cursory look, to combine the strength of iron with as little labour as was possible in putting it together. He felt the truth of Professor Robinson's remarks about the necessity of there being really skilled labour in the construction. He endorsed what the Professor had said, but at the same time he differed from him in a certain degree. That there must be skilled supervision to generally look after the construction he quite admitted, but he did not think that it required skilled labour right down to every man employed upon it. For instance, he would not like to put a man in charge of the work who had not a very good notion of what mixing concrete was, but he thought that with one, or possibly two, the whole thing would go along very satisfactorily when they took into account the very small amount of concrete necessary for a building of the size that they had had instanced in Swansea. The amount of labour, after all, was very small, and one or two men of skill to supervise would do everything that was required.

At Plymouth he was impressed with the great ease with which the piles were put in. He had had practical experience of the soil. It was a mixture of clay with enormous boulders in it, and it was very difficult to get piles into it, and the ease with which the concrete piles were put in astonished him, although failure had been prophesied by some. It seemed to him that very satisfactory results were produced. He held no brief for M. Mouchel; had never had the pleasure of speaking to him, although he had seen him at Plymouth. He must, however, say that there ought to be a very great future for the system, so far as he could see.

Mr. E. J. SILCOCK said that he had had no experience of the Hennebique system, but he had had a little experience in dealing with expanded metal, and several years ago when expanded metal was first put on the market he carried out experiments to see the effect of putting it in concrete slabs. He found that, roughly speaking, for ordinary slabs about $2\frac{1}{2}$ or 3 inches thick, a sheet of expanded metal would increase the strength of a flag when used as a beam from two to three times, and in constructing floors since he had adopted it with very successful results where the floors had been loaded with heavy loads. In addition to that he had used expanded metal in what he believed were quite novel circumstances; at all events he had never heard of anyone doing the same thing. He was constructing a malt kiln, and he wanted to get barges inside the building and to construct semicircular arches over the waterway through which the barges were brought in. The semicircular arches had no spandrel load or filling at all; and they were loaded in the centre. Obviously the tendency would be for the crown of the arch to sink and for the haunches to spread, and it occurred to him that putting expanded metal in between the rings of brickwork would give a tensional strength in the arch which would resist that tendency. He constructed it in that way. It was with some qualms that he gave instructions for the centring to be removed, but the arches stood perfectly well and never showed the slightest sign of opening at the joints. He could not give any calculations to show how he arrived at the result, but the effect was all right.

He should like the author in his reply to address himself to the question of the effect of the stirrups of the Hennebique system in taking up the shearing stresses of a beam. The author stated in the paper that the stirrups were one of the leading features of the system, and that their object was to take up the shearing stresses, and he showed diagrams in which the method of construction of beams was illustrated. The shearing stresses must be dealt with by something which was vertically continuous throughout the length of the beam, or by some system of triangulation, and in the latter case the stresses must be applied at the apices of the triangulation. In the girders shown in the diagrams there was no triangulation in the centre-third of the girder, and he failed to see how the vertical stirrups could in any degree deal with the stresses of shearing in that portion. With regard to the other two-thirds of the girder, it was said that the bracing formed by the stirrups and the horizontal tension member and the raking tension member made a statical triangle; but there was no connection between the members of that triangle as had already been pointed out, and further than that, the triangle must have at least one compression member, and he took it that the shearing stress if applied to the top of the stirrups would put them in compression, and if they were in compression they seemed to be of no greater value than the concrete itself. It was very doubtful indeed whether the stirrups had any effect whatever on the shearing stresses, and he must say that he thought some further experiments on that subject would be of value. The author stated that two beams had been tested, one of which had stirrups and one of which had not. He would suggest that an experiment on only two beams was insufficient to settle such a vital point.

Mr. GEORGE GREEN said that he had not intended to say anything, for two reasons—first, that he had not read the paper, and, secondly, that when he arrived it was half read. He had been very much interested in it, and he should read it afterwards. In July last he had the pleasure of seeing several buildings built in the style described in Paris; he was not sure whether they were ferro-concrete or not, but they were iron-framed buildings, filled in and roofed with concrete; they were built to a very considerable height and of considerable dimensions. He was very much struck with the number of such buildings existing in Paris. The French seemed to be very much in front of us in regard to this method of construction. He agreed with Mr. Sherley-Price in regard to the buildings in Swansea; he could not say that they were beautiful, but such buildings must be very much quicker and cheaper in construction, or he could not see why they should be built. Buildings of an engineering nature, where strength and lightness probably were the chief things and where appearance did not matter at all, or mattered very little, gave opportunities for the method of construction under discussion.

Mr. R. J. GIFFORD READ said that he had had some experience of this class of construction, although it only came to his notice a short time ago. It was a new thing, but he did not disregard it for that reason. He looked into it, and the more he looked into it the more he saw how rational it was, and for that reason he had recommended it on two or three occasions for work upon which he had been engaged. One of those works was now carried out. It was a foundation for a warehouse on bad ground, and in the ordinary way it was proposed to sink trenches 15 feet deep and build up concrete walls, but there was a danger of disturbing the neighbouring premises. He therefore recommended the Hennebique system of concrete floor over the whole area, and on this the walls of the warehouse were built. Whether the work was absolutely right or not he did not yet know, but he might say that the cost compared to the work that the architect was going to do was about one-half. It was carried out by Mr. Jackaman as contractor in a fortnight.

Another foundation for a building that he had to deal with was in bad ground near the Albert Docks. In this instance he had proposed to carry the walls on girders resting on the top of cast-iron screw piles. However, he obtained an estimate for a foundation on the Hennebiqne system, consisting of concrete piles and sills made up of iron and concrete, and the cost of that compared to the system which he had previously contemplated was less than half. He had now orders to put the work in hand, but as there was danger of shaking the adjoining building it was proposed that one pile should be first driven in order to see what the vibration would be like. If there was too much they might cancel the contract and adopt another system.

He had found a very great difficulty in explaining the nature of the work to his clients. Everyone said at first, as he did himself, "It is too thin, it will never stand." That was the objection taken to the first work he had named, but the district surveyor who at first opposed it finally said, "If this succeeds, all I can say is it will mean a revolution in foundations down in this part of the country." He (the speaker) had had to make several reports on the nature and strength of ferro-concrete, and he had consequently looked up authorities quite independent of M. Mouchel, who was the Hennebique representative in this country.

The best work that he had come across, so far, dealing thoroughly with the subject, was that of M. Resal, chief engineer of the Ponts et Chaussées in Paris, and also the notes of M. Considère, government engineer and head of the department that the author had spoken of. From their investigations he could see pretty clearly the reasons for the merits of the system.

Mr. Devonshire spoke of examples of the Hennebique system breaking down. There were such examples, and he had also come across a striking example of concrete and expanded metal breaking down. That was the foundation of a large oil tank, constructed of concrete 2 feet thick with a layer of expanded metal in the bottom, the whole resting on the top of hollow cast-iron piles. The weight of the tank when filled broke the concrete over the piles, for the simple reason that there was no expanded metal in the upper surface, where the greatest tensional stress came. It clearly showed that the concrete and metal were good, but that they were wrongly combined.

In order to test the mathematical calculations and the theories that he had read on this subject he had had four experimental beams made:—one of concrete, without any iron in it; another, with a single armature of iron in the bottom; a third, with iron in top and bottom; and the fourth, a Hennebique beam, having a large flanged top of concrete and iron rods in the bottom. He hoped to test these in a few weeks, and would be pleased to communicate the results to the Society.

Mr. GALBRAITH, in replying upon the discussion, said that he agreed in the main with Mr. Sheild's remarks, and he would deal generally with the issues raised. Professor Robinson spoke of the conservatism of the British engineer. He (the speaker) was a British and an Irish engineer, and he admitted that there was perhaps a little too much conservatism about British engineers, but it was a good point in many instances. Professor Robinson had also remarked upon the care required in designing ferro-concrete structures. They all knew that the greatest care was required to ensure success and avoid disastrous failures. Mr. Devonshire's remarks were very interesting, and he should certainly give them every attention. With regard to what Mr. Sherley-Price said as to the Swansea flour mill, he (the author) could only say that it was the first building in England erected under the Hennebique system, and the accepted tender for the Hennebique process was about fifteen per cent. higher than the lowest acceptable tender for the same building in ordinary construction, but against that had to be set all the advantages of the composite structure. Referring to Mr. Silcock's inquiry as to the working of the stirrups in a Hennebique beam, his (Mr. Galbraith's) opinion was, that the method of working of the stirrup was that it withstood the tensile stresses working in the direction of the tension curve, and that in vertical arrangement it raised the bearing strength of the beam by the prevention of the premature entrance of cracks, but to what extent he quite agreed that there was plenty of room for a series of comparative tests to show. He was very pleased to hear that Mr. Gifford Read contemplated carrying out some experiments in that direction, and he agreed with that gentleman that the Hennebique piles were peculiarly well adapted to foundations in bad ground. With regard to the best works for the study of this particular construction, he did not know a better reference on the subject than the notes which M. Considère had communicated from time to time to the French Institution of Civil Engineers.

The author wishes it to be understood that he has no interest whatever in the Hennebique system of ferro-concrete construction beyond the professional interest he has naturally taken in the invention owing to his having experience of works in which that system has been adopted. His paper has been prepared solely with a view to bring before the Society what he conceives to be an interesting and promising innovation in engineering construction in this country.

The following communication was received from Mr. J. W. BROUGH, Assoc. M. Inst. C.E., of Raevels, Belgium, subsequently to the reading of the paper.

Ferro-concrete undoubtedly holds a leading position in modern engineering construction. I consider Mr. Galbraith's paper to be of great value, contributing as it does to the literature of this very important subject. More experimental investigations should be conducted in order to obtain reliable data concerning the ultimate resistance and moduli of resistance, the influence of almost unknown internal forces, and the behaviour of columns in flexure.

The presence of rolling scale on the steel increases the sliding resistance, and the surface should be, if possible, roughened before the application of a coat of cement grout. Great attention should be paid to the time of setting of the cement employed. I make a point of employing cement the initial setting time of which is not less than two hours, with good results. In concrete composed of a good mortar, the exposure to dry air should be guarded against, excessive drying causing contraction and increased tensile strain. Immersion of this concrete causes an expansion which is less dangerous, the resistance of cement to compression being ten times the tensile strength. It is, therefore, of great importance in exposed positions, in elevation, that the mortar employed should not contain too much cement, and the proportion should never exceed from 50 to 55 lb. cement to one cubic foot of clean sharp sand.

The different systems of ferro-concrete construction may be divided into two classes, namely, ferro-concrete constructions proper, in which the steel and concrete employed are calculated to resist respectively the tensile and compressive strains that the structure is called upon to bear, and structures in which the application of the ferro-concrete system is employed for the purpose of rendering them fire-proof.

There is a great future for ferro-concrete in the possibility of constructing almost any engineering work in mountainous districts where the transport of girders and heavy sections is impossible.

SECTION A.B.   SECTION C.D.

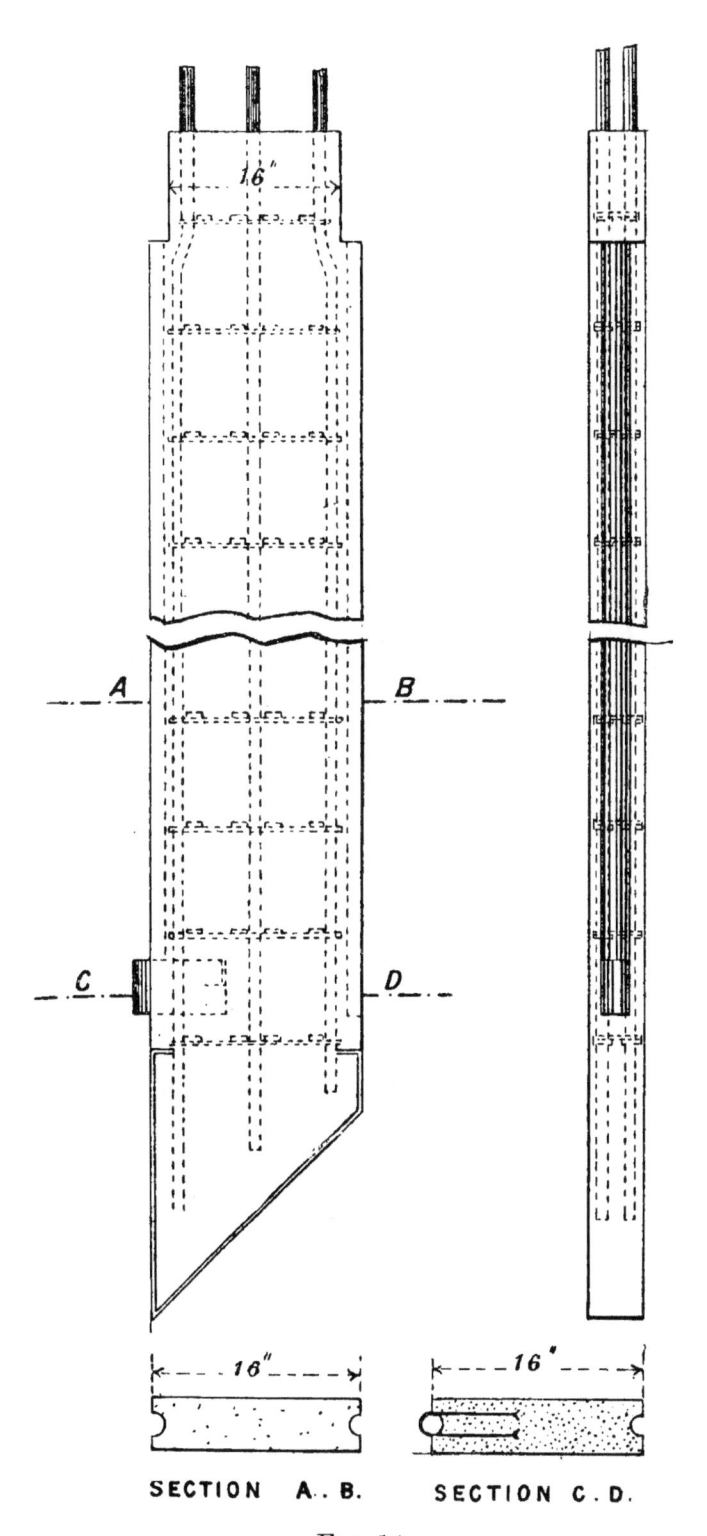

Fig. 14.

THE HENNEBIQUE SYSTEM OF
BY A

November 3rd, 1902.

PERCY GRIFFITH, PRESIDENT, IN THE CHAIR.

EFFECT OF SEGREGATION ON THE STRENGTH OF STEEL RAILS.*

BY THOMAS ANDREWS, F.R.S.

TRANSVERSE SEGREGATION.

THE occasional occurrence of segregation of the combined carbon and other elements which the author has recently observed in steel rails, rendered it important to trace, if possible, the distribution and location of the segregation in steel rails, and thus to indicate the weakest point in a segregated rail, and if possible to suggest a remedy for this evil.

SEGREGATION OF COMBINED CARBON IN STEEL RAILS.

Segregation in steel is productive of uneven physical structure, and also of initial internal stresses liable to promote fracture under concussional or vibratory stress. It is therefore of considerable importance in examining rails to be on the look out for indications of segregation.

Rail No. A 2302 (see Table I.), being a typical case of transverse segregation, a large number of analytical determinations were made (about 140 in all), to trace the distribution of the combined carbon throughout the rail. This investigation at the same time indicated the approximate location in the finished rail of the central part of the original ingot. The results are exceedingly interesting and show the points of greatest weakness in rails where segregation has occurred; this would probably be at or near the point where the most concentrated segregation is found. The rail would, of course, also be weakened towards the centre of the head, and especially at the junction of the head with the web. This is clearly proved on referring to the results of the physical tests on Tables III. and V.

* The President's Gold Medal was awarded to the Author for this paper.

Fig. 1, Plate I. shows the varying percentages of the combined carbon, in decimal parts of 1 per cent., in various parts of the transverse section of the rail; and the central portion of the rail, enclosed in the dotted lines, may be regarded as approximately indicating the part where the segregation was most pronounced. This also at the same time shows the probable location and distribution in the finished rail of the central portion of the original ingot. This search for the centre of the ingot has been something like seeking for the North Pole.

In Table I. are given the results of the general chemical analyses of the rail No. A 2302.

TABLE I.

Percentage Results.

	Rail Head.	Rail Foot.	Approximate Percentage of Segregation.
Combined carbon by coloration test	0·38 to 0·54	0·390	42
Silicon	0·074	0·065	14
Manganese	1·160	1·120	3
Sulphur	0·060	0·050	20
Phosphorus	0·051	0·049	4

The results given in Table I. having clearly shown the rail to be segregated, a more minute general analysis was therefore made after the completion of the investigation on the segregation of the carbon given on Fig. 1. This was done to trace with greater accuracy the local segregation of the elements other than carbon. The results are given in Table II., which show that the segregation was of an extensive character.

As between the top of the rail head and the central area of segregation near the centre of the rail head the carbon had segregated to about 59 per cent., the silicon had segregated to the extent of 47 per cent., the manganese to the extent of 8 per cent., the sulphur as much as 60 per cent., and the phosphorus equal to 46 per cent.

In Table III. are given the results of the physical tests of the segregated rail No. A 2302, portions being tested from the top of the rail head, and another test was made near the bottom of the rail head adjacent to the junction with the web, which latter situation would be near the area of greatest segregation. These results account for the very inferior physical properties found in this part of the rail. Tests were also made from the rail bottom, which being comparatively free from the main area of segregation, see Fig. 1, gave fairly satisfactory results. The results of the numerous tests given on Fig. 1 of

TABLE II.—RESULTS OF CHEMICAL ANALYSIS, showing the extent of the segregation of combined carbon and other elements in the rail head No. A 2302.

Percentage Results.

	Drillings from the Centre of the Area of Segregation.	Drillings from near the Rail Face.	Approximate Percentage of Segregation of the various Elements.
Combined carbon by coloration test	0·49 to 0·54	0·34 to 0·39	59
Silicon	0·088	0·060	47
Manganese	0·937	0·864	8
Sulphur	0·080	0·050	60
Phosphorus	0·095	0·065	46

TABLE III.—PHYSICAL TESTS OF RAIL No. A 2302, showing the varying physical properties in different parts of the rail induced by segregation of the chemical elements.

Description.	Original Dimensions. Size in inches.	Area, square inches.	Distance between Gauge Points.	Elastic Stress per square inch.	Ratio of Elastic to Maximum Stress.	Maximum Stress per square inch.	Elongation per cent.	Reduction of Area per cent.	Remarks
From top of rail head from point A, Fig. 2.	·564	·250	in 2	tons. 22·32	54·7	tons. 40·84	18·0	27·6	25 per cent. silky fibrous, 75 per cent. finely granular.
From bottom of rail head at point B, Fig. 2.	·564	·250	2	25·88	55·2	46·88	5·5	5·2	Finely granular.
From bottom of rail at point C, Fig. 2.	·564	·250	2	20·76	53·0	39·16	23·5	36·4	25 per cent. finely granular, 75 per cent. silky fibrous.

the distribution and extent of the segregation, clearly show the causes of the varying physical properties in the different parts of the rail.

From the results given in Table III. it will be seen that the maximum stress per square inch varied from 39·16 tons per square inch, to as high as 46·88 tons per square inch, or a variation of about 20 per cent. Whereas the elongation ranged from 23·5 per cent. in the rail bottom, to as low as only 5·5 per cent. in the bottom of the rail head, at the point of concentrated segregation; and all this within a length of 7 inches in the same rail.

Examination of various Segregated Steel Rails.

The general chemical analysis of another segregated rail, No. A 2052, is given in Table IV., and the physical tests of rail No. A 2052 are given in Table V., which latter table affords another illustration of the great variation in the physical properties of the same steel rail induced by segregation.

Table IV.

Percentage Results.

	Rail Head.	Rail Foot.	Approximate Percentage of Segregation of the various Elements.
Combined carbon by coloration test	T. ·38 M. ·57	0·390	50
Silicon	0·046	0·044	5
Manganese	0·979	0·918	7
Sulphur	0·065	0·040	62
Phosphorus	0·046	0·046	—

In Table IV., T ·38 is the percentage of combined carbon at the top of the rail head, and M ·57 is the percentage of combined carbon in the centre of the rail head.

On further careful analyses on numerous other drillings from the central area of the segregation at point B (see Fig. 3), the combined carbon was found to run as high as 0·70 per cent., so that in this one rail there was a range of combined carbon from as low as 0·30 per cent. to as high as 0·70 per cent., this representing a total variation in the carbon constituents of about 133 per cent.

ON THE STRENGTH OF STEEL RAILS. 213

The test pieces for this rail (Table V.) were cut severally from the positions A, B and C, shown on Fig. 3.

TABLE V.

Description.	Original Dimensions. Size in inches.	Area, square inches.	Distance between Gauge Points.	Elastic Stress per square inch.	Ratio of Elastic to Maximum Stress.	Maximum Stress per square inch.	Elongation, per cent.	Reduction of Area, per cent.	Remarks.
From top of rail head near point A, Fig. 3.	·564	·250	inches. 2	tons. 23·88	54·8	tons. 43·60	17·0	21·6	15 per cent. fibrous, 85 per cent. granular.
From centre of rail head near point B, Fig. 3.	·798	·500	2	23·66	58·4	40·54	5·0	6·8	Granular.
From bottom flange of rail near point C, Fig. 3.	·564	·250	2	21·44	52·7	40·64	23·5	36·4	50 per cent. fibrous, 50 per cent. finely granular.

These further results given on Tables IV., V., VI. and VII., fully confirm those recorded on Tables I., II., III. and on Figs. 1 and 4.

Another instance of segregation is afforded by Table VI. which gives the analyses of another rail.

From the results in Table VI. it will be seen that the carbon was very much segregated in this rail.

TABLE VI.

Percentage Results.

—	Drillings from near Junction of Rail Head with Web.	Drillings from the Rail Foot.	Approximate Percentage of Segregation of the various Elements.
Combined carbon by coloration test	0·42 to 0·58	0·450	38
Silicon	0·066	0·065	—
Manganese	0·576	0·547	5
Sulphur	0·090	0·075	20
Phosphorus	0·069	0·064	8

TABLE VII.—PHYSICAL TESTS OF RAIL OF THE CHEMICAL COMPOSITION given on Table VI.

Description.	Original Dimensions. Size in inches.	Original Dimensions. Area, square inches.	Distance between Gauge Points.	Elastic Stress per square inch.	Ratio of Elastic to Maximum Stress.	Maximum Stress per square inch.	Elongation, per cent.	Reduction of Area per cent.	Remarks.
			inches.	tons.		tons.			
From rail head	·564	·250	2	16·96	46·3	36·60	17·0	21·6	10 per cent. fibrous, 90 per cent. finely granular.
From rail foot	·564	·250	2	17·84	47·0	38·00	25·0	30·8	fibrous, with a trace finely granular.

FURTHER OBSERVATIONS ON THE SEGREGATION OF THE COMBINED CARBON IN STEEL RAILS.

In order more minutely to study the phenomena of segregation in steel rails, no fewer than 116 separate determinations of the combined carbon were made in the segregated rail No. A 2052, which was a rail of a section of 92 lb. per yard. This rail was a very typical sample of segregation in rails, the segregation of the combined carbon and other elements being in places very pronounced.

The general results of the numerous analyses of rail No. A 2052 are graphically delineated in Fig. 4, which gives the exact position in the rail whence drillings were taken for examination, the numbers on Fig. 4 giving the percentages of combined carbon, in the various parts of the rail, in decimal parts of 1 per cent. The space enclosed by the dotted line (Fig. 4) shows the chief area of segregation in the rail, and indicates also the approximate location in the finished rail of the central parts of the rail ingot. The main area of the segregation is enclosed within the dotted line, which may fairly be regarded as approximately indicating the distribution, in the finished rail, of the central portions or core of the ingot from which the rail was rolled.

It is important to observe that the most congested and concentrated portion of the segregation in this rail was found located at the junction of the rail head with the web, in the central part of which the combined carbon was found as high as 0·70 per cent. This shows that a maximum segregation to the extent of more than 133 per cent. existed in this part of the rail compared with the combined carbon found near the outer, and other portions, of the top and bottom part of the rail. The situation of this nucleus of segregation, at the junction of the head with the web (a place where great strength should be found), indicates the serious action of segregation in promoting variable physical properties resulting in a material reduction of the resisting strength of segregated steel rails.

This is further accentuated by the fact that this rail only endured the first impact from a tup weighing 1 ton which fell on it from a height of 7 feet, the rail being placed on bearings 3 feet 6 inches apart. It broke at the second drop from a height of 20 feet. The results of the general analyses of the rail head and rail bottom (Table IV.) show how the other chemical constituents had segregated, and emphasises the danger existing in segregated rails.

The results of the tensile tests of rail No. A 2052, given in Table V., show that a wide difference of physical properties existed between the top of the rail head, the bottom of the rail, and the part near the junction of the head with the web, i.e. at the place where the greatest amount of segregation existed, the chemical analyses fully confirming the physical tests.

Further Observations on Segregation in Steel Rails.

Another typical instance of segregation in steel rails is afforded by the results of an examination which the author made of a new steel rail (Index No. A 2193) sent to him for investigation from one of the chief railways in England. This rail, about 5 feet from the end, was found to be in a state of considerable segregation, as will be seen on reference to the analyses in Table VIII.

The extent to which the segregation had affected the physical properties of the rail is shown by the results of the physical tests given in Table IX.

Test marked A was taken from top of rail head. Test marked B was taken from rail bottom. Test marked C was taken from near junction of rail head with web. A consideration of the results given in Table IX. shows to what a considerable extent the physical properties of this rail had been affected by the segregation of the chemical elements.

TABLE VIII.—CHEMICAL ANALYSES OF SEGREGATED RAIL, about 5 feet from the rail end. Index No. A 2193.

Percentage Results.

	Analysis of Rail Head.	Analysis from Drillings near junction of rail head with web.	Analysis of Rail Foot.	Approximate Maximum Percentage of Segregation.
Combined carbon	0·440	0·600	0·440	36½
Silicon	0·058	0·061	0·058	5
Manganese	1·166	1·260	1·087	16
Sulphur	0·070	0·140	0·060	133
Phosphorus	0·065	0·110	0·062	77
Iron by difference	98·201	97·829	98·293	—
	100·000	100·000	100·000	—

TABLE IX.—PHYSICAL TESTS OF SEGREGATED STEEL RAIL, from various parts of the vertical section of the rail. Index No. A 2193.

	Original Dimensions. Size in inches.	Original Dimensions. Area, square inches.	Distance between Gauge Points.	Maximum Stress per square inch.	Elongation per cent.	Reduction of Area per cent.	Remarks.
			inches	tons			
A	·564	·250	2	43·96	15·0	18·4	Granular, with a trace fibrous. Slightly cracked on the surface.
B	·561	·247	2	43·04	24·0	41·3	Silky fibrous with a trace finely granular.
C	·564	·250	2	44·20	2·5	5·2	Granular.

SEGREGATION INDUCES DEFECTIVE WEARING FACES IN STEEL RAILS.

Excessive segregation not infrequently produces an evil, consisting of a rough and irregular and unsound condition of the rail face. This is due to the uneven physical structure and irregular formation induced by the local segregation of the chemical elements. The last mentioned rail (Index No. A 2193) affords a typical example of this. It was noticed that there were numerous longitudinal and transverse flaws and fissures on the surface of the rail face which constituted very considerable sources of danger in this rail. There were numerous areas of micro-segregation of the carbides, and of the impurities such as sulphides, which had produced lines of weakness in the internal crystalline structure of the metal. The micrographs afford a good illustration of the influence of segregation on the ultimate micro-crystalline structure of steel, and of the injurious effects produced by such an abnormal physical condition.

High-Power Microscopical Examination of Segregated Steel Rail. Index No. A 2193.

A careful examination of sections prepared from near the junction of the rail head with the web, showed that in this area of segregation there existed a number of open fissures running in some places vertically and transversely across the thickness of the web. As these internal flaws are typical of those due to segregation, the author made a careful microscopical examination of their nature and character. On Figs 5, 6 and 7 are delineated the appearance of these typical transverse fissures, as seen in vertical section at a magnification of 300 diameters, at the junction of the web with the head, which will convey an idea of the danger resulting from a segregated condition in steel rails. There was practically no metallic contact across the whole width of the web in several places.

A suitable micro-section was prepared, polished and etched, and a careful microscopical examination made of the ultimate crystalline structure and internal micro-flaws. Fig. 8 is an illustration of the normal micro-crystalline structure of the rail in an area comparatively free from segregation. Fig. 9 is an illustration of the micro-crystalline structure in an area where the combined carbon was 0·60, and where it had segregated to the extent of about 36½ per cent. Fig. 10 shows the ultimate crystalline structure in a segregated area, and illustrates both the segregated condition of the combined carbon and the micro-segregation of the sulphide of manganese and iron, and other impurities which have apparently produced the internal micro-flaws delineated on this figure. Figs. 9 and 10 also further illustrate the manner in which the excess of manganese has apparently interfered with the uniform crystallisation of the carbide of iron areas.

In order to further investigate the matter another series of tests was made from a portion of the same rail (Index No. A

2193), cut transversely from near the middle (longitudinally) of the rail. The results of the chemical analyses of this rail are given in Table X., and the physical tests are given on Table XI.

TABLE X.—CHEMICAL ANALYSIS OF MIDDLE (LONGITUDINALLY) OF THE RAIL. Index No. A 2193.

	Analysis of Rail Head.	Analysis of Rail Foot.
Combined carbon	0·440	0 440
Silicon	0·059	0·061
Manganese	1·067	1·059
Sulphur	0·065	0·070
Phosphorus	0·063	0·064
Iron by difference	98·306	98·306
	100·000	100·000

TABLE XI.—LOSS OF STRENGTH IN VARIOUS PARTS OF THE VERTICAL SECTION OF THE RAIL, due to segregation. Index No. A 2193.

	Original Dimensions. Size in inches.	Area, square inches.	Distance between Gauge Points.	Maximum Stress per square inch.	Elongation, per cent.	Reduction of Area, per cent.	Remarks.
A	·564	·250	inches. 2	tons. 45·12	8·5	8·4	Granular.
B	·562	·248	2	46·53	14·0	17·7	Granular, with a trace fibrous. Cracked on the surface.
C	·564	·250	2	44·80	10·0	11·6	Granular. Slightly cracked on the surface.

It will be seen that in this part the rail was comparatively free from segregation, so that the segregated condition was local, and practically confined to the junction of the rail head with the web. As this localisation of the segregation was near one end of the rail, and seeing that the centre of the rail was comparatively free from segregation, the fair inference was that a sufficient portion had not been cut off from the top of the ingot from which the rail was rolled, and hence a considerable deficiency of physical properties had resulted in the rail as a whole.

Test marked A was taken from top of rail head. Test marked B was taken from rail bottom. Test marked C was taken from near junction of rail head with web.

Another illustration of the bad effects of segregation in steel rails, which the author recently met with in course of his practice as consulting metallurgical engineer and metallurgical chemist, is afforded by the following examination of another segregated rail.

The chemical analysis of this rail is shown in Table XII., which indicates to what a considerable extent the carbon and other elements had segregated in this rail near the junction of the rail head with the web.

TABLE XII.—CHEMICAL ANALYSIS OF ANOTHER SEGREGATED RAIL.

—	Analysis of Rail Head.	Analysis of Drillings near Junction of Rail Head with Web.	Analysis of Rail Foot.	Approximate maximum Percentage of Segregation.
Combined carbon	0·390	0·600	0·420	54
Silicon	0·084	0·093	0·084	10
Manganese	1·015	1·116	0·972	15
Sulphur	0·120	0·150	0·065	130
Phosphorus	0·069	0·107	0·065	64
Iron by difference	98·322	97·934	98·394	—
	100·000	100·000	100·000	—

The deleterious effects of this on the physical properties of the rail will be seen on referring to the results on Table XIII., which show that a considerable want of uniformity of physical

TABLE XIII.—LOSS OF STRENGTH IN PARTS OF THE VERTICAL SECTION OF THE RAIL DUE TO SEGREGATION.

—	Original Dimensions. Size in inches.	Area, square inches.	Distance between Gauge Points, inches.	Maximum Stress per square inch, tons.	Elongation per cent.	Reduction of Area per cent.	Remarks.
A	·564	·250	2	42·48	20·0	27·6	40% fibrous. 60% granular.
B	·564	·250	2	47·00	12·0	15·2	Granular.

strength existed between different portions in the vertical section of the rail. This condition of things indicates the presence of considerable sources of initial weakness in the rail as a whole.

Test marked A was taken from the top of the rail head. Test marked B was taken from near junction of the rail head with web.

These illustrations clearly show several aspects of the danger which lurks in segregated rails; as, for instance, the internal weakness revealed by the open fissures (see Figs. 5, 6 and 7) running transversely across the web at its junction with the rail head. Again, Fig. 8, compared with Figs. 9 and 10, affords visible and tangible evidence of the locally segregated condition of the combined carbon which, aided by the excess of manganese, has apparently produced the irregular massed micro-structure of the grey carbide of iron seen in the illustration. The carbide of iron areas have massed together and the rail is, in such segregated parts, liable to brittleness owing to the want of a uniformly surrounding and interlocking meshwork of ferrite. Moreover, considerable internal stress exists in steel possessing an irregular micro-structure of the above character.

Further, on Fig. 10 is seen the lurking insidious presence of numerous micro-segregations of micro-flaws (probably sulphide of manganese, silicide or phosphide of iron, etc.), or germs of metallic disease, so massed or segregated together as to produce in the ultimate structure considerable lines of internal weakness, which would readily develop into fracture under the influence of vibratory stress.

It may be remarked, that the examples given of segregation in steel rails in this paper are not isolated ones, but they are fairly typical of the evil induced in steel rails whenever segregation exists.

Longitudinal Segregation.

The author having demonstrated the position in transverse vertical section of the local segregation which is sometimes found in steel rails, a typical illustration of which is given in Figs. 1 and 4, it seemed desirable to make a further investigation, to trace the longitudinal extent of such areas of local segregation, and further, if possible, to ascertain the primary cause, and to suggest a remedy.

The segregated state of the chemical constituents sometimes observed in steel rails appears chiefly due to the primarily segregated condition occurring in the top part of the rail ingot. During the slow cooling of ingots, the impurities, such as sulphide, phosphide and silicide of iron and manganese, which freeze or consolidate at lower temperatures, remain liquid or semi-plastic for a longer period of time than the metal itself, and in course of the consolidation of the metallic mass the above impurities, being of less specific gravity, gradually liquate towards the top of the ingot and there chiefly and finally consolidate. Local segregation of the carbon contents also frequently occurs. The impurities appear to liquate towards the central longitudinal axis or pipe of the ingot, near the summit of the ingot, and disseminate themselves amid the intercrystalline spaces of the primary crystals of the steel, and in the vertical transverse section of the finished rail these parts of the ingot are mostly found at the lower part of the rail head, adjacent to the junction of the rail head with the web.

It will therefore be seen that the top of the ingot is likely to be the place where the greatest local segregation of impurities occurs. There the impurities are more massed together, and in this situation it is probable that the greatest segregation or dissociation of the chemical elements from mass uniformity takes place. The abnormal local concentration, or local aggregation of the chemical elements is thus accounted for. On the rolling out of the ingot into a finished rail this concentrated aggregation, or zone, becomes extended along the interior longitudinal axis of the rail, and in that end of a rail representing the top of the ingot the greatest tendency to internal weakness in segregated rails will be found. The term longitudinal segregation therefore appositely describes such a condition in the composition and structure of a segregated rail.*

Careful examinations and chemical analyses were made of numerous rails, commencing at one end of each rail and continuing the examination towards the other. The rails examined were of 85 and 96 lb. per yard section. Analyses were made of drillings from the end of the rails representing the top of the rail ingot, and other comparative analyses were made on sections at various distances in the length of the rail, with the results given in Tables XIV., XVI., XVIII. and XIX. The analyses were taken from the respective situations in the length of the rail shown on Figs. 11 and 12, and as described on the last-mentioned Tables. Physical tests were also made, showing the

* Segregation is likely to be most pronounced at that end of rails representing the tops of ingots. Some idea of the prevalence of segregation in steel rails is afforded by the fact that out of a total number of 379 analyses of steel rails, selected promiscuously from large bulks, the chemical constituents of 68 rails were found by the author to be more or less locally segregated at the end of the rail presumably representing the top of the ingot. This indicates approximately that about 18 per cent. of the total number of rails analysed were in a condition of local segregation at that end of the rail examined.

TABLE XIV.—LONGITUDINAL SEGREGATION IN STEEL RAILS. Examination of Rail No. 1714.

	Analysis of Rail in Transverse Section. Position A at end of rail representing Top of Ingot.			Analysis of Rail in Transverse Section. Position B, 5 ft. from rail end A.			Analysis of Rail in Transverse Section. Position C, 8 ft. from rail end A.			Analysis of Rail in Transverse Section. Position C1, 9 ft. from rail end A.		
	General Average Analysis of the Rail.	Analysis of Drillings near Junction of Head with Web.	Approximate Maximum Percentage of Segregation.	General Average Analysis of the Rail.	Analysis of Drillings near Junction of Head with Web.	Approximate Maximum Percentage of Segregation.	General Average Analysis of the Rail.	Analysis of Drillings near Junction of Head with Web.	Approximate Maximum Percentage of Segregation.	General Average Analysis of the Rail.	Analysis of Drillings near Junction of Head with Web.	Approximate Maximum Percentage of Segregation.
Combined carbon	0·457	0·560	22½	0·560	0·700	25	0·500	0·580	16	0·470	0·540	14¾
Silicon	0·075	0·078	4	0·082	0·126	53½	0·075	0·078	4	0·066	0·066	—
Manganese	0·965	1·008	4½	0·945	1·008	6¾	0·971	1·001	3	0·965	0·965	—
Sulphur	0·071	0·105	48	0·090	0·150	66¾	0·075	0·120	60	0·065	0·090	38½
Phosphorus	0·057	0·067	17¾	0·064	0·098	53	0·060	0·087	45	0·054	0·059	9¼
Iron by difference	98·375	98·182	—	98·259	97·918	—	98·319	98·134	—	98·380	98·280	—
	100·000	100·000		100·000	100·000		100·000	100·000		100·000	100·000	

The positions A, B, C and C1 named in the above Table will be understood on referring to Fig. 11.

TABLE XIV. (*continued*).—LONGITUDINAL SEGREGATION IN STEEL RAILS. Examination of Rail No. 1714.

	Analysis of Rail in Transverse Section. Position D, 11 ft. from rail end A.			Analysis of Rail in Transverse Section. Position F, 18 ft. from rail end A.			Analysis of Rail in Transverse Section. Position H, 24 ft. from rail end A.			Analysis of Rail in Transverse Section. Position J, 30 ft. from rail end A.		
	General Average Analysis of the Rail.	Analysis of Drillings near Junction of Head with Web.	Approximate Maximum Percentage of Segregation.	General Average Analysis of the Rail.	Analysis of Drillings near Junction of Head with Web.	Approximate Maximum Percentage of Segregation.	General Average Analysis of the Rail.	Analysis of Drillings near Junction of Head with Web.	Approximate Maximum Percentage of Segregation.	General Average Analysis of the Rail.	Analysis of Drillings near Junction of Head with Web.	Approximate Maximum Percentage of Segregation.
Combined carbon	0·510	0·510	—	0·480	0·490	2	0·480	0·480	—	0·470	0·490	4¼
Silicon	0·078	0·078	—	0·078	0·078	—	0·076	0·076	—	0·080	0·073	9½
Manganese	0·994	0·994	—	0·972	0·972	—	0·987	0·979	1	0·943	0·969	3
Sulphur	0·090	0·090	—	0·075	0·072	4	0·075	0·075	—	0·075	0·090	20
Phosphorus	0·065	0·073	12	0·054	0·061	13	0·054	0·070	29	0·062	0·061	1½
Iron by difference	98·263	98·255	—	98·341	98·327	—	98·328	98·320	—	98·370	98·317	—
	100·000	100·000		100·000	100·000		100·000	100·000		100·000	100·000	

The positions D, F, H and J named in the above Table will be understood on referring to Fig. 11.

effect of the local segregation in the vertical section of some of the rails at various places in the length of the rail. The results are given on Tables XV., XVII. and XX. (see also Fig. 15).

Analyses of the various sections of the rail head were made at the positions in the vertical section of the rails as shown on Figs. 13 and 14. The drillings for the general average analysis were taken from the positions marked A on Figs. 13 and 14. The drillings from the segregated areas, near junction of rail-head with web, were taken from the positions marked B on Figs. 13 and 14.

Examination of Segregated Rail, Index No. 1714.

The following rail examined was indexed No. 1714. It was of a section of 85 lb. per yard, and the length of the finished rail was about 30 feet. A certain portion had been sawn off in course of manipulation in the ordinary manner, and the first analysis was made from end A, which was the sawn end just as the rail was received from the makers. Other analyses were made of this rail in the respective longitudinal positions marked B, C, C1, D, F, H and J, the latter being the end furthest removed from the end of the rail representing the top of the ingot. The detailed results are given on Table XIV., and are graphically shown on Fig. 11.

In summarising the results, it will be observed that the local segregated area, near the junction of the rail head with the web, practically extended longitudinally from end A to point B, or for a distance of about 11 feet. From thence to the opposite end of the rail there was practical freedom from segregation of the chemical constituents, except to some extent in the case of phosphorus. In this rail sulphur was the element which showed the greatest tendency to local segregation, phosphorus taking the next place, and the carbon being practically next in order. In position B the local segregation of the silicon was considerable. The other elements had not been materially influenced by segregation. At the opposite end J of the rail there were, however, distinct indications of a reappearance of local segregation of all the chemical elements, though not of an extensive character. The results with this rail showed that the area of local segregation extended longitudinally for an approximate distance of about 11 feet from the end A of the rail, and its point of greatest concentration was about 5 feet from this end of the rail. Beyond the distance of about 11 feet from end A there was no appreciable local segregation, except to some extent just at the end J of the rail, as previously mentioned.

The physical tests (see Table XVI.) showed variable results,

ON THE STRENGTH OF STEEL RAILS. 225

TABLE XV.—LOSS OF STRENGTH IN STEEL RAILS FROM SEGREGATION, SHOWN BY PHYSICAL TESTS. Rail Index No. 1714.

Position.	Relative Position of Test Pieces longitudinally, see Table XIV. Fig. 11. Distance from End of Rail, representing Top of Ingot.		Position in Vertical Section of Rail, see Fig. 15.	Maximum Stress per sq. in. Tons.	Elongation per cent.	Reduction of Area per cent.	Remarks.
A	End of rail representing top of ingot.	A	Top of rail head.	42·68	18·0	24·4	25 % fibrous, 75 % granular, slightly cracked on surface after test.
		B	Junction of head with web.	48·56	6·0	8·4	Granular with a trace fibrous.
B	5 feet	A	Top of rail head.	44·04	16·0	18·4	Granular with a trace fibrous, slightly cracked on surface after test.
		B	Junction of head with web.	36·76	3·0	5·2	Granular.
C	8 feet	A	Top of rail head.	43·52	11·0	11·6	Granular with a trace fibrous, slightly cracked on the surface.
		B	Junction of head with web.	40·44	4·5	5·2	Granular.
D	11 feet	A	Top of rail head.	45·04	10·5	11·6	Granular with a trace fibrous, cracked on the surface after test.
		B	Junction of head with web.	40·88	4·5	5·2	Granular with a trace fibrous.
J	Opposite end of rail	A	Top of rail head.	45·80	14·0	18·4	15 % fibrous, 85 % granular.
		B	Junction of head with web.	46·72	13·5	18·4	15 % fibrous, 85 % granular.
		C	Rail bottom.	47·00	14·0	18·4	15 % fibrous, 85 % granular.

Q

and indicated the local loss of strength in the section of the rail consequent on the normal physical conditions having been disturbed by the existence of local segregation.

EXAMINATION OF SEGREGATED RAIL. Index No. 1524.
85 lb. per yard section.

The rail was analysed in a somewhat similar manner to rail 1714, the drillings being taken from the different positions A and B in the vertical section, as shown in Fig. 13. Sections were cut for examination from various parts in the length of the rail at points marked A, B, C, D, F, H and J, shown in Fig. 12. End A was the end of the rail representing the top of the ingot. The results of the several analyses are given on Table XVI., and they are graphically delineated on Fig. 12.

The physical tests (see Table XVII.) showed variable results, and indicated the local loss of strength at the point of greatest segregation longitudinally, in the section of the rail consequent on the normal physical conditions having been disturbed by the existence of local segregation. At end A of the rail the segregation was comparatively insignificant. At point B, 4 feet from end A, the local segregation of the carbon was considerable, and at its maximum, the sulphur and phosphorus were here also very much segregated. The sulphur had segregated to the greatest extent, the percentage of segregation of the phosphorus closely approaching that of the sulphur in this portion of the rail. It will be noticed that in this rail the area of concentrated local segregation was situated about 4 feet from end A of the rail. At point C, 8 feet from point A, the local segregation had practically died out, the remaining 22 feet in the length of the rail showing but slight indication of local segregation though its existence in a very modified form was detectable.

FURTHER EXAMINATION OF LONGITUDINAL SEGREGATION IN STEEL RAILS.

For these experiments a series of rails of heavy main line section, 96 lb. per yard, were rolled, care being exercised to ensure that the end of the rail representing the top of the ingot was specially stamped for subsequent identification in each case. The rails were rolled in one length, to finish to about 72 feet, which was sawn in two. Each rail examined in connection with this research represented at one end, A, the top of the ingot from which it had been rolled, the total length of each finished rail examined being 36 feet. Careful chemical analyses were made in transverse section at the end representing the top of the ingot and at different distances removed longitudinally therefrom, as stated in detail in Tables XVIII. and XIX. The drillings for the general average analysis were taken from the positions marked A on Fig. 14. The drillings from the segregated areas, near junction of rail head with web, were taken from the position marked B on Fig. 14.

Rail No. 1144.—A general summary of the results shows in the case of this rail that there was no segregation at the end of the rail representing the top of the ingot from which it was made. Nine feet removed from this there was a distinct segregation of carbon, sulphur and phosphorus, but not to any considerable extent. At the middle of the rail, 18 feet removed from the end of the rail representing the top of the ingot from which the rail was made, there was slight segregation of the sulphur, silicon and manganese, but at the other end of the rail there were no indications of local segregation.

Rail No. 1244.—At the end of the rail representing the top of the ingot from which the rail was made, the carbon and sulphur were much segregated locally and the phosphorus excessively, the silicon and manganese but slightly so. Nine feet removed longitudinally from this end there were but slight evidences of segregation, which at this distance, and again at a distance of 18 feet from the end of the rail, representing the top of the ingot from which the rail was made, had practically died out. At the opposite end of the rail, which would practically represent the centre longitudinally of the ingot, there was no local segregation; see results on Table XVIII.

Rail No. 1344.—At the end of this rail, representing the top of the ingot, segregation of the carbon, silicon, manganese and phosphorus was observed, the greatest percentage of segregation being in this instance with the carbon. At a distance of 9 feet from the above point the existence of local segregation had practically ceased, though there were still traces left. At 18 feet the local segregation of the sulphur and phosphorus had very considerably increased, but at the opposite end of the rail, representing the centre of the ingot, there were no traces of the existence of local segregation.

Rail No. 1444.—At the rail end, representing the top of the ingot, there was very heavy local segregation of all the chemical constituents, especially of the carbon, silicon, sulphur and phosphorus. At a distance of 9 feet from this end of the rail traces of segregation had almost ceased. At 18 feet there were indications of slight increased sulphur segregation and slight traces

228   EFFECT OF SEGREGATION

TABLE XVI.—LONGITUDINAL SEGREGATION IN STEEL RAILS. Examination of Rail Index No. 1524.

	Analysis of Rail in Transverse Section. Position A, at end of rail, representing Top of Ingot.			Analysis of Rail in Transverse Section. Position B, 4 ft. from rail end A.			Analysis of Rail in Transverse Section. Position C, 8 ft. from rail end A.			Analysis of Rail in Transverse Section. Position D, 11 ft. from rail end A.		
	General Average Analysis of the Rail.	Analysis of Drillings near Junction of Head with Web.	Approximate Maximum Percentage of Segregation.	General Average Analysis of the Rail.	Analysis of Drillings near Junction of Head with Web.	Approximate Maximum Percentage of Segregation.	General Average Analysis of the Rail.	Analysis of Drillings near Junction of Head with Web.	Approximate Maximum Percentage of Segregation.	General Average Analysis of the Rail.	Analysis of Drillings near Junction of Head with Web.	Approximate Maximum Percentage of Segregation.
Combined carbon	0·395	0·380	4	0·440	0·600	36	0·435	0·430	1	0·450	0·410	9¾
Silicon	0·068	0·066	3	0·069	0·069	—	0·066	0·066	—	0·066	0·067	1½
Manganese	0·778	0·778	—	0·788	0·792	—	0·828	0·828	—	0·778	0·778	—
Sulphur	0·085	0·075	13	0·120	0·190	58	0·120	0·140	16¾	0·102	0·122	19½
Phosphorus	0·052	0·054	4	0·055	0·085	54½	0·046	0·049	6½	0·053	0·051	4
Iron by difference	98·622	98·647	—	98·528	98·264	—	98·505	98·487	—	98·551	98·572	—
	100·000	100·000		100·000	100·000		100·000	100·000		100·000	100·000	

The positions A, B, C, D, referred to in the above Table, will be understood on referring to Fig. 12.

TABLE XVI. (*continued*).—LONGITUDINAL SEGREGATION IN STEEL RAILS. Examination of Rail Index No. 1524.

	Analysis of Rail in Transverse Section. Position F, 17 ft. from rail end A.			Analysis of Rail in Transverse Section. Position H, 23 ft. from rail end A.			Analysis of Rail in Transverse Section. Position J, 30 ft. from rail end A.		
	General Average Analysis of the Rail.	Analysis of Drillings near Junction of Head with Web.	Approximate Maximum Percentage of Segregation.	General Average Analysis of the Rail.	Analysis of Drillings near Junction of Head with Web.	Approximate Maximum Percentage of Segregation.	General Average Analysis of the Rail.	Analysis of Drillings near Junction of Head with Web.	Approximate Maximum Percentage of Segregation.
Combined carbon	0·450	0·420	7	0·450	0·430	4½	0·420	0·420	—
Silicon	0·068	0·067	1½	0·068	0·068	—	0·068	0·070	3
Manganese	0·778	0·785	1	0·785	0·781	—	0·785	0·785	—
Sulphur	0·105	0·110	5	0·110	0·115	4¾	0·115	0·122	6
Phosphorus	0·052	0·050	4	0·050	0·048	4	0·048	0·052	8⅓
Iron by difference	98·547	98·568	—	98·537	98·558	—	98·564	98·551	—
	100·000	100·000		100·000	100·000		100·000	100·000	

The positions F, H and J, referred to in the above Table, will be understood on referring to Fig. 12.

of carbon and phosphorus segregation. At 36 feet there were no indications of local segregation; see results on Table XIX.

TABLE XVII.—LOSS OF STRENGTH IN STEEL RAILS FROM SEGREGATION, SHOWN BY PHYSICAL TESTS. Rail Index No. 1524.

Relative Position of Test Pieces longitudinally, see Table XVI. Fig. 12.		Position in Vertical Section of Rail, see Fig. 15.		Maximum Stress per sq. in. Tons.	Elongation per cent.	Reduction of Area per cent.	Remarks.
Position.	Distance from End of Rail, representing Top of Ingot.						
B	4 feet	A	Top of rail head.	38·20	19·0	21·6	15 % fibrous, 85 % granular, slightly cracked on the surface after test.
		B	Junction of head with web.	38·96	6·0	8·4	15 % fibrous, 85 % granular.

Rail No. 1544.—At the rail end answering to the top of the ingot there was considerable local segregation of the carbon; the sulphur and phosphorus were also much segregated, as also the silicon. At a length of 9 feet from the end of the rail, representing the top of the ingot from which the rail was made, there were no signs of any practical local segregation. At a distance of 18 feet the sulphur was found to be heavily segregated locally, as also the phosphorus, but the latter not to a great extent. At 36 feet from the end of the rail, representing the top of the ingot from which the rail was made, no local segregation was found.

Rail No. 1644.—Local segregation had occurred extensively at the end of this rail representing the top of the ingot. This was chiefly the case with the carbon, sulphur and phosphorus. At a distance of 9 feet from this end there was no local segregation. At 18 feet from the end there were traces of segregation of the sulphur, and, to a very slight extent, of the other elements. At 36 feet, answering to the centre of the ingot, there were no signs of any local segregation.

The adverse manner in which the segregation had locally affected the physical strength and properties of the several rails, No. 1144 to No. 1644 inclusive, is shown on reference to Table XX. Careful chemical analyses and physical tests were made of the preceding rails, indexed 1144, 1244, 1344, 1444, 1544 and 1644, but for brevity the full details are given only for rails indexed 1244 and 1444; see Tables XVIII. and XIX. The whole of the results, however, show that the longitudinal segregation of some of the chemical elements was greatest at that end of each rail representing the top of the ingot from which it had been made.

Further Observations on the Influence of Segregation on the Loss of Physical Strength in Steel Rails.

The author has demonstrated, by a series of tensile tests made in various parts of the vertical section of segregated steel rails, that a considerable loss of tensile strength occurs in that part of the section of the rail near the junction of the rail head with the rail web. The author has made a further series of physical tests to confirm this circumstance, taking the test pieces from the positions indicated by the letters given on Fig. 15. The results of these physical tests show the effects of the local segregation on the loss of strength in steel rails (the results are given on Table XX.). The results of the chemical analyses, showing the extent of the segregation of the rails are given on Table XXI., so that a comparison may be made between the relative amount of segregation and the extent to which that condition had affected the physical properties of the rails experimented upon.

Physical tests were made of twenty-two other rails, which confirm the loss of strength from segregation illustrated in the typical samples of segregated rails given on Table XX.

The Relative Extent of the Segregation of the various Chemical Constituents in Steel Rails.

This investigation has practically demonstrated the approximate longitudinal extent of the segregated area in steel rails, and has further indicated the desirability of cutting off a greater length than has hitherto been deemed necessary from the end of the rail representing the top of the ingot, or, otherwise, of cutting off an equivalent portion from the top of the ingot itself, to ensure soundness and freedom from both segregation and piping voids in the finished rail. An additional point of interest has been the indication afforded of the relative extent of the segregation of the respective chemical constituents in steel rails, near the end of a rail representing the top of an ingot.

A reference to the Tables XIV., XVI., XVIII., XIX. and XXI. of chemical analyses shows that sulphur and phosphorus

TABLE XVIII.—LONGITUDINAL SEGREGATION IN STEEL RAILS. Examination of Rail, Index No. 1244.

	Analysis of Rail in Transverse Section. Position A at end of rail representing Top of Ingot.			Analysis of Rail in Transverse Section. Position D, 9 ft. from rail end A.			Analysis of Rail in Transverse Section. Position C, 18 ft. from rail end A.			Analysis of Rail in Transverse Section. Position B, 36 ft. from rail end A.		
	General Average Analysis of the Rail.	Analysis of Drillings near Junction of Head with Web.	Approximate Maximum Percentage of Segregation.	General Average Analysis of the Rail.	Analysis of Drillings near Junction of Head with Web.	Approximate Maximum Percentage of Segregation.	General Average Analysis of the Rail.	Analysis of Drillings near Junction of Head with Web.	Approximate Maximum Percentage of Segregation.	General Average Analysis of the Rail.	Analysis of Drillings near Junction of Head with Web.	Approximate Maximum Percentage of Segregation.
Combined carbon	0·450	0·660	47	0·415	0·430	3½	0·465	0·450	3⅓	0·467	—	No segregation in this part of the rail.
Silicon	0·092	0·097	5½	0·100	0·101	1	0·100	0·097	3	0·100	—	
Manganese	0·954	0·979	2¼	0·994	0·994	—	0·954	0·936	2	0·969	—	
Sulphur	0·085	0·132	55	0·065	0·070	7¾	0·100	0·100	—	0·075	—	
Phosphorus	0·063	0·121	92	0·064	0·066	3	0·061	0·058	5	0·062	—	
Iron by difference	98·356	98·011	—	98·362	98·339	—	98·320	98·359	—	98·327	—	
	100·000	100·000		100·000	100·000		100·000	100·000		100·000		

ON THE STRENGTH OF STEEL RAILS.   233

TABLE XIX.—LONGITUDINAL SEGREGATION IN STEEL RAILS. Examination of Rail, Index No. 1444.

	Analysis of Rail in Transverse Section. Position A at end of rail representing Top of Ingot.			Analysis of Rail in Transverse Section. Position D, 9 ft. from rail end A.			Analysis of Rail in Transverse Section. Position C, 18 ft. from rail end A.			Analysis of Rail in Transverse Section. Position B, 36 ft. from rail end A.			
	General Average Analysis of the Rail.	Analysis of Drillings near Junction of Head with Web.	Approximate Maximum Percentage of Segregation.	General Average Analysis of the Rail.	Analysis of Drillings near Junction of Head with Web.	Approximate Maximum Percentage of Segregation.	General Average Analysis of the Rail.	Analysis of Drillings near Junction of Head with Web.	Approximate Maximum Percentage of Segregation.	General Average Analysis of the Rail.	Analysis of Drillings near Junction of Head with Web.	Approximate Maximum Percentage of Segregation.
Combined carbon	0·430	0·610	42	0·415	0·430	3½	0·450	0·430	4½	0·427	—	No segregation in this part of the rail.
Silicon	0·081	0·130	60	0·092	0·090	2¼	0·088	0·088	—	0·086	—	
Manganese	0·972	1·037	6¾	0·958	0·965	¾	0·979	0·979	—	0·972	—	
Sulphur	0·070	0·110	57	0·070	0·070	—	0·080	0·070	14¼	0·070	—	
Phosphorus	0·065	0·123	89	0·068	0·063	8	0·062	0·060	3⅓	0·062	—	
Iron by difference	98·382	97·990	—	98·397	98·382	—	98·341	98·373	—	98·383	—	
	100·000	100·000		100·000	100·000		100·000	100·000		100·000		

TABLE XX.—LOSS OF STRENGTH IN STEEL RAILS FROM SEGREGATION, SHOWN BY PHYSICAL TESTS.

Rail Index No.	Position in Vertical Section of Rail. See Fig. 15.	Maximum Stress per sq. in. Tons.	Elongation per cent.	Reduction of Area per cent.	Remarks.
1755	A. Top of rail head	42·28	20·0	27·6	40 per cent. fibrous; 60 per cent. granular. Several flaws on test piece.
	B. Rail head and web junction	48·36	4·0	5·2	Granular. There is a longitudinal flaw in the test piece, extending inwards about $\frac{1}{8}$ inch from the surface.
1294	A. Top of rail head	43·80	16·0	21·6	Granular, with trace fibrous. Cracked on surface after test.
	B. Rail head and web junction	41·20	3·0	5·2	Granular.
1855	A. Top of rail head	41·40	17·5	21·6	25 per cent. fibrous; 75 per cent. granular. Cracked on surface after test.
	B. Rail head and web junction	50·84	7·0	8·4	Granular.
1494	A. Top of rail head	44·08	18·0	27·6	15 per cent. fibrous; 85 per cent. granular.
	B. Rail and web junction	29·72	1·5	1·6	Granular.
1026	A. Top of rail head	41·64	20·0	27·6	15 per cent. fibrous; 85 per cent. granular.
	B. Rail head and web junction	46·32	10·0	11·6	Granular, with a trace fibrous.
1894	A. Top of rail head	43·70	15·5	20·3	Granular, with a trace fibrous.
	B. Rail head and web junction	42·00	4·0	5·2	Granular.
1526	A. Top of rail head	39·92	19·0	24·4	15 per cent. fibrous; 85 per cent. granular. Slightly cracked on surface after test.
	B. Rail head and web junction	40·28	8·0	8·4	Granular, with a trace fibrous.
1236	A. Top of rail head	41·32	19·0	21·6	Granular, with a trace fibrous. Slightly cracked on the surface after test.
	B. Rail head and web junction	40·16	6·5	7·7	Granular, with a trace fibrous.

TABLE XX. (continued).—LOSS OF STRENGTH IN STEEL RAILS FROM SEGREGATION, SHOWN BY PHYSICAL TESTS.

Rail Index No.	Position in Vertical Section of Rail. See Fig. 15.	Maximum Stress per sq. in. Tons.	Elongation per cent.	Reduction of Area per cent.	Remarks.
1336	A. Top of rail head B. Rail head and web junction	42·44 47·24	20·5 7·5	30·8 8·4	30 per cent. fibrous; 70 per cent. granular. Granular.
1546	A. Centre of rail head B. Rail head and web junction	37·92 42·28	26·0 12·0	36·4 15·2	50 per cent. fibrous; 50 per cent. granular. 10 per cent. fibrous; 90 per cent. granular.
1386	A. Top of rail head B. Rail head and web junction	37·84 35·32	17·5 7·5	24·4 8·4	20 per cent. fibrous; 80 per cent. granular. Cracked on surface after test. Granular, with a trace fibrous. Badly cracked on surface after test.
1244	A. Top of rail head B. Rail head and web junction	42·44 44·76	11·0 2·5	11·6 4·4	Granular. Granular.
1344	A. Top of rail head	42·32	11·0	11·6	Granular.
1444	A. Top of rail head B. Rail head and web junction	45·52 45·80	12·5 6·5	11·6 6·8	Granular. Granular.
1514	A. Top of rail head B. Rail head and web junction	43·24 43·92	11·0 5·0	11·6 4·4	Granular, with a trace fibrous. Slightly cracked on the surface after test. Granular.
1644	A. Top of rail head	44·36	8·0	8·4	Granular.

NOTE.—The position of rail head and web junction mentioned in the foregoing Table is approximately the area of greatest segregation; see illustration of this on Figs. 1 and 4.

TABLE XXI.—EXTENT OF LOCAL SEGREGATION IN STEEL RAILS SHOWN BY CHEMICAL ANALYSES.

	Rail Index No. 1755.			Rail Index No. 1855.				Rail Index No. 1026.				Rail Index No. 1526.			
	Analysis of Rail in Transverse Section at Positions given below.		Approximate Percentage of Segregation.	Analysis of Rail in Transverse Section at Positions given below.		Approximate Percentage of Segregation.	Analysis of Rail in Transverse Section at Positions given below.		Approximate Percentage of Segregation.	Analysis of Rail in Transverse Section at Positions given below.		Approximate Percentage of Segregation.
	Rail Head.	Junction of Rail Head with Web.		Rail Head.	Junction of Rail Head with Web.			Rail Head.	Junction of Rail Head with Web.			Rail Head.	Junction of Rail Head with Web.	
Combined carbon	0·440	0·670	52	0·430	0·600	40		0·427	0·560	31		0·400	0·480	20
Silicon	0·066	0·066	—	0·080	0·080	—		0·080	0·087	8¾		0·099	0·095	4¼
Manganese	1·082	1·080	—	0·963	0·961	—		0·886	0·970	9½		0·788	0·886	12½
Sulphur	0·070	0·120	71½	0·060	0·130	117		0·063	0·115	82½		0·093	0·140	50
Phosphorus	0·049	0·098	100	0·060	0·104	73		0·057	0·083	45½		0·062	0·083	34
Iron by difference	98·293	97·966	—	98·407	98·125	—		98·487	98·185	—		98·558	98·316	—
	100·000	100·000		100·000	100·000			100·000	100·000			100·000	100·000	

Complete chemical analyses were made of 49 other segregated rails, which confirmed the results of the typical segregated rails given on the above Table. For brevity, however, the details of the chemical analysis are not given.

appear to have the greatest tendency to segregation, and afterwards carbon, though in some instances carbon is capable of heavily segregating. Silicon does not so frequently appear to have segregated heavily, though in some instances it has been observed to have done so, as in the cases of rails indexes 1714 and 1444. It does not, however, appear that an absolute rule can be established, as local and other special circumstances may induce varied segregated formations.

TABLE XXII.

	Index No.	Length of Test Piece and Distance from End of Rail.	Weight of Ball 1 ton. Rail placed on Bearings 3 feet 6 inches apart.	
			First Fall 7 feet.	Second Fall 20 feet.
			Deflection in inches.	Deflection in inches.
		feet.		
First test from segregated end "A" of rail.	1044	5	1	Broke
	1294	,,	1	,,
	1494	,,	$1\frac{1}{8}$	
	1894	,,	$1\frac{1}{16}$	,,
	1925	,,	$1\frac{1}{8}$	,,
Second tests (unsegregated) from point "C" or middle of rail.	1044	,,	$\frac{7}{8}$	$3\frac{5}{16}$
	1294	,,	1	$3\frac{5}{8}$
	1494	,,	$1\frac{1}{8}$	$3\frac{11}{16}$
	1894	,,	$1\frac{1}{8}$	$3\frac{7}{8}$
Third tests from the opposite (unsegregated) end "D" of rail.	1044	,,	1	$3\frac{3}{16}$
	1294	,,	1	$3\frac{5}{8}$
	1494	,,	$1\frac{1}{16}$	$3\frac{3}{4}$
	1894	,,	$1\frac{1}{16}$	$3\frac{3}{4}$

* Impact tests were only made at the segregated end of rail 1925.

INFLUENCE OF LOCAL SEGREGATION IN REDUCING THE IMPACT RESISTANCE OF STEEL RAILS.

A test was made to show the relative resistance to impact of segregated and normal portions of the same new steel rail, index No. 1644 (see Fig. 16), of a section of 97 lb. per yard. The comparative results of this test are given in detail in Fig. 16. The results show the extent to which the resistance to impact had been locally reduced in the different portions of the section of the rail, and add confirmation to similar results derived from the other physical tests which have been made.

This rail had a segregated chemical composition, and its physical properties are given on Table XX.

TABLE XXIII.

Index No.	Section Weight in Pounds per yard.	Length of Test Piece and Distance from End of Rail.	First Fall 7 feet. Deflection in inches.	Second Fall 15 feet. Deflection in inches.	Third Fall 10 feet. Deflection in inches.
	lb.	feet.			
1536	85	5	$1\frac{3}{8}$	Broke	—
1486	,,	,,	$1\frac{1}{2}$	,,	—
1027	,,	,,	$1\frac{7}{16}$	,,	—
1137	,,	,,	$1\frac{3}{16}$	,,	—
10137	,,	,,	$1\frac{3}{16}$	,,	—
1337	,,	,,	$1\frac{7}{16}$	,,	—
1367	,,	,,	$1\frac{3}{8}$	,,	—
10367	,,	,,	$1\frac{5}{16}$	,,	—
1667	,,	,,	$1\frac{1}{4}$	$3\frac{3}{8}$	Broke
1877	,,	,,	$1\frac{1}{4}$	Broke	—
1287	,,	,,	$1\frac{5}{8}$	,,	—
1687	,,	,,	$1\frac{5}{16}$	,,	—
10687	,,	,,	$1\frac{5}{16}$	,,	—
1008	,,	,,	$1\frac{7}{16}$	,,	—
1388	,,	,,	$1\frac{1}{4}$	,,	—
100388	,,	,,	$1\frac{5}{16}$	,,	—
*1636	96	,,	$1\frac{1}{16}$	,,	—
*1107	,,	,,	$1\frac{1}{8}$	,,	—

Total number of rails tested, 285; total number of rails broken, 18. Percentage of failures, 6·31.

* In the tests of rails Nos. 1636 and 1107 the second fall was 20 feet instead of 15 feet in the case of the others.

Further confirmation of the adverse influence of local segregation on the impact resistance of steel rails is afforded by additional large scale experiments made on rails of the 96 lb. bull-head section, numbered 1044, 1294, 1494, 1894 and 1925. The detailed results of the complete chemical analyses and physical tests have been omitted for brevity. On Table XXII. will be found the results of the drop tests obtained respectively with each rail.

The rails should have withstood the two impacts without sign of fracture, whereas each broke at the second blow during

TABLE XXIV.—CHEMICAL ANALYSES OF STEEL RAILS WHICH BROKE UNDER DROP TESTS. (See Table XXIII.)

	Rail Index No. 1636.			Rail Index No. 1636.			Rail Index No. 1486.			Rail Index No. 1027.		
	General Average Analysis of the Rail.	Analysis of Drillings near Junction of Head with Web.	Approximate Maximum Percentage of Segregation.	General Average Analysis of the Rail.	Analysis of Drillings near Junction of Head with Web.	Approximate Maximum Percentage of Segregation.	General Average Analysis of the Rail.	Analysis of Drillings near Junction of Head with Web.	Approximate Maximum Percentage of Segregation.	General Average Analysis of the Rail.	Analysis of Drillings near Junction of Head with Web.	Approximate Maximum Percentage of Segregation.
Combined carbon	0·490	0·560	14¼	0·470	0·620	32	0·470	0·660	40½	0·416	0·560	35
Silicon	0·071	0·080	12¾	0·082	0·106	29¼	0·080	0·083	3¾	0·070	0·074	8½
Manganese	0·886	0·933	5¼	1·021	1·059	3¾	0·922	0·951	3	0·936	1·030	10
Sulphur	0·090	0·120	33	0·060	0·110	83⅓	0·060	0·130	116⅔	0·050	0·100	100
Phosphorus	0·058	0·069	19	0·052	0·065	25	0·054	0·098	81¼	0·063	0·103	63
Iron by difference	98·405	98·238	—	98·315	98·040	—	98·414	98·078	—	98·465	98·133	—
	100·000	100·000		100·000	100·000		100·000	100·000		100·000	100·000	

TABLE XXV.—PHYSICAL TESTS OF STEEL RAILS WHICH BROKE UNDER DROP TESTS. (See Table XXIII.)

Rail Index No.	Position in Vertical Section of Rail.	Maximum Stress per sq. in. Tons.	Elongation per cent.	Reduction of Area per cent.	Remarks.
1536	Centre of rail head	43·56	14·0	18·4	Granular, with a trace fibrous; cracked on surface after test.
	Rail bottom	44·20	18·0	21·6	10% fibrous, 90% granular; slightly cracked on surface after test.
1636	Centre of rail head	39·92	4·0	5·2	Granular, with a trace fibrous.
	Rail bottom	45·68	22·0	38·7	Fibrous, with a trace finely granular.
1486	Centre of rail head	41·20	10·0	11·6	Granular, with a trace fibrous.
	Rail bottom	41·00	24·0	39·2	Silky fibrous.
1027	Top of rail head	42·14	11·5	14·5	Granular.
	Centre of rail head	43·52	11·0	11·6	Granular.
	Rail bottom	42·54	22·5	41·5	Silky fibrous, with a trace very finely granular.
1137	Top of rail head	45·36	5·0	8·4	Granular.
	Centre of rail head	45·68	6·0	8·4	Granular.
	Rail bottom	45·60	5·5	8·4	Granular.

TABLE XXV. (continued).—PHYSICAL TESTS OF STEEL RAILS WHICH BROKE UNDER DROP TESTS. (See Table XXIII.)

Rail Index No.	Position in Vertical Section of Rail.	Maximum Stress per sq. in. Tons.	Elongation per cent.	Reduction of Area per cent.	Remarks.
1367	Centre of rail head	43·81	5·0	5·9	Finely granular.
	Rail bottom	45·11	11·0	12·9	Finely granular, with a trace fibrous; slightly cracked along the surface after test.
1877	Centre of rail head	47·32	15·0	18·4	Granular, with a trace fibrous.
	Rail bottom	47·24	22·0	39·2	50 % fibrous, 50 % finely granular.
1287	Centre of rail head	39·44	15·5	18·4	Granular, with a trace fibrous; cracked on the surface after test.
1388	Centre of rail head	46·16	9·0	11·6	Granular.
	Rail bottom	45·64	12·5	15·2	Granular, with a trace fibrous; cracked on the surface after test.
1107	Centre of rail head	43·84	15·5	21·6	20 % fibrous, 80 % finely granular; slightly cracked along the surface after test.
1337	Centre of rail head	43·88	17·0	24·4	25 % fibrous, 75 % finely granular.
1667	Centre of rail head	49·78	15·0	18·5	Granular.
1687	Centre of rail head	46·08	16·0	24·4	10 % fibrous, 90 % finely granular.
1008	Centre of rail head	42·68	21·0	33·6	20 % finely granular, 80 % fibrous.

R

the tests of the segregated end of the rail. In the centre and other end of the rails, remote from local segregation, the rails stood the drop tests satisfactorily.

Additional investigations were made on these four rails, 1044, 1294, 1494 and 1894, and impact tests were made on other portions taken from the middle and also from the opposite end of each rail. These portions of the rails satisfactorily stood the test imposed. The results of these second and third sets of impact tests are also given on Table XXII.

For purposes of comparison the author made complete chemical analyses of the ends of the last-mentioned rails, representing the tops of the rail ingots, and the chemical analyses taken respectively at distances of 9 feet and 18 feet from the segregated ends, and also from the opposite ends, about 36 feet removed from the segregated ends. In these positions there were no practical indications of local segregation. On comparing the results of the chemical analyses with the results of the drop tests it was at once seen that the reduction of impact resistance was local and consequent on the locally segregated condition of the rails.

Another series of tests were made, in course of which a total of 285 rails, representing a very large bulk of rails received from many of the chief English rail manufacturers, were submitted to the usual impact test. Out of this number 18 rails failed, or a percentage of 6·31 of those tested. The chemical analyses and physical tests showed that in these cases the failure was due to the locally segregated condition of the carbon, sulphur or phosphorus, and also, in most instances, to the high percentages of combined carbon and manganese present in the rails.

A complete chemical analysis was made of each of the 285 rails used in this part of the investigation. The chemical composition was normal, and there was no appreciable local segregation of any of the chemical components in the 267 rails which satisfactorily stood the test. There was, however, considerable local segregation of most of the chemical elements observed in each of the 18 rails which broke under the drop test. For brevity the chemical analyses only of some typical ones are given on Table XXIV., and the physical tests of the segregated rails which broke are given on Table XXV.

Internal Flaws induced by Local Segregation and Piping in Steel Rails.

A typical illustration of internal flaws arising from a segregated state of the chemical elements is afforded by rail No. 1444, see Fig. 17. The chemical composition of this rail and its want of uniformity of physical properties are seen on Tables XIX. and XX. Owing to the locally segregated condition of this sample, and the apparent piping of the ingot, it seemed desirable to search for internal flaws in the top of the web, adjacent to the junction of the web with the rail head, near the end of the rail. On cutting sections the author discovered a long internal flaw running longitudinally in the centre of the web of the rail for a distance of more than about $2\frac{1}{2}$ inches. The flaw varied in width from about $\frac{1}{16}$ inch, and extended downwards, from the junction of the web with the rail head, for some distance into the web. It will thus be seen that the rail web was practically split in two, as there was no cohesion between the metal on either side of the web centre, which was for some distance divided by the internal flaw above mentioned.

The appearance of the flaw, as seen in horizontal section near the summit of the rail web at its junction with the rail head, is shown on Fig. 17. The flaw had towards its longitudinal termination proceeded in a transverse direction towards one side of the web. The risk of this flaw is accentuated by the fact that it runs longitudinally across one of the fish-plate bolt-holes. The author considers the flaw is due to the locally segregated state of the rail near this part, and to the apparent piping which had taken place in the ingot, and it is an illustration of the lurking danger attending a segregated condition in steel rails. A typical instance of internal flaws existing near the end of a rail from piping of the ingot is shown on reference to Figs. 18 and 19. These are from a rail of 85 lb. section, which had fractured after only three years' wear, the rail having been laid on gravel ballast on the level in a curve. The rail had worn down in three years to a weight of $81\frac{1}{4}$ lb. per yard, representing an average loss in weight of $1 \cdot 25$ lb. per yard per annum.

The author cut sections from near the end of the rail and discovered a long internal flaw running longitudinally in the centre of the web of the rail for a distance of above 7 inches. This flaw extended downwards from the junction of the web with the rail head for a distance of about $1\frac{1}{4}$ inches. From this it will be seen that the rail web was practically split in two as there was no cohesion between the metal on either side of the web centre, which was divided by the long internal flaw. The width of this flaw was in some places nearly $\frac{1}{16}$ inch, and varied in width downwards. The appearance of this flaw, as seen in section on the summit of the rail web at its junction with the rail head, is shown on Fig. 18. This flaw bifurcated and formed other ramifications extending upwards into the rail head, as seen in Fig. 19. Further, on attempting to machine a portion from the rail head, for purposes connected with the investigation, the rail head actually split up into about half a dozen separate pieces; the lines of fracture being along the line of the centre of the ingot, and its ramifications extending into the rail head, as shown on Figs. 18 and 19. These various indications, and especially the presence of the large central longitudinal flaw alluded to, which practically divided the web and also weakened the lower part of the rail head (see Figs. 18 and 19), point to the probability that sufficient length had not been sawn off from the end of the finished rail, which at that end (evidently representing the top of the rail ingot) showed manifest signs of unsoundness traceable to the foregoing cause.

External Segregation.

The author has sometimes detected, in the course of his examinations and analyses of steel rails, the existence of a local external segregation of some of the chemical components; in addition to internal transverse segregation and internal longitudinal segregation of some of the chemical constituents of steel rails. The phenomenon is allied to that of the internal transverse and longitudinal segregation which the author has recently investigated, but its effect on the strength of a rail is somewhat different.

With regard to the cause of external local segregation in steel rails, he has demonstrated that local internal segregation mostly exists in those rails which represent the top part of an ingot. During the cooling of large steel ingots the author thinks that in addition to local segregation, near the top of the ingot, there may sometimes be a further transverse migration of the carbon and other elements from the central axial part towards the exterior of the ingot. This tendency to external migration of the chemical elements he considers is most likely to occur near the bottom or wider part of the ingot. Now, when such an ingot is rolled into a rail say 90 feet long (which is finally cut in 30-feet lengths), that rail representing the bottom of the ingot is likely from the above cause to manifest a condition of local external segregation, as distinct from interior axial longitudinal segregation; whereas, the rail representing the top portion of the ingot would probably afford indications only of internal local axial segregation. The author further thinks that the keeping of steel ingots for any lengthened period in soaking pits tends to promote local segregation of some of the chemical constituents.

If either internal or external local segregation arising from any cause is present to any considerable extent, it becomes a serious source of danger leading to the sudden fracture of rails. The author has recently met with a practical illustration of the danger to rails from external segregation of some of the chemical constituents of the above nature. This example was afforded by the fracture of a rail on a bridge of one of the chief English railways, which nearly produced a serious accident to an express train. The author made a careful investigation into the cause of that accident, which was distinctly traced to an extensive external segregation of some of the chemical constituents of the rail at the point of fracture. The author has been requested to examine other cases of the fracture of steel rails which have occurred in main line service on various railways, and on careful examination he found, in many instances, the cause of the fracture was mainly due to the local segregation of some of the chemical constituents.

The present research on segregation has demonstrated the nature and approximate extent of the evil, and has enabled the author to point out a remedy with some approach to reliability. The investigation has also shown the desirability as far as practicable of preventing segregation in steel rails, so as to minimise the risks arising therefrom.

The author gratefully desires to express his indebtedness to several railway companies, who are his clients, for their great kindness in affording every facility and assistance for the carrying out of the experiments of the research.

DISCUSSION.

The PRESIDENT said that it was hardly necessary for him to explain that the paper which had been read represented but a small proportion of the mass of information which the author had available on the subject. From that mass he had selected only such portions as would enable the meeting to grasp the most important results of his very elaborate investigations. As engineers, they must highly appreciate the fact of a scientist, a member of their profession, devoting himself to a detailed study of any branch of the profession as the author of the paper had done with regard to the segregation of steel rails. Railway engineers must feel particularly grateful to Mr. Andrews, because his investigations bore especially upon their branch of the profession. Speaking for himself and for the Members generally, he must confess that they owed a great debt of gratitude to the author for giving the Society the first use of his very valuable results, and the opportunity of publishing them. Exhaustive as the paper appeared to be, it only barely represented the magnitude and extent of the investigations which the author had made. It was a matter of regret to the Council that they were unable to present to the Members the whole of the material which the author had offered. The regret, however, was mitigated by the fact that the author had successfully abbreviated the material, so as to bring it within a reasonably small compass without seriously interfering with the value and usefulness of the paper. He was sure that the meeting would accord to the author a very hearty vote of thanks for his most valuable paper.

The vote of thanks was accorded by acclamation.

Mr. J. E. STEAD said that he heartily endorsed every word that the President had said with reference to the colossal amount of work which Mr. Andrews had performed. Mr. Andrews was continually giving the world the result of his researches. It was, however, to be regretted that he had not given them the benefit of other people's researches in the same direction.

Mr. Alexander Pourcel read a paper on 'Segregation' at the Engineering Congress at Chicago, in which he stated that it was near the top of ingots where it was most pronounced, and that rails made from that part were weaker than those made from the lower parts of the same ingots. He referred to a great mass of literature on the same subject published by Professor Howe, of New York, and drew attention to Tchernoff's researches in 1878, in which he spoke of the segregation in ingots having been observed in 1866. Pourcel also referred to segregation in guns, shafts, etc., and stated that in the latter it was not dangerous in consequence of the segregate being confined to the central axis. He (Mr. Stead) referred to these points lest it should be imagined that metallurgists were not well aware of segregation in steel rails. Many years ago, Mr. T. W. Hogg, of the Newburn Steel Works, Newcastle-on-Tyne, read a paper before the Society of Chemical Industry which contained the results of a very careful research he had made with reference to segregation in ingots.

He (Mr. Stead) had been intimately connected for about thirty-two years with the manufacture of wrought-iron, steel rails, plates and other material, and it had been well known to him for at least twenty-five years that segregation was usually found exactly in the position shown by Mr. Andrews. One might be inclined to conclude from Mr. Andrews' statement that the segregation which he admitted to be present in about one-fifth of the total number of rails produced, must be regarded as liable to cause or lead to fracture.

A most admirable report on the subject of the wear of steel rails had been published by the Board of Trade which had been compiled by many gentlemen of metallurgical prominence, and at the end of that document there was a valuable report by Sir Lowthian Bell. Sir Lowthian Bell had, perhaps, devoted more time to the study of iron and steel than any other living person, and they could not but have great consideration for what he said. In concluding his report, he stated that there were "15,000,000 rails on British railways, and of that number only 600 rails broke per annum." That was equal to one rail in 25,000. Yet it was a fact that there were at least 3,000,000 segregated rails in use at the present time, for every ingot contains a segregated part, and that part made one rail, the remainder giving three or four rails. Last year he believed that there had not been a death by the breaking of a rail in the whole of the United Kingdom. That was a good record. But if they were to judge from the remarks of Mr. Andrews, they ought to believe that one rail in every five was liable to break on the track. It was, however, only one in every 25,000 which actually did break. Were they then justified in assuming that, as a rule, it was segregation which led to those fractures? He did not say that there were no isolated instances in which it might; but he asserted, judging from his experience in the examination of rails which had been found fractured on the railway track, that, generally, it was not segregation that was responsible.

He noticed that Mr. Andrews sometimes classed segregation and piping together as the causes of weakness, which would lead to the conclusion that he was not certain which it was. Was it not more likely to be piping rather than segregation which was responsible for those evils? He knew as a fact that occasionally piping occurred at the point where there was the greatest segregation. Steel-makers did cut off a considerable portion of the top of the ingot and endeavoured to get rid of the piped portion. They were all aware that it was likely to be there, but even "in the best regulated families" occasionally there might be a mistake, and, do as they might, occasionally an ingot would come out with a pipe in the middle of it at the very part where there was the segregation. He was inclined to attribute many of the fractures to which Mr. Andrews had referred to the actual separation through the piping and not to segregation. What was the remedy? He did not exactly glean from Mr. Andrews' paper what the remedy was. Did Mr. Andrews propose to reject one-third or one-fifth of the whole ingot? Generally steel-makers were very anxious to know everything they could, and were glad to receive suggestions; and he was sure that, if Mr. Andrews could give them any practical remedies for avoiding segregation, they would act upon them. He (Mr. Stead) happened to be intimately connected with the testing of the material which went into the iron and steel works of this country. Many thousands of tons had passed through his hands, and he knew that, if steel-makers could get any means by which they could still further increase the purity of their steel, or avoid segregation, they would be very glad to use them.

He would now pass to the question of external segregation. He thought that all who had had experience in examining steel rails would agree that external segregation was an exceedingly rare thing. He (Mr. Stead) had analysed and microscoped and looked at hundreds of rails, and he must confess that he had not found a single case of external segregated rail.

With regard to central segregation, many metallurgists held that it had a tendency to strengthen the rail and not to make it weak. They could conceive that that would be so, for such a rail practically had a rod of hard steel right through the centre of it, and unless the rail was actually bent it would not break, but act as a strong support to the perfectly good portions of steel surrounding it. With the vibration caused by the comparatively slight shocks that fell upon the rail, it was not likely that the internal part would break. In fact he had never found a case in which the fracture had actually been proved to have started in the central axis, though he had examined many rails which had fractured on the track. Was it not likely that segregation, instead of being an actual evil, was an advantage? He did not say that it was so, but a large number of people believed that it strengthened a rail instead of weakening it.

He must now discuss what Mr. Andrews called "sulphur flaws." What were they? They really consisted of sulphide of manganese, and sometimes silicate of manganese, or of both, but never contained either silicide or phosphide of iron. If the flaws were in the little cigar-shaped forms, it was very difficult to see how they could affect and cause the steel directly in contact with them to fracture at right angles to their longitudinal axes. He quite agreed that the Yorkshire iron so often recommended was an excellent material, and was not liable to fracture with fair use, but if they took that iron and submitted it to the microscopic test they would find that it had enormous flaws, for the layers and threads of cinder had more right to the term "flaw" than the torpedo micro-specks of sulphide of manganese in steel. It did not require a power of 300 diameters to see them as it did in steel. If, then, the so-called sulphide flaws lead to fracture in steel, one would expect that the exact equivalent of such flaws (but in a much more exaggerated form) in wrought iron would make such iron more liable to fracture than steel.

He was afraid that Mr. Andrews' paper was largely based on hypotheses, and not upon a sufficient number of practical facts. There were fractures; there were micro-flaws (so-called); a rail broke down; therefore it was caused by the sulphide flaws. That was apparently the way Mr. Andrews argued. But he would return to the report of the Board of Trade previously referred to. Sir Lowthian Bell had there given a record of the examination of 37 rails which broke in actual use, and he (Mr. Stead) had co-related his results, and found that the evidence tended to show that "high sulphur rails were less likely to break on the permanent way than those containing little sulphur, for, of 37 rails broken in use, 24 were low in sulphur and 13 high, a difference of 11 in favour of the high sulphurous rails." Those high sulphurous rails contained as much sulphur as many of Mr. Andrews' segregates, and as many so-called sulphur flaws. They must, therefore, conclude that the sulphur flaws were not responsible for the fracture or the breaking down of the rails in Mr. Andrews' examples. It was also shown in this report that the average life of high sulphurous rails before breaking was about 4 years longer than that of low sulphurous rails. A third point proved beyond doubt was that the sulphurous rails after long use resisted percussive shocks, and were mechanically better than rails which were low in sulphur. The conclusion drawn from Sir Lowthian's data was that the so-called flaws were beneficial microbes and not, as Mr. Andrews believed, germs of disease. His own experience—and in this he was confirmed by many steel-makers—was that, provided a rail was perfectly sound and was not redshort in consequence of the presence of sulphur, it would stand the ball test equally well if not better than a rail containing less sulphur. What he was stating was a fact, not a hypothesis. The only objection to sulphur in rails, so far as he could find by the evidence afforded by the Board of Trade report and his own long experience, was that such rails oxidised a little more rapidly than those comparatively free from sulphur. There was no doubt that sulphide of manganese on the surface of a rail assisted corrosion: the sulphide oxidised, producing sulphate and, rusting, crept along the surface of the rail. With regard to that point, Mr. Andrews and he (Mr. Stead) agreed perfectly.

Before concluding, he wished to refer to the work of Brinell. Mr. Brinell made some very interesting researches to ascertain the effect of sulphur on the mechanical properties of steel. He purposely introduced large percentages of sulphur into steel, much larger quantities than Mr. Andrews had dealt with, and he found that such steel when rolled into bars stood shock better than material with less sulphur. Why was that? It was for the same reason that wrought iron was not so liable to fracture by shock in its fibrous condition as it was after melting, for such sulphurous steel was slightly fibrous. The sulphide of manganese when in large quantity separated the threads or elongated crystals of steel from each other, just in the same way that cinder separated the threads of iron in wrought iron. If a minute fracture began, it soon travelled to one of the cinder or sulphide areas, and its growth was checked; whereas if no cinder were present it more readily continued to travel. Some of Mr. Andrews' lantern slides appeared to confirm that conclusion. That was his (Mr. Stead's) explanation. He had tried to get at the bottom of the matter, and to find out how the fractures originally were formed in steel rails, and he must confess that there was very great difficulty in many cases in finding out what actually caused a fracture. One might find a crack right through a rail, and on examining each side of the fracture one might or might not find sulphide or segregation, and it was not wise to conclude because on each side of the fracture there might or might not be sulphide, that that was the cause of the fracture. There might be an initial flaw along which the cleavage travelled.

He would like to know where the fractures occurred in the rails which Mr. Andrews examined. Were they right in the middle, or were they generally near one end? His own experience was that the greater number of breaks occurred at a point very near to where the fish-plates were located. That would point to the necessity of introducing a stronger fish-plate, one which would go right under the rail as well as on the side of it, so as to prevent tension at any time being placed on the head of the rail. A steel rail after being in use under heavy traffic for only one or two years became brittle on the surface of the head, and often broke under a ton weight falling 10 feet, but if $\frac{1}{8}$ of an inch were planed off the head of the rail, it would stand such a test without breaking. The crushing of the rolling stock on the head of the rail produced hardening. It eliminated all power of extension. Consequently, if a sleeper sank, or the rails were not well supported at the ends, the surface of the head was put in tension. He had seen repeated cases in which fracture could only be explained in that way, for everything was right with the steel excepting that the surfaces of the head were brittle, a state of things common to all rails after they had been in use.

He must compliment Mr. Andrews upon trying his very best to get at the bottom of these mysteries. He had also tried to do so in his own researches, and had frequently formed hypotheses which he had had to abandon for better, as more and more truth was brought to the surface; and he believed that Mr. Andrews, when he examined many of the millions of rails which were segregated and which were doing good service on railways, would modify his opinion as to the effect of segregation in rails.

Mr. C. H. RIDSDALE said that he endorsed what Mr. Stead had said as to the very great amount of pains that Mr. Andrews had taken in preparing the paper. At the outset he would admit the broad fact of the occurrence of segregation. It was, as Mr. Stead had said, not a new and disquieting state of things that had recently arisen. It was an inevitable law of nature. Segregation was always found. At the North-Eastern Steel Company's works, with which he was connected, they only made basic steel, although he gathered that Mr. Andrews referred chiefly to acid steel; still their experience might be interesting. Fourteen years ago they made an exhaustive series of investigations with many hundreds of analyses at different points of the ingots, many of which were split in two and examined, and also of rails at different points, representing the top and a very few inches down. That exhaustive series of investigations was made to determine the extent of segregation, and particularly how far it was of moment in practice, and how best to prevent harm resulting from it. He proposed to refer a little later to the results of that series of experiments. He would make, first of all, a few remarks more particularly as to the paper itself.

It was of no use to get unduly alarmed at segregation. In the first place, Mr. Andrews in his paper, so far as he could see, had brought forward no proofs of any serious practical ill effect resulting from segregation—he did not think that there was any example of serious or bad results; and, although by his (Mr. Andrews') showing, a large proportion of the rails in regular use had the faults he described, he did not mention one accident or one fracture at the junction of the head directly traceable to that cause, and he (Mr. Ridsdale) had never heard of one. He had, of course, come across a number of fractures, but they had always been traceable either to a small flaw such as Mr. Stead mentioned—probably a red-shortness flaw which had been accidentally overlooked, or, perhaps, to corrosion from sea water on a line near the coast. He was sure that the remarkable immunity from accident which railways enjoyed bore witness to the same thing. He believed during the whole of 1901 not one life was lost through accident to a train on English railways.

Again, the amount of impurities described in the segregates was only what the whole rails frequently averaged in other countries than England. In England there was a much purer standard. He had referred to that in a paper read before the Iron and Steel Institute in 1901, on 'The Correct Treatment of Steel.' He had summarised several of the instances. One was from the Rails Commission Report. In the summary he then gave he mentioned that the conclusion to be drawn from the report was that, on the whole, the harder or impurer rails wore better than the purer or softer. Since that he had gone carefully into another set of tables in the same report dealing with 66 rails.

He ought to say that it was through seeing the results of Mr. Stead's summarising of the sulphur in those same tables that made him think he would look up the phosphorus results. Therefore he did so on precisely the same lines; so in that respect it was a copy of Mr. Stead's work. Of the 66 rails, the first group, of 8 rails, contained 0·124 per cent. of phosphorus; the next group, of 24 rails, had only a medium amount of phosphorus, about 0·071 per cent.; and the last group, of 34 rails, had only 0·044 per cent. Those with the 0·12 of phosphorus averaged 17½ years' service against 14½ years for the pure ones, and the wear in the rails with 0·12 per cent. phosphorus was only 0·26 lb. per yard per annum. In the purer rails it was as much as 0·33 lb.—about 27 per cent. more. And yet, notwithstanding the higher phosphorus, after their long service of 17½ years in the case of the impure ones, and only 14½ years of service in the case of the pure ones, the impure ones stood 34¾ ton-feet—that was an impact test—and the purer rails only stood 27 ton-feet.

Another point he had mentioned in his paper was the experience of Dormus, an Austrian railway engineer, who made a very long and elaborate investigation, and found that pure rails only had a life of eleven years with 80,000,000 tons of traffic passing over them, and many of them had broken, whereas impure rails, with 0·11 to 0·15 per cent. of phosphorus, had been in use for twenty years with 102,000,000 tons passing over them, and not a single rail had broken. On the strength of that he had deliberately selected an impure composition (with about the same amount of impurities as that contained in the segregates Mr. Andrews had mentioned) for his railway. He (Mr. Ridsdale) was speaking, not of carbon, but of phosphorus and other impurities.

The experiments at the North-Eastern Steel Company's works bore out that such impure steel would stand severe practical tests. They had often cut out the segregated impure portions, as shown in Fig. 16, and subjected them to bending tests, together with the purer portion (cut parallel with them), and they had found them no worse, and generally better and tougher, than the latter. Some of those tests he had described in the discussion on Baron H. Jüptner von Jonstorff's paper on 'The Influence of Phosphorus on Cold-Shortness' at the Iron and Steel Institute, as reported in Vol. I. (1897) of the *Journal* of that Institute. The greater toughness was, in spite of the elongation, often less. He quite admitted that the elongation of the segregated portion was less, but when they came to the bending test the result was much better. He had not had much time before coming to that meeting to get samples to show this, but he got a piece of rail, and, first of all, had taken a slice about $\frac{3}{8}$ inch thick off the top of the head. Then he cut up the next $\frac{3}{4}$ inch, or inch, into three slices. Of course there were two outside slices and the centre one. It was just in that portion that Mr. Andrews had shown that the segregate occurred. That [showing it] was the central portion, and it had bent through 68 degrees and had not broken yet. The outer purer portion [which he also exhibited] he had bent with the skin inside (because, of course, it would not be fair to bend it with the skin outward), and it broke after bending through 4 degrees less—only 64 degrees. Then he had cut off that "dangerous" portion just at the top of the web, and it bent through 92 degrees [as shown]. Of course he did not mean to say that the segregated portion would always act in that way, but it showed that there was nothing very terrible to be feared from it.

With regard to the low tensile tests quoted, the strains a rail had to bear were crushing and bending strains, the segregate never had to bear tensile strains. Segregation might be regarded in some respects as a good thing (as he pointed out in a paper some years ago), because it automatically purified the outer portions which had to bear all the strain due to cooling, wear, and so on, and it drove the impurities into the central portion, which was the best possible position, as they could do less harm there and were protected from the strains mentioned. In any event the impure portions were only perhaps one-tenth of the whole.

Coming now to flaws, photographs such as had been shown gave an enormously exaggerated and false idea of their importance. They averaged only about one five-hundredth of an inch long. On the same scale as printed a cross section of a rail-head would have to be 40 to 50 feet wide; so he thought that it was absurd to talk of such flaws as "lurking and insidious sources of danger." Faults like Fig. 17 were not properly described as due to segregation, but were due to laying ingots whilst molten in a horizontal position, or to bleeding of the ingot. They could quite understand that if a little steel ran out of the ingot, or if it ran forward into the cavity at the top, a bubble would flow back, and it caused the pipe to extend a long way back. As an example of modern steel-works practice, and to show what the steel-makers were trying to do to avoid trouble from any such causes, he might mention that the North-Eastern Steel Company adopted, ten years ago, the course of never letting an ingot get into a horizontal position till set, the furnaces also being vertical and not horizontal. Again, at many works, no doubt, not enough margin of weight was allowed, and for fear of getting "shorts" not enough was cropped off. The North-Eastern Steel Company made ample allowance, however, and rolled the ingots top foremost so as to see any hollowness when shearing, and carefully inspected the blooms at the cogging shears, the men having strict instructions to cut well into the solid. After this there was at least 5 feet of crop from the rail. Lastly, in order to be sure that the worst part was good enough, the North-Eastern Steel Company tested the top crop with a 1-ton weight falling 20 feet. Then, again, since they had adopted the Darby recarburising process, they had obtained greater regularity in the carbon content. There could be no doubt that fractures which occurred were not due to the segregation which Mr. Andrews had described, but it was much more likely that they were due to such causes as strains set up by irregular cooling and distortion, requiring excessive cold straightening. The North-Eastern Steel Company had recently set up probably the finest rail-bank in the world, to allow of the rails cooling perfectly flat and under proper control, so as to minimise straightening.

How successful the results of the various means taken had been in avoiding the dangers treated of by Mr. Andrews, might be gathered from a letter by Sir Lowthian Bell in the *Engineer* of October 26, 1900, showing that of a total of 352,000 tons of rails in use on the North-Eastern Railway two-thirds—230,000 tons—were of basic make (very largely made by the North-Eastern Steel Company under the conditions described). The breakage statistics showed that where one rail of acid make had broken per 750 tons, only one rail of basic make had broken per 3700 tons—five times better.

Mr. J. W. JACOMB HOOD said that Mr. Andrews had given engineers an opportunity of following the investigations which Sir William Roberts-Austen put forward about two years ago, and which opened the eyes of many engineers to a line of investigation that might hereafter give them the results they were all looking for, namely, some means of discovering how they were going to get a rail that could be relied upon. Segregation was not a new thing. Everyone who had to deal with rails knew that the subject of segregation had been a subject of suspicion for many years. It was not, as Mr. Stead had pointed out, a novelty. They would like to know from Mr. Andrews whether, from the investigations he had carried out, he could tell them how they were to guard against the evils arising from segregation. They had seen evidence of them again and again. With very great respect to Mr. Andrews, he (Mr. Hood) would say that Mr. Andrews had not indicated what line they would have to adopt in order to discover how segregation arose and how they could overcome the danger caused by it.

Mr. L. ARCHBUTT said that he had much pleasure in tendering his thanks to Mr. Andrews for having brought the results of his investigations before the Society. It was quite true, as had been pointed out by previous speakers, that the knowledge that segregation occurred was not new. It had been known for years, and yet he thought that Mr. Andrews had done valuable service in bringing before them so prominently as he had done the fact that segregation occurred in steel rails to the extent it did. He (Mr. Archbutt) might say that he had a large number of analyses to make of rails and other structural steel, and during the last few years he had analysed a number of sets of drillings taken from the head, web and foot of defective rails, as near as possible to the point of fracture. In the majority of cases there were no differences in the analytical results which pointed in the slightest degree to segregation. In three cases there were small differences, but not sufficient to attract his attention very much at the time. But a case that had been submitted to him within the last few weeks was certainly the most striking that he had met with, although it was not, perhaps, more striking than some of the instances which Mr. Andrews had brought forward in his paper. In the particular case that he referred to a piece of rail only 12 inches in length, cut from a rail 30 feet long, gave the following analytical results from drillings taken at different points:

—	No. of Tests.	Head of Rail. Results.	Average.	No. of Tests.	Web. Results.	Average.	No. of Tests	Foot of Rail. Results.	Average.
Carbon	7	·284 to ·470	·356	7	·414 to ·572	·474	6	·284 to ·482	·327
Silicon	5	·087 „ ·099	·091	5	·089 „ ·103	·095	5	·084 „ ·096	·086
Manganese	6	·838 „ ·904	·863	6	·920 „ ·947	·935	6	·826 „ ·873	·840
Sulphur	5	·018 „ ·042	·031	6	·050 „ ·066	·055	6	·010 „ ·027	·021
Phosphorus	6	·057 „ ·074	·066	7	·108 „ ·122	·114	7	·050 „ ·060	·054

The rail in question was an 85-lb. rail, laid in 1884 in a tunnel, and taken out unbroken in 1902, after eight years' wear. When being cleaned from rust, previous to weighing, a longitudinal crack, 21 inches in length, was discovered in the web, commencing 4 feet 6 inches from one end of the rail. The weight of the rail when taken out was 74 lb. per yard, showing an average loss per annum of 1·37 lb. per yard. What astonished him most was the extraordinary extent to which that piece of rail varied in composition in different parts of such a short length. He thought the possibility of such variation had not been sufficiently taken into consideration as an explanation of the differences which sometimes occurred between the results obtained by different analysts, working with what was supposed to be the same sample.

He was not prepared to enter into a discussion of whether segregation in steel rails was or was not such a serious matter as Mr. Andrews, from his far larger experience, had concluded it to be; but he thought it would require a great deal of argument to convince him that such an amount of segregation as had been proved to exist in some steel rails was a desirable thing. Segregation was liable to occur, not only in rails, but also in tyres, and even boiler plates. He thought Mr. Andrews had done great service in calling attention to the subject. No doubt there was a temptation to manufacturers not to cut off so much of the ends of rails or of ingots as they should do. Apart from the question of whether the variation in the composition of the steel was an evil, they knew that the metal in the top end of the ingot was very unreliable. Consequently, it seemed to him, that it was very necessary that proper precautions should be taken to remove the unreliable metal before the rails were sent out.

He hoped that Mr. Andrews, in his reply, would give them a better indication than he had given them in his paper of the steps which he proposed should be taken in the future with regard to steel rails, by which danger due to segregation would be guarded against.

Mr. EDWARD A. HARMAN asked, "Who shall decide when doctors disagree?" Experts on one side affirmed that segregation was disastrous, while experts on the other side affirmed it was beneficial. The effect of segregation, concentration or collection on the strength of steel rails, at first sight appeared to be distinctly the subject of a specialist, and came only within the range of metallurgy.

After listening to Mr. Andrews' lucid paper upon the subject one was bound to perceive that, so far from the subject being an abstract one, it was of the most momentous importance to all engineers. In the presence of so many manufacturers and users of steel rails it might be wondered where gas-works engineering, with which he was identified, came in. Certainly the consideration of the segregation of steel rails was more particularly connected with railway, tram and travelling-crane lines, which, however, were now becoming an integral part of all engineering works, and were of as vital importance as the machinery therein.

The segregation of steel rails was one of vast importance in connection with all structural appliances for all kinds of engineering—for the moment his thoughts turned to it as applied to gas-works apparatus, such as to gas-holders having steel stanchions with rails running up their front, and also descending to the bottom of the tank, and for girders, bridges, roofs, purifiers, and all other kinds of gas-works plant for which steel or steel rails were employed.

The tables of chemical analyses and physical tests set forth in the paper would be most valuable for reference. Indeed, the list of interesting figures stated were such that the paper itself would take some time to digest and thoroughly appreciate.

In most of the tests one read with interest the column entitled "reduction of area per cent.," which was in some instances remarkable. In one case it was shown that the reduced area at the top of the rail-head under consideration amounted to not less than 27 per cent., and at the bottom to 36 per cent. The strength, therefore, due to such an enormous reduction of area appealed to one in a very significant way. The percentage of loss in area did not represent fully nor adequately the loss of life of rail. It was interesting also to notice that at the place where the greatest amount of segregation existed the chemical analyses fully confirmed the physical tests. To engineers the loss of strength in the various parts of the section of the rails demanded the most careful consideration.

It was exceedingly good of Mr. Andrews to give the results of his varied experience and extensive operations as an engineer and chemist—experience which it was impossible for other than a specialist to have obtained, or, indeed, to have had the opportunity of investigating and theorising upon in the thorough manner he had, and had, moreover, so kindly placed before the Society, in concentrated form, his views and ideas. One was unable to appreciate the enormous amount of work in the paper, the tables of which had been so concisely and clearly placed before the Members.

Mr. Andrews stated that the excessive segregation frequently produced showed the very unsound condition of the rail which indicated lines of weakness in the internal structures of the metal, but it was, however, not only to the external appearances one had to look for danger, but to the internal, which could only be determined from testing or examination of sections.

He noticed that Mr. Andrews stated that the segregated state of the chemical constituents was due to the slow cooling of the ingots wherein the impurities, such as the sulphides, having various temperatures for their consolidation, remained liquid, or otherwise, for a longer period of time than the metal itself. In the course of the consolidation of the metal the impurities, being of less specific gravity, rose towards the top of the ingot while others chiefly consolidated. That account was, to his mind, a straightforward explanation of the results obtained. He might observe that, in connection with gas-works, in violation of all the laws relating to the diffusion of gases, he had found a stratification of gas to take place in the holders, gas of a high illuminating power and heavier specific gravity going out of the holder during the emptying process and leaving a lighter specific gravity gas on the top. Of course the question of temperatures came into consideration with the stratification of gases, as it did also in the subject under consideration, as mentioned by Mr. Andrews in reference to the cooling of the various composites of metals and impurities. Probably the question of temperatures was of more importance than at first appeared, remarks concerning which were made by Mr. Andrews where he dealt with the conditions of segregation.

One important point in connection with rails was that they had to meet not only a compressive strain but a tensile strain. In the spaces between the sleepers there was not only compressive strain but a strong tensile strain as well. Those spaces formed the rail into a series of miniature bridges. But, at all events, he thought that whatever their opinions might be on the subject of segregation, they had no reason for alarm or apprehension of danger in travelling upon a railway on account of segregation, taking into consideration the rare occurrence of a rail being broken in use.

There were many other points that he should like to deal with, such as the effect of electrolysis, magnetism, exhaustion, and other matters, but time did not permit. He should like to express his obligations to Mr. Andrews for the excellent paper he had given them. Not the least important function of the paper would be to give a stimulus to the examination of cases of segregation with respect to steel and steel rails generally, and that would doubtless afford satisfaction to Mr. Andrews, who, although not a steel rail manufacturer, had evinced his great interest in it, and had given the Society the benefit of his research work in so interesting a manner.

The PRESIDENT said that as the time of the meeting was practically exhausted, he would ask those present who had any remarks to make upon the paper to submit them in writing. He would ask Mr. Andrews to make a few observations now, and to reply fully upon the discussion in writing.

Mr. ANDREWS assured the Members that it had been a very great pleasure indeed to him to be present. One reason was that he had nearly reached his silver wedding with the Society of Engineers, for it was about twenty-five years since he first joined. During that period he had had the pleasure of listening to papers and of contributing several himself. He thanked those present for their kind attention to his somewhat detailed and elaborate paper. As they had observed, it raised a controversy between Mr. Stead, himself, and one or two other metallurgical experts. Just at present they appeared to be in that state in which they could not agree. They saw things in a different light, as they looked at them from different aspects. But he was bound to say, with all deference to his old friend Mr. Stead, that he could scarcely go with him so far as to say that not only segregation was no source of danger at all, but that it was really an improvement to have segregated rails on their lines. He failed to see that, by increasing the impurity of the steel, they got a better and more reliable material for wear.

Those papers on segregation which Mr. Stead had referred to were well known to him (the speaker), and he had gained considerable information by reading the works of others, and he continued to do so now. He never intended in the course of the paper to claim priority in the discovery of the phenomenon of segregation. He merely desired to extend the knowledge of that important question.

There were many other matters of technical inquiry and suggestion in Mr. Stead's remarks which he should have much pleasure in replying to in writing, and he hoped before long that he and those who differed from him would be more united in deciding upon the cause of rail fractures. He hoped to help engineers to ameliorate the condition of segregation before very long. Mr. Stead had raised the question of sulphur flaws, but he could not accept all his conclusions. However he would not detain the meeting longer, but would reply fully in writing, as suggested by the President.

The following was communicated by Mr. Arthur W. Richards, who was present at the meeting, but for want of time was prevented from taking part in the discussion:—

The question of segregation in steel for rails is one to which manufacturers have for many years paid close attention, and investigated with a view of its prevention as far as possible. Knowing that irregular and high temperatures of casting the steel, want of regularity and speed in stripping and delivering the ingots to the reheating furnaces, heating, etc., are the chief causes, Messrs. Bolckow, Vaughan and Co. have spent large sums of money in improving the manufacture of steel rails.

Starting with the pig iron, it is most carefully selected and passes through high capacity reservoirs or mixers, so that a regular iron in analysis is obtained. The Bessemer department can blow and cast this iron at a regular temperature, and the ingots are very solid in the moulds. The firm have adopted the system of casting on cars, and the ingots are more easily and uniformly stripped and delivered to the soaking pits. The blooms are well cropped, especially the top end, both at the blooming shears and at the finishing saw. Besides this, the first crop from the saw from each blow number is always subjected to a severe falling weight test, and it stands the test excellently. The firm have every confidence, therefore, that what segregation takes place will not affect the rails to make them break in the road, and they certainly see no necessity to cut a large portion of the ingots away for scrap, unless the different railways are prepared to pay for it. Further, the firm manufacture about 4000 tons of steel rails per week, and have never had a breakage due to segregation. The few breakages that have occurred have been admitted by the various railway engineers to be due to deficient packing of the sleepers, which has caused the heads of the rails to be in tension, whilst the remainder are the results of incipient flaws due to overheating.

The following is Mr. Andrews' communicated reply upon the discussion on his paper :—

In reply to the remarks of Mr. Stead, I cannot understand his apparent attempt to convey the impression that I was not conversant with the earlier works of others on segregation in steel. This is not the case, neither have I infringed on the labours of others. I have always regarded as most valuable the works of others on the general subject of segregation in steel, but these did not afford assistance in the line of my present research, hence it was not essential to refer to them, as they are already well known.

My paper is simply a record of the results of a series of original experiments, undertaken, in the interests of public safety, with the hope of increasing our knowledge of the serious nature of segregation in steel rails; and the facts I have brought forward are incontestable. My paper is based on solid facts, and not on hypotheses. Equally with Mr. Stead I claim a knowledge of modern steel works practice, and also of the most recent requisites of the permanent way.

I have indisputably proved, by these large scale chemical,

physical and microscopical experiments, that the physical uniformity and mass strength of rails, of a segregated chemical composition, is impaired. That segregation is the chief factor in such physical deterioration is shown by the fact that the rails of a normal unsegregated composition, tested under the same conditions as the segregated ones, showed no such sign of non-homogeneity of physical structure or weakness. Mr. Stead has not by any experiments disproved this fact, which is based on too large a series of careful experiments to be disturbed.

I have moreover observed, in course of a wide experience, that the majority of rails which fracture in service, prior to the termination of a reasonable time life, have had a segregated chemical composition, and that the percentages of sulphide, or other impurities, were also in such cases abnormally high; and further, that the fractures mostly occurred at the place where the greatest local segregation of the impurities was found.

Mr. Stead doubtless knows that there is a regular breakage of rails occurring on the permanent way. During the half-year ending June 30, 1902, no fewer than 171 rails broke in service on British railways, any one of which fractures might have caused a serious accident. I have recently examined a number of rails which fractured in main line service on various railways, some of them in tunnels and on bridges, and others on the open track—and in which segregation was the chief cause of their destruction. This is shown by the fact that the rails broke in the segregated area, and after careful chemical, physical, and microscopical examination, there appeared no other apparent cause than segregation to account for their fracture. Mr. Stead, however, in spite of these facts, disputes that segregation is a source of weakness, and he even suggests that it is a source of strength; it is nevertheless very remarkable that the place of fracture occurred in these rails in the area of greatest local segregation.

It is the one suspicious rail, amongst the vast number of good ones, which is most likely to give trouble, and every tendency to risk should be carefully avoided. In the face of such facts it is surprising that Mr. Stead should attempt to make light of so serious an evil as segregation in steel rails.

In segregated rails the percentage of sulphur not infrequently is locally as high as and even higher than $0 \cdot 21$ per cent., which forms a sulphide of manganese representing $1 \cdot 17$ per cent. of the mass volume of the steel. Hence $1 \cdot 17$ per cent. of the mass volume of the segregated part of the rail consists of manganese sulphide, a substance which is simply an impurity. In addition to this large mass volume percentage of impure material, there is in segregated rails also the increased percentage of locally segregated phosphide or other impurities to be taken into account. It is therefore difficult to conceive that the mass strength of a rail will not suffer by reason of such a physically non-homogeneous structure as is produced by segregation.

That eminent metallurgist, Mr. Henry Marion Howe, M.A., of New York, to whom Mr. Stead has referred, states that, "Heterogeneous composition implies heterogeneous strength and ductility: and the strength of a heterogeneous substance is usually nearer the strength of the weakest component or part than the average of all the parts: the piece tends to break down piecemeal. So with ductility." The comparative series of impact tests on segregated and unsegregated rails given in my paper fully confirm this, but Mr. Stead appears to have carefully avoided any reference to the results of the impact tests in my paper.

It having been shown in my paper that segregation and consequent physical weakness chiefly occur at that end of the rail representing the top portion of an ingot, I fail to see why there should be objection to having a greater length cut off that end of a rail representing the top of the ingot. This would add to the prime cost, but there would be a consequent increase in safety.

It may be mentioned, that much greater portions are cut from the ingots used for naval, gun or other marine forgings, to ensure greater safety, and surely the safety of the permanent way is of equal consequence, and the question of cost should not be the prime factor where safety is concerned.

With regard to the percentage of sulphur in steel rails, what object can Mr. Stead have in attempting to ignore the universally admitted evil effects of sulphur in steel rails? Railway engineers, and also the best makers of steel rails, endeavour, in the interests of public safety, to obtain as pure and physically homogeneous steel rails as practicable.

The comparison which Mr. Stead has attempted to make between the micro-structure of wrought iron and steel is irrelevant to the present case, as the impurity referred to in wrought iron is slag and not sulphides.

A reference to the Report of the Royal Commission on 'The Loss of Strength in Steel Rails through use on Railways,' which I have carefully read and studied, and which, I may remark, contains various references to my previous investigations on the strength of steel rails, shows that the great majority of the railway companies specify a percentage of sulphur and phosphorus as low as it is possible to obtain. Metallurgists and engineers are not so ignorant of the evils of excess of sulphur in steel rails as Mr. Stead's remarks would lead one to suppose.

Mr. Stead is now apparently advocating that a high percentage of sulphur is harmless in steel rails. This is opposed to the best and most recent practice. It practically means, that makers would be allowed to use inferior and common pig-iron, high in percentage of sulphur or other impurities, in order to make rails of such high percentage of sulphur as Mr. Stead appears to approve of, instead of using the purer metal now universally required to fulfil the best and most recent specifications of railway companies. Not many, either railway engineers or metallurgists, I think, are likely to support Mr. Stead in defending his belief in favour of a higher percentage of sulphur in rails. It is more reasonable to assume that the purer the raw material is the more reliable will be the finished rail.

In the Board of Trade Report on 'The Loss of Strength in Steel Rails through use on Railways' (p. 74), Mr. E. Windsor Richards, one of the Royal Commissioners, proposes the following chemical specification for ensuring safety in rail service:

	Minimum.		Maximum.
Carbon	·35	to	·5
Silicon	·05	,,	·1
Sulphur	·04	,,	·08
Phosphorus	—	,,	·08
Manganese	·75	,,	1·00

Here it will be noted that the percentage of sulphur and phosphorus is to be kept within low limits. Further, in the same report, the average chemical specification for steel rails of ten of the leading British railway companies was as follows:

CHEMICAL SPECIFICATION OF STEEL RAILS.

—	Carbon.	Manganese.	Silicon.	Sulphur.	Phosphorus.
Average Maximum	·46	·89	·093	·067	·072
Average Minimum	·40	·80	·065	·041	·057

Here again it will be seen, that in the interests of safety the percentage of sulphur and phosphorus is recommended to be low; so that Mr. Stead's advocacy of a higher percentage of sulphur is opposed to the recommendations contained in the Board of Trade report. Moreover, it was shown by Sir Lowthian Bell, that 1 in 3500 rails were found broken, and in some cases Mr. Footner found that 1 rail in 2000 broke on the permanent way. This does not agree with Mr. Stead's figures, that only 1 in 25,000 rails was fractured in service. Fractures occur sometimes in more than one place in a rail.

I have always carefully discriminated between defects arising from piping in the ingot, and those consequent on local chemical segregation, although these two sources of weakness are often intimately associated. External segregation in steel rails is of far more rare occurrence than internal segregation, though I have occasionally found cases of steel rails which have fractured in service through an area of considerable external segregation.

Segregation, like other evils, is one of degree, and it is fairly correct to assume that about 18 per cent. of the rails tested chemically and physically from a total of 285 rails were found to be in a more or less varied degree of segregation; the percentage of segregated rails, which were, however, so far defective as not to pass the ordinary drop test, has been shown in my paper to be about 6 per cent. of the whole number of the rails tested.

It is, moreover, a singular coincidence, as mentioned in my paper, that a number of the segregated rails examined showed external flaws and local unsound places which would cause them to wear down rapidly in service. It would in fact have been positively dangerous, from the above cause alone, to have used some of these rails. Mr. Stead does not appear to admit that a longitudinal surface flaw in a steel rail is capable of developing, under stress, a number of fine transverse fissures or cracks. This, however, has been demonstrated in my report on the fractured St. Neots rail, and I have made numerous additional experiments to demonstrate this.

Mr. Brinnel's experiments on rolled bars, referred to by Mr. Stead, are not applicable to the point in question, viz. steel rails. It is irrelevant to compare wrought-iron with steel rails. I agree with Mr. Stead that stronger fish-plates than those ordinarily used are desirable.

My experience of the part played by internal flaws in causing the fracture of a rail does not quite agree with Mr. Stead's remarks. I have observed numerous cases in which flaws have developed in the central axis or segregated part, near the railhead and web junction, both in rails which have fractured in service, and also in new rails which have been submitted to the drop test, and I can therefore only conclude that the segregation formed a considerable factor in inducing such internal flaws. See Figs. 5, 6, 7, 17, 18 and 19 in my paper.

A praiseworthy effort is being made by railway companies to obtain rails of the best chemical composition and physical structure, but Mr. Stead's remarks appear to support a retrograde movement in favour of retaining a higher percentage of sulphide impurities, and also in attempting to ignore the evils which I have indisputably demonstrated as existing in segregated steel rails.

I regret I cannot accept Mr. Stead's conclusions, as I have knowledge of quantities of rails of the type advocated by Mr.

Stead which have worn badly in main line service on several railways, some of which have fractured in main line service, and I have met with new rails of this class which have broken in pieces on simply unloading from the truck in the ordinary manner.

It is curious why Mr. Stead should ask for a remedy for segregation, when the tenor of his remarks implies that segregation and high sulphur in steel rails are beneficial and not injurious. I fear Mr. Stead's zeal in favour of sulphur and segregation in rails has outrun his discretion, and, notwithstanding his arguments, I am convinced that the presence of segregated sulphides, or other impure elements, in rails, will not conduce to public safety on our railways.

In reply to Mr. Ridsdale's remarks, I would first observe that he does not appear to appreciate the dangers arising from segregation of the impurities in steel rails, and he seems to consider that local segregation may be beneficial rather than injurious to steel rails, and that it may be a source of strength rather than weakness in rails. I fail, however, to see how an abnormal local concentration of some of the chemical impurities in any part of a rail can add to its mass strength. The experiments described in my paper have demonstrated the opposite fact, that the mass strength of rails is considerably weakened by segregation. This is confirmed by the opinion of Mr. Henry Marion Howe, quoted in my reply to Mr. Stead.

According to Mr. Ridsdale's theory, the segregated portion in the section of a segregated rail should be the best part of it. I have recently examined by chemical, physical and microscopic methods, a rail which fractured in main line service, which had a considerably segregated chemical composition. To accord with Mr. Ridsdale's theory, this rail ought therefore to have been strengthened and not weakened by such a condition. It fractured, however, in service through the area of greatest segregation, and there was in consequence a narrow escape from a serious calamity. Another segregated rail which fractured in main line service, and which I thoroughly examined, contained a higher percentage of sulphur and phosphorus than ordinary.

I have also recently examined, for another railway company, a rail which fractured in main line service after a very short life. This rail was found to be in a condition of considerable internal local segregation, and the chemical, physical and microscopical examination showed that the rail had fractured through the local area of greatest segregation.

I may mention another segregated steel rail which I carefully and fully examined, and which fractured in service after a life of only three years: the fracture of this rail also occurred through the area of greatest local segregation. I have also lately examined two other rails which have fractured in main line service, on two different railways, which showed a chemical composition with a high percentage of phosphorus and sulphur, such as Mr. Ridsdale and Mr. Stead appear to approve of, and which also showed considerable local segregation of the chemical constituents. Unfortunately both these rails broke in service, and there was a very narrow escape of a serious accident in each case.

Further, I have made a careful examination of another rail which recently fractured in main line service, and which was high in percentage of phosphorus, and which also had a segregated chemical composition. The fractured rails just referred to contained a percentage of phosphorus of about 0·10 and higher, and the percentages of sulphur varied from about 0·10 to 0·18, and, moreover, they had a segregated chemical composition. I have also knowledge of another rail of the type advocated by Mr. Ridsdale, which very recently fractured in main line service. A piece of this rail broke clean out, leaving a gap in the continuity of the rail. The piece which broke out was caught at the time of fracture in the driving-wheel of the engine of a passing express. All these fractures of segregated rails in main line service have happened recently, and are not raked up for the occasion from ancient history.

I have, moreover, met with new rails high in percentage of phosphorus, and of the type apparently admired by Mr. Ridsdale, which have broken in pieces on simply unloading from the truck in the ordinary manner.

These rails, according to Mr. Stead's and Mr. Ridsdale's theory—that a high percentage of sulphur and phosphorus is advantageous—ought to have had the best endurance in main line service, but, unfortunately for this theory, they fractured in main line service after very comparatively short lives, and, as the Scotsman said, "facts are things that winna ding."

I might mention other rails, of a segregated chemical composition, which have broken in main line service, but the breakages of segregated rails in main line service to which I have alluded are sufficient to disprove Mr. Ridsdale's theory that segregation, which implies a locally high percentage of phosphorus and sulphur, rather improves than deteriorates the quality of steel rails.

Mr. Ridsdale says that I have not referred in my paper to a single example of a segregated rail which has fractured in service. This is not correct, as on the last page of my paper I have referred to a number of typical segregated rails which I had examined, and which had fractured in main line service on various railways.

I have now referred to many segregated rails which have

fractured in main line service, the causes of fracture in these instances having apparently been mainly due to local segregation of some of the chemical constituents. I moreover showed micrographic lantern illustrations of the structure of some of these typical fractured segregated rails during the reading of my paper.

In course of the discussion which followed, Mr. Archbutt, of the Midland Railway Company, gave another example of a rail which was flawed, and which he showed had a segregated composition, and that the internal flaw occurred in the area of local segregation of the chemical constituents.

It is at least a significant fact that the rails fractured in main line service which I have recently examined have mostly had segregated chemical compositions, and that they broke through the area of the greatest local segregation. My paper was intended to deal chiefly with new rails. I expect, however, before long to record various other instances of the fracture of segregated steel rails which have occurred in main line service.

It is impossible to imagine that a segregated rail is strengthened because it has a brittle central segregated core. The whole of the experiments recorded in my paper have shown that such a segregated core is brittle, and that it has no better physical properties than common cast metal, and that the elongation is often as low as three per cent. Moreover, this central segregated core contains in excess the concentrated essence of all the impurities. How then can such a core strengthen a steel rail? Mr. Ridsdale's hypothesis that rails are strengthened rather than weakened by a local segregated condition of the chemical constituents cannot therefore be accepted.

Mr. Ridsdale, when praising those old rails high in phosphorus and sulphur, omits to remark that many of them were probably made near thirty years ago, in the earlier days when ingots received more work and when they were hammered previous to rolling. The present method of manipulation is somewhat different to the older methods, and I think that the influence of impurities was possibly considerably modified by the extra work which was put upon the rails made in earlier times by the older method of hammering the rail ingots.

With regard to high percentage of sulphur in rails, Professor Arnold has shown that sulphur is the "deadly enemy" in steel, and he considers that $0 \cdot 03$ per cent. of sulphur produces quite as great a danger as $0 \cdot 1$ per cent. The case is analogous to the presence of a fine crack; it matters little whether the crack be $0 \cdot 001$ or $0 \cdot 0001$ wide so long as the metallic continuity is broken.

Mr. Ridsdale attempts to minimise the evil effect of sulphide flaws in steel by referring to their small size. It is, however, incorrect to limit the size of sulphide flaws in steel rails to $\frac{1}{500}$ inch in size. I have frequently met with many micro-sulphide flaws as large as $\frac{1}{10}$ inch in size, and they are not invariably longitudinal in direction. I have noticed dangerous transverse ramifications sometimes develop from longitudinal flaws (see Figs. 5, 6 and 7 in my paper). Moreover, these micro-flaws often concentrate in large clusters, forming segregated areas. They prevent metallic homogeneity (as seen in Figs. 5, 6, 7 and 10 of my paper), and they encourage the development of lines of weakness when rails are under the vibratory stress of service.

These segregates of micro-sulphides are almost invariably microscopically observable at, or adjacent to, the point of fracture in fractured segregated rails, hence they must constitute a considerable factor in weakening the mass strength of such rails. I cannot agree with Mr. Ridsdale's suggestion that rails never experience tensile strain; the fact is, that rails laid on chairs form a continuous girder, and they are consequently alternately exposed to tensile and compression stress. I have seen fractured rails which have broken into several pieces, the fractures having occurred in some parts in compression and others in tension.

In reply to the remark of Mr. Ridsdale that "the impure portions were only perhaps one-tenth of the whole," I think it is sufficient to refer to the sectional area of the segregated portion inclosed in the dotted lines on Figs. 1 and 4 in my paper, which clearly demonstrate that the sectional area of segregation is more than one-tenth, and practically amounts to about half of the sectional area of the rail web, and occupies nearly the whole of the area of the rail head and web junction. The largeness of the segregated area is visibly and tangibly demonstrated in some other samples of sections of segregated rails I have recently prepared. The extent of the segregated area in segregated rails is very much greater than that suggested by Mr. Ridsdale.

Mr. Ridsdale's remarks appear to be simply an advocacy of rails of impure chemical composition, and I hope before long to show that rails of such impure chemical composition do not wear well, and that such rails more frequently break in service.

With reference to Mr. Ridsdale's remarks on rails high in phosphorus, the old rails, high in impurities, referred to by him must have been made nearly thirty years ago, and as I have previously remarked, the method of manufacture in those days was different to present methods, therefore the comparison of these old rails with modern rails is not quite admissible. It should, however, be remembered that the results of my present large scale experiments are based upon the examination of above 285 rails of modern manufacture, and the results are therefore more extensive, and they were obtained on a wider series of rails made by the most recent practice.

Mr. Ridsdale states that some of the rails containing a high percentage of phosphorus appeared to have reached a time-life of about seventeen years. I may, however, observe that I have lately examined a number of rails which have endured without failure or fracture the main line traffic of several English railways for periods of about twenty to twenty-four years. These rails had a normal unsegregated chemical composition, and were comparatively free from impurities. This circumstance also shows that such rails are better than rails of segregated chemical composition.

With regard to the effect of phosphorus on steel, the opinion of Mr. Howe appears to be opposed to that of Mr. Ridsdale. Mr. Howe remarks that "Phosphorus, however, usually raises the elastic limit and thus the elastic ratio, an index of brittleness. The effect of phosphorus on the elastic ratio, as on elongation and contraction, is very capricious. Phosphoric steels are, however, liable to break under very slight tensile stress if suddenly or vibratorily applied, or shock-like. Phosphorus diminishes the ductility of steel under a gradually applied load. The influence of $0·01$ per cent. of phosphorus is perceptible, that of $0·2$ per cent. is generally fatal to ingot metal."

Mr. Ridsdale admits that makers often do not cut off sufficient length at the rail ends to eliminate the segregated portion. If his suggestion that segregation is beneficial be correct, why should he regard this omission as a fault with makers?

One might suppose from Mr. Ridsdale's remarks that I had singled out basic steel rails as being specially liable to segregation, and perhaps this may be so, but I am not aware that I have made any special allusion to this class of rails, some of which, however, were included in some of the experiments. My observations were made on rails generally, without special regard either to the makers or the methods of manufacture.

Mr. Ridsdale appears to have carefully avoided any reference to the results of the impact tests recorded in my paper. These impact tests were made on a series of 285 rails, representing a very large bulk of rails received from many of the chief English rail manufacturers, which were submitted to the usual impact test. Out of this number 18 rails failed, or a percentage of $6·31$ of those tested. The chemical analyses and physical tests showed that in these cases the failure was due to the locally segregated condition of the carbon, sulphur, or phosphorus in the rails which failed to stand the drop test.

A complete chemical analysis was made of each of the 285

rails used in this part of the investigation. The chemical composition was normal, and there was no appreciable local segregation of any of the chemical components in the 267 rails which satisfactorily stood the test. There was, however, considerable local segregation of most of the chemical elements observed in each of the 18 rails which broke under the drop test. These large scale experiments have, therefore, incontestably and conclusively proved that rails of a segregated chemical composition will not endure impact shock so well as normal rails, free from segregation.

I am not disposed to accept Mr. Ridsdale's inference, that railway engineers have hitherto been wrong in requiring a pure material for their rails, neither can I accept his assumption that the quality of rails is improved by an increase of the impurities, sulphur and phosphorus.

I was very pleased to learn from the remarks of **Mr. J. W. Jacomb Hood**, of the London and South-Western Railway, that railway engineers are beginning to appreciate the value of the chemical and microscopic methods of examining steel rails. With regard to the remedy for segregation, owing to the length of my paper it was only possible briefly to mention that I have under consideration remedial methods for alleviating the evils of segregation.

I notice that **Mr. Archbutt**, of the Midland Railway, has recently met with an instance of a segregated rail which had failed in actual service. This is confirmatory of my own experience, as mentioned on the last page of my paper, and helps to dispose of Mr. Ridsdale's remark that no example of actual failure of segregated rails had been given in the paper. I fully appreciate the value of Mr. Archbutt's remarks, and I hope ere long to be able to make further communications concerning remedial methods for reducing segregation in steel rails.

I am obliged to Mr. Edward A. Harman for having clearly pointed out that rails are alternately subject to compressive and tensile strains, owing to the fact that a rail resting on chairs constitutes a series of miniature bridges.

I have also perused with pleasure the interesting account which Mr. Arthur W. Richards has given of the care exercised by makers in the manipulation of steel rails. I have noticed similar precautions in various rail mills.

In conclusion, I can only repeat that my labours in connection with this research have been undertaken in the interests of public safety, and I shall be satisfied if any humble efforts of mine may help to a clearer understanding of the insidious danger lurking in rails of a segregated chemical composition.

FIG. 7.

SEGREGATED STEEL RAIL. Transverse internal flaws and fissures near junction of rail head and web, as seen at magnification of 300 diameters. Carbon average 0·44 per cent. Manganese 1·12 per cent Carbon, sulphur and phosphorus badly segregated. Highest carbon 0·60 per cent. Manganese 1·26 per cent.

FIG. 10.

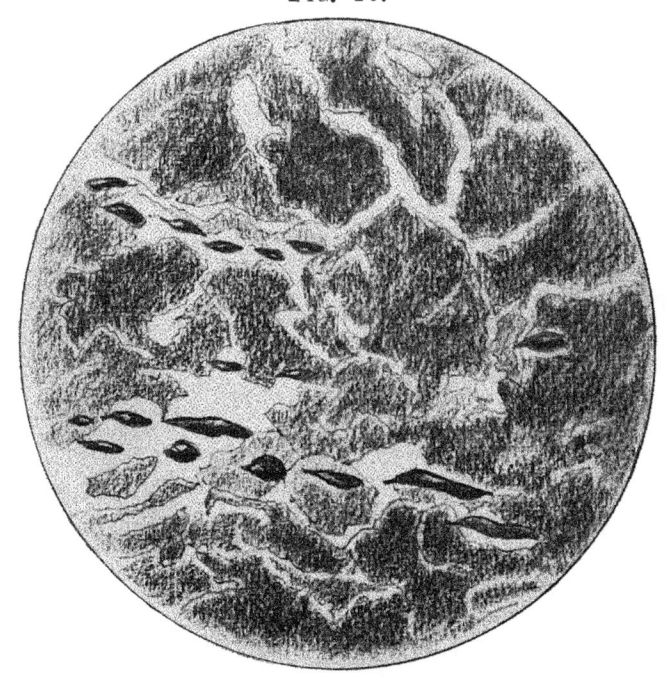

SEGREGATED STEEL RAIL. Micro-crystalline structure in segregated carbon area, also showing micro-flaws, as seen at magnification of 300 diameters. Carbon 0·60 per cent. Manganese 1·26 per cent. Carbon, sulphur and phosphorus badly segregated.

Fig. 5.

Fig. 6.

Fig. 7.

SEGREGATED STEEL RAIL. Transverse internal flaws and fissures near junction of rail head and web, as seen at magnification of 300 diameters. Carbon average 0·44 per cent. Manganese 1·12 per cent. Carbon, sulphur and phosphorus badly segregated. Highest carbon 0·60 per cent. Manganese 1·26 per cent.

SEGREGATED STEEL RAIL. Transverse internal flaws and fissures near junction of rail head and web, as seen at magnification of 300 diameters. Carbon average 0·44 per cent. Manganese 1·12 per cent. Carbon, sulphur and phosphorus badly segregated. Highest carbon 0·60 per cent. Manganese 1·26 per cent.

SEGREGATED STEEL RAIL. Transverse internal flaws and fissures near junction of rail head and web, as seen at magnification of 300 diameters. Carbon average 0·44 per cent. Manganese 1·12 per cent. Carbon, sulphur and phosphorus badly segregated. Highest carbon 0·60 per cent. Manganese 1·26 per cent.

Fig. 8.

Fig. 9.

Fig. 10.

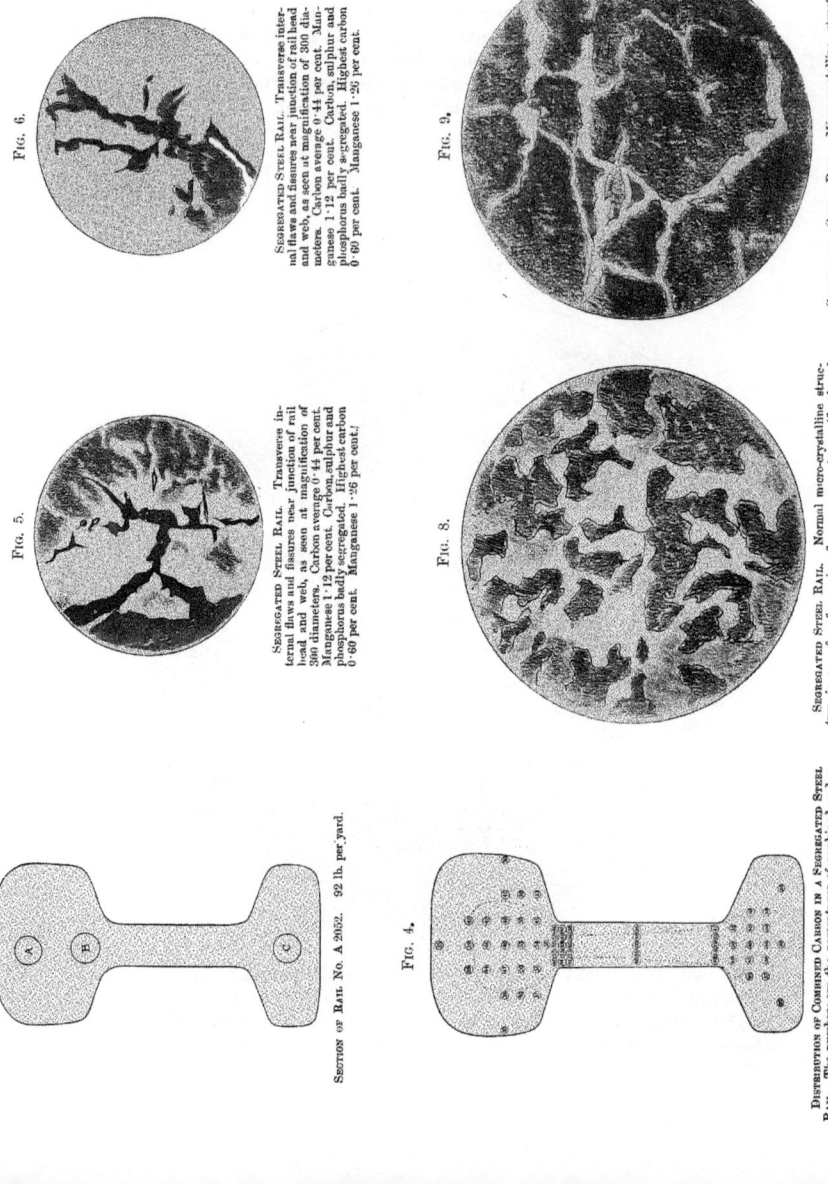

SEGREGATED STEEL RAIL. Normal micro-crystalline structure, in area free from micro-flaws, as seen at magnification of 300 diameters. Carbon average 0·44 per cent. Manganese 1·12 per cent. Carbon, sulphur and phosphorus badly segregated.

SEGREGATED STEEL RAIL. Micro-crystalline structure in segregated carbon area, as seen at magnification of 300 diameters. Carbon 0·60 per cent. Manganese 1·26 per cent. Carbon, sulphur and phosphorus badly segregated.

SEGREGATED STEEL RAIL. Micro-crystalline structure in segregated carbon area, also showing micro-flaws, as seen at magnification of 300 diameters. Carbon 0·60 per cent. Manganese 1·26 per cent. Carbon, sulphur and phosphorus badly segregated.

SECTION OF RAIL No. A 2052.   92 lb. per yard.

Fig. 4.

DISTRIBUTION OF COMBINED CARBON IN A SEGREGATED STEEL RAIL. The numbers are the percentages of combined carbon in decimal parts of 1 per cent. Section 92 lb. per yard. The dotted line encloses the area of the greatest segregation. The ingot centre is approximately in a line through highest numbers.

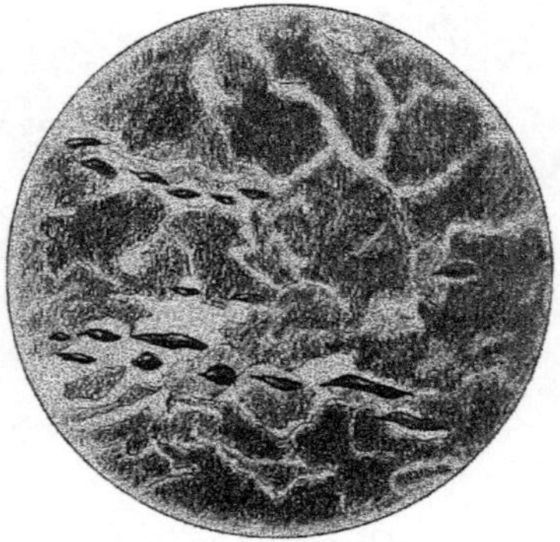

se Curves.
ments locally segregated, near Rail Head and Web

ie per Cent.

F　　　　　　　　　　H　　　　　　　　　　J

irves.
e per Cent.

F　　　　　　　　　　H　　　　　　　　　　J

17 Ft　　　　　　　23 Ft　　　　　　　30 Ft

epresenting the Top of the Ingot.
Curves "N" represent the Normal percentages.
see Fig 13 from each place in the length of the Rail as lettered above.

EFFECT OF SEGREGATION ON STEEL RAILS. BY T. A.

Fig. 11.
Longitudinal Segregation in Steel Rails. Rail Index 1714.
Carbon, Silicon, and Manganese Curves.

The dotted curves show the relative percentages of the Chemical Elements locally segregated, near Rail Head and Web Junction, compared with the normal chemical composition of the Rail.

Curves 'S' represent the percentages in the segregated marks. Curves 'N' represent the normal percentages. The drillings for Analyses were taken in transverse vertical section, see fig 13, each place in the length of the Rail as lettered above.

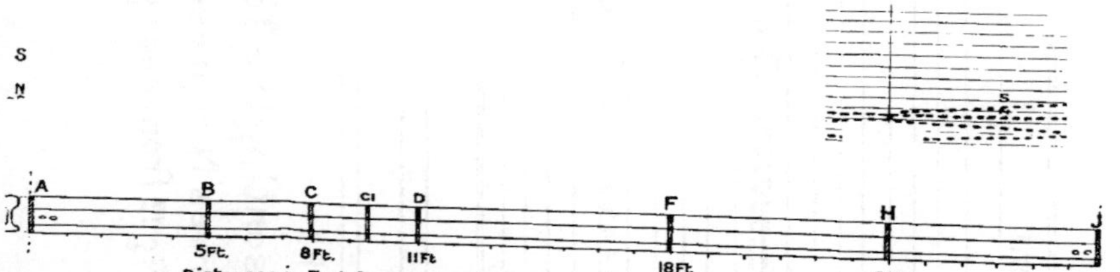

Distances in Feet from End of the Rail representing Top of the Ingot.

Fig. 12.
Longitudinal Segregation in Steel Rails. Rail Index 1524.
Carbon, Silicon, and Manganese Curves.

The dotted curves show the relative percentages of the Chemical Elements locally segregated, near Rail Head and Web Junction, compared with the normal chemical composition of the Rail.

Curves 'S' represent the percentages in the Se Distances in Feet from The drillings for Analyses were taken in transver

End of Rail representing Top of Ingot.

Section of Rail 85 lbs per yard.

STEEL RAILS. BY T. ANDREWS—PLATE III.

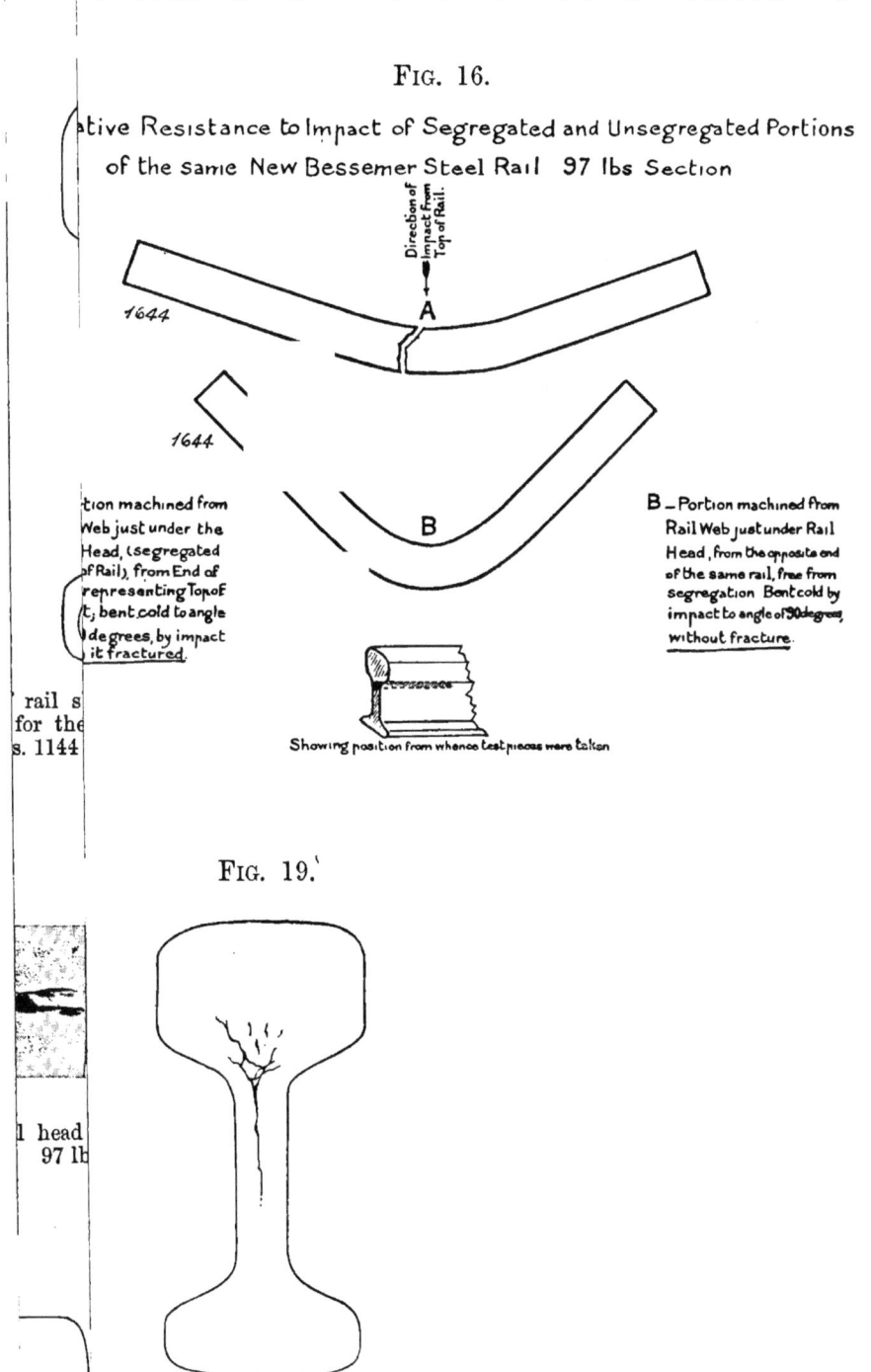

FIG. 16.

ative Resistance to Impact of Segregated and Unsegregated Portions of the same New Bessemer Steel Rail 97 lbs Section

Direction of Impact from Top of Rail.

1644

A

1644

B

tion machined from Web just under the Head, (segregated of Rail), from End of representing Top of t; bent cold to angle degrees, by impact it fractured.

B — Portion machined from Rail Web just under Rail Head, from the opposite end of the same rail, free from segregation. Bent cold by impact to angle of 90 degrees, without fracture.

Showing position from whence test pieces were taken

FIG. 19.

of rail near the end showing internal flaws.

EFFECTS OF SEGREGATION ON STEEL RAILS. BY T. ANDREWS—PL

Fig. 13.

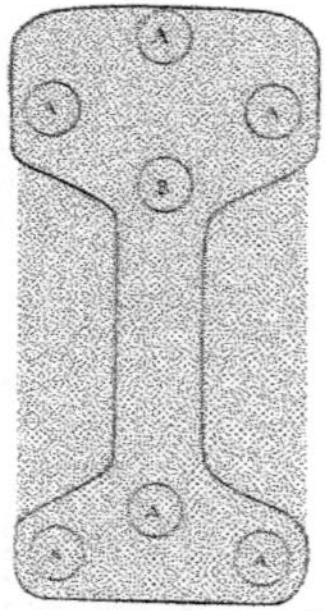

Section of rail showing the positions from which drillings were taken for the chemical analyses of rails Nos. 1714 and 1324.

Fig. 14.

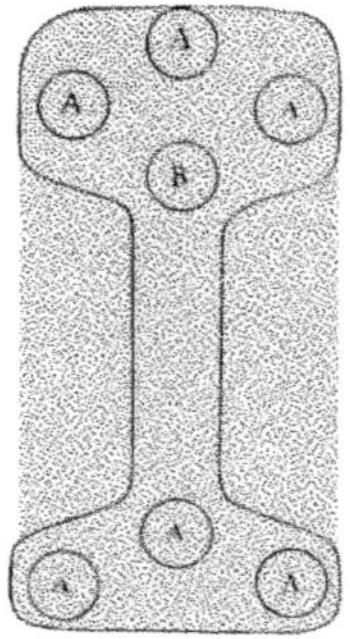

Section of rail showing the positions from which drillings were taken for the chemical analyses of rails of the 96 lb. section. Nos. 1144 to 1644 inclusive.

Fig. 15.

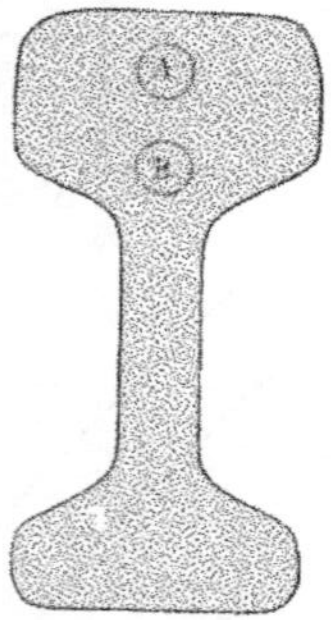

Section of rail showing the positions from which portions were taken for physical tests.

Fig. 16.

Relative Resistance to Impact of Segregated and Unsegregated of the Same New Bessemer Steel Rail, 97 lbs Section

A. Portion machined from Rail Head just under the Head and that's Part of Rail representing Segregated Segment bent cold to angle of 46 degrees by impact was 2 footlocks

B.

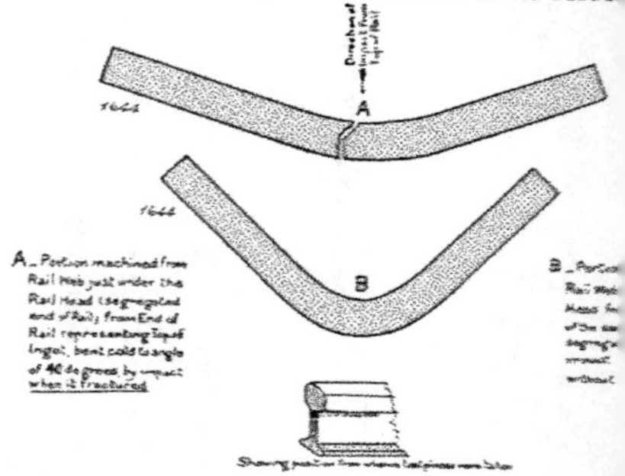

Fig. 17

Large internal longitudinal flaw, near junction of rail head with web, owing to segregation. New Bessemer steel rail, 97 lb. per yard section. Horizontal section at AB. Actual size.

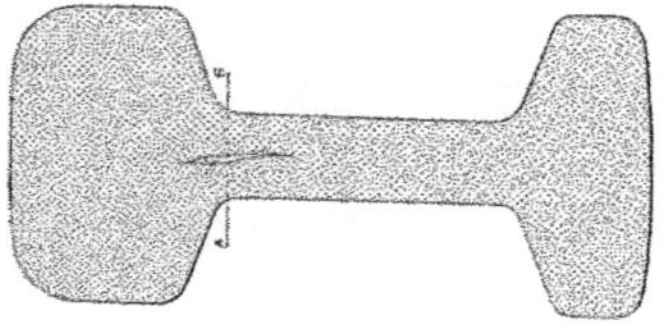

Fig. 18.

INTERNAL FLAWS.

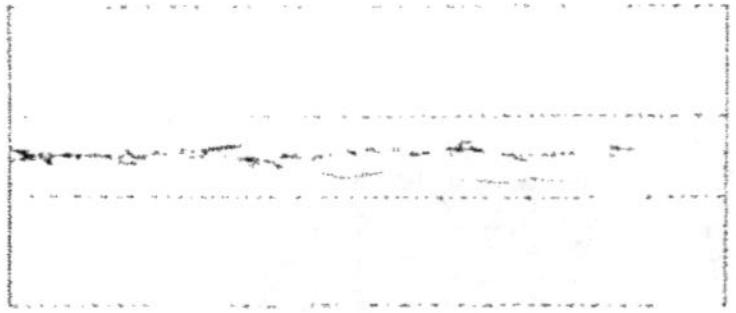

Section at AB

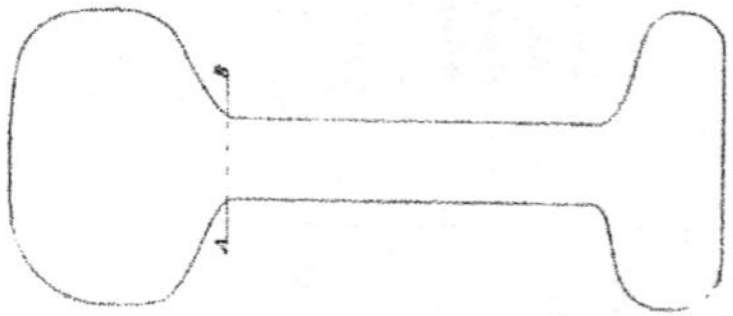

Steel Rail fractured after 3 years service

Fig. 19.

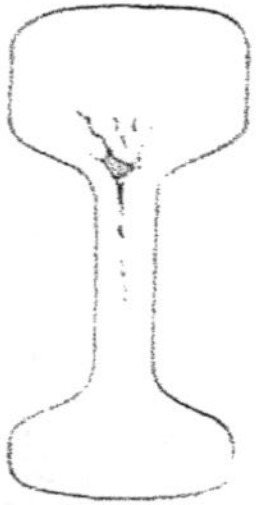

Section of rail near the end showing internal flaws.

December 1st, 1902.

PERCY GRIFFITH, PRESIDENT, IN THE CHAIR.

DEPRECIATION OF PLANT AND WORKS UNDER MUNICIPAL AND COMPANY MANAGEMENT.

BY CHARLES H. W. BIGGS,

IT is necessary to offer some excuse for bringing a subject of a non-technical—though highly controversial—character before a technical Society, but for this the author must refer the critic to the paper itself, trusting that sufficient justification will be found in the manner in which the subject is treated. At the present time a good deal of desultory discussion is taking place concerning municipal and company affairs, but in this paper the author does not propose to take part in any such discussion. His aim is rather to attempt to define and explain one or two points from whence controversialists can start, for unless there exist common ground at the start all discussion is valueless.

Connection of Engineers with the Subject.—It will be asked, what have engineers to do with the question of depreciation? but the author will endeavour to state their connection with it as briefly as possible. Logically, the reason ought to be evolved as a moral, and a dogmatic assertion avoided, yet the latter will serve his purpose in this case. Think for a moment of iron or steel used for structural purposes. The engineer in his design calculates the stresses or strains a piece of iron or steel has to withstand. Say the load on a beam varies from a minimum to a maximum. The engineer uses a factor of safety and uses a beam which in its initial life will carry a much greater load than the one it is designed to carry. Why does he do this? Simply because he knows that deterioration of material begins from the very moment of its use, and the life of his material depends absolutely upon the rate of deterioration. As soon as the weakening influences have got rid of the material representing the factor of safety, the structure becomes dangerous, its useful life is ended, and it should be immediately consigned to the scrap heap. Hence the rate of depreciation in such a case is determined by the rate of deterioration of the factor of safety. Classified by their idiosyncrasies in regard to the factor of safety, engineers are of three kinds:—

1. Those permanently safe, who refuse to be guided by complicated mathematical formulæ, preferring the certainty of safety and long life to everything else. Such gentlemen have been typical British engineers in the past.

2. The highly mathematically trained gentlemen who will split hairs about the strength of a cobweb and who preach lightness and elegance, say in bridge work, using a smaller factor of safety which they claim to be quite adequate, dubbing all notions to the contrary old-fashioned; and

3. The residuum whose designs are guided by what they imagine their clients will spend, and whose god is cheapness.

Many of the second and third class, when taxed with producing short-lived works, are content to reply, " We work for the present, not for the future; futurity has done nothing for us, why should we trouble about futurity?" Such an argument may be smart or witty, but when the case is examined the benefit to the present generation is doubtful, in that the rate of depreciation is much greater than is necessary. No doubt many will remark that in the example selected, length of life depends not only upon the factor of safety but very largely upon the methods adopted to preserve the material. That is so; but given similar and suitable methods of preservation the principal element in length of life, and therefore rate of depreciation, is the factor of safety adopted, and the author claims that the rate of depreciation depends very much upon the engineer.

Municipal Capital Account.—According to a recently published blue-book of the Local Government Board, the total outstanding loans for all local authorities for the year 1899–1900 amounted to 293,864,224*l*. Against this there existed a credit of about 11,000,000*l*., reducing the balance of debt to about 283,000,000*l*. Ten years earlier the outstanding loans amounted to 198,671,312*l*., and allowing for a corresponding credit it will be seen that the municipal indebtedness has been increasing lately at the rate of about 10,000,000*l*. a year. By this time the debt cannot be far short of 300,000,000*l*., and this at $3\frac{1}{2}$ per cent. means 10,500,000*l*. a year interest. This amount then may be taken as the capital in the hands of, and used by the various local authorities. It cannot, however, be discussed as a whole, but must be divided into two parts.

1. Capital used in (so-called) productive undertakings; and
2. Capital used in non-productive undertakings.

The proportion of capital used in these different classes of undertakings does not matter for the purpose of this paper, although it can easily be ascertained from the blue-book.

Repayment of Capital.—The whole of this capital is obtained under the condition that a certain definite part must be repaid annually. The rate of repayment is not settled by the borrowers but by an independent tribunal, the Local Government Board, and varies according to the purposes for which it is obtained. There is no need to discuss the time of repayment, but there is little doubt that the fundamental principles underlying the whole of the transactions is that all human works decay, or by the progress of discovery require to be superseded, and that no capital expenditure should exist at the period of extinction, whether that be due to natural decay or to supercession. The policy is certainly a wise one, and in carrying it into effect it is but natural that this period of repayment should rather be underestimated than otherwise. The question whether a Government department should be given to stereotyping its rules and slack to consider new departures is quite beyond the scope of the present paper. In trying to be on the safe side, no doubt a heavier cost is imposed upon the present generation than upon those that follow. The ideal method of repayment would of course be to pay an annual sum exactly equivalent to the deterioration in value of the work.

Non-Productive Works.—Let us, however, consider a little more closely this question of repayment in connection with the work for which the capital is required. Suppose it be for an unproductive undertaking, such as a sewage scheme. Put aside any return in the shape of farm produce or manure, and look upon the capital as expended unproductively. Interest and repayments fall solely upon the rates, as also does the upkeep of the undertaking. The author holds that the repayment of capital in such a case is merely a system of depreciation, and is true depreciation, even though the sewers exist and are efficient when the total capital is repaid. If the sewers are efficient and are utilised after the period of repayment, that is merely a proof that the rate of depreciation was too great and that one generation of ratepayers has borne a charge which ought in equity to have been borne by a subsequent generation of ratepayers. The author has never heard it suggested that besides the upkeep of these works and the repayments of loan, there ought to be a third charge upon the rates to be called by the name "depreciation." The pertinence of this will be seen from what follows, as there ought to be valid reasons for acting differently when the other class—the so-called productive works—are to be considered.

Productive Works.—Up to this point there is probably no very great difference of opinion, and it is only in cases where municipal authorities embark in the other class of undertakings that opinions differ. The author has ventured to call this class "so-called productive undertakings," because there exists no necessity that they should really be productive. They may be carried on by the authorities for the benefit of a part of the community without any desire to obtain profit, and of course with as little desire to risk a loss. It is a question perfectly open to discussion whether any municipality should undertake work for the benefit of a portion only of the community, with a chance of the work becoming a burden upon the whole of the ratepayers should the work not be productive enough to meet the various charges due to construction and maintenance. The discussion of the wisdom or otherwise of embarking upon these undertakings is interminable, and more suited for another place. Hence it will here be left severely alone. Thus, for the sake of argument, the author will assume an undertaking sufficiently productive to pay all the charges upon it—say a tramway. The first question to determine is, what are these charges?

Charges on an undertaking.—Here opinions differ; but take the most obvious. We have to consider:—

1. Interest.
2. Repayment of Capital.
3. Depreciation.
4. Maintenance—i.e. the cost necessary to keep the whole undertaking in a state of perfect efficiency; and
5. Working.

The author would here emphasise the point that everything said here relates to municipal undertakings, and not to those in the hands of companies.

Items 1 and 5 are open to no criticism. What will be the effect of item 4 upon items 3 and 2? There are two ways of considering item 4; firstly, that tramway undertakings are of a permanent character, likely to remain at any rate for a century or so; secondly, that they are merely a temporary stopgap, bad at the best. Personally the author is inclined to the latter opinion. If, however, tramways are a permanent institution, and are kept in a state of perfect efficiency out of revenue, the rate of depreciation, if depreciation is necessary at all, and the rate of repayment of capital, should be very small, as, whatever the rate may be, it is a tax upon present users in favour of future users. The law, however, demands repayment of capital, and the Local Government Board decides the period over which it extends. It is therefore evident that when item 4 (undertaking kept perfectly efficient) is combined with item 2 (repayment through a term of years), the repayment is not in this case proportionate to natural deterioration unless item 4 be charged to capital account. This, however, neither is, nor ever has been done, nor has it been suggested, but if it were, municipal capital would remain constant. Accepting the present position of affairs, that efficiency must be maintained out of revenue, we see that repayment of capital is depreciation at a rate only to be tolerated upon the assumption that tramways are temporary in character.

Coming now to item 3, the most violently criticised of all the charges, it is held by a very influential and numerous class that, in addition to maintenance and repayment, depreciation is absolutely necessary to put these undertakings upon a sound financial basis.

The rate of depreciation suggested is that which, when the initial capital expenditure becomes extinct, should equal in amount the original expenditure. The contention is that when the capital is paid off, there should be in hand from this depreciation account a sum sufficient to renew the whole undertaking. The author considers this contention to be economically unsound, and it is upon this point that he asks for clearly expressed opinions. Assume for a moment that the views just expressed are carried into effect, and that the length of life of a tramway and its equipment be twenty years—of course this figure is merely for purposes of argument—at the end of twenty years we get the following situation, viz.:—

1. Initial capital account closed by annual repayments.
2. An efficient undertaking maintained in perfect working order out of revenue.
3. A sum of money at the bank or elsewhere equal in amount to the repaid capital.

The author's interpretation of such a state of affairs is that this system involves a threefold depreciation, each equal in amount to the original capital, viz.:—(1) the original capital is depreciated out of existence; (2) the maintenance cost must amount to the same as the original capital; and (3) there is cash in hand equal in amount to the original capital. Those, then, who insist that municipalities must depreciate as well as pay back capital, ought to have very strong reasons to warrant such an extreme position. What are these reasons? The author has never heard of any reasons being given.

Before discussing company work the author will elaborate a little the results of the combination of repayment, maintenance and depreciation. The users of the undertaking during the first twenty years of its existence will not only have to pay the working cost and the interest upon capital, but an amount equivalent to three times the capital in addition.

Assuming that at the end of the twenty years it is not necessary to further increase the sum obtained by depreciation, the users will then have to pay the working and maintenance costs. In other words, the initial users will have been mulcted of large sums in order to benefit subsequent users.

As regards the double payment of capital, namely by repayments and by maintenance, the author sees no way out of the difficulty, and that, willy nilly, the initial users will have to pay; but in his opinion the insistance upon depreciation as well as repayment of loans is absurd, and the only possible motive for the suggestion is an insidious attempt to handicap a municipal undertaking because it is a municipal undertaking.

Summary.— Depreciation, then, as applied to municipal undertakings, is founded upon no logical principle. Repayments are over an arbitrary period, dictated by an irresponsible Government department, but usually the period is too short, and the rate of repayment therefore too fast. As at present carried out, it imposes far too heavy a tax upon the first users for the benefit of subsequent users.

The questions the author submits for discussion are :—

1. Is efficient maintenance—as is absolutely necessary in productive undertakings such as water-works, electrical works, tramways, etc.—a capital or a revenue charge? If the latter, is not that one method of depreciation?

2. Is the repayment of capital depreciation? If not, what is it?

3. And, most important, If capital is repaid, is further depreciation necessary, and why?

Companies' Capital.—It will be necessary to discuss this part of the subject in two ways, first assuming an ideal which does not exist, and secondly the actual conditions under which companies must work. First the ideal, whereby a company would obtain the same amount of capital as an authority, that is, the total money obtained would be expended on the work done without any diminution for promotion or watering, thus leaving the total capital available for the cost of the actual works. One important question arises here, does it cost a company or a municipality the greater sum to carry out similar work? The answer is an important one, and an evening's discussion might well be devoted to it; but whatever the answer, if there be any difference, it means that one or the other party has a less capital outlay for depreciation. For the moment—because the question is too large to grapple with now—the author will assume that the conditions are even, and that similar work can be accomplished at like cost in either case. The author will further assume that the revenue in each case is identical, and that the working costs are the same in both cases. The question then arises, does the management cost differ? There can be little doubt that the expenses of company management are largely in excess of those of a municipality, hence the amount of revenue divisible in the shape of interest, depreciation, etc, is lessened by the difference in the cost of management. Although the author has insisted upon the fact that he is discussing an ideal state, he wishes to guard against the obvious retort that the management in one case is possibly more efficient than in the other, and so what is paid in excess on the one hand is saved by the greater efficiency on the other hand. Passing over this point, however, what is done with the divisible residue? In the first place the undertaking must be kept in a state of efficiency; then come the questions of reserve, depreciation, and dividends. For simplification, assume cost of management and efficiency of management equal in both cases. This means an equal amount for disposal. Concerning dividends, if the dividend of a company be more than $3\frac{1}{2}$ per cent. (the interest on the capital of an authority), then we may safely contend that the extra percentage over the $3\frac{1}{2}$ per cent. is merely another form of repaying capital, and that any company paying more than $3\frac{1}{2}$ per cent. is in this way doing exactly what municipalities do. If the shareholders were to invest the return over $3\frac{1}{2}$ per cent. at compound interest, they would at the end of a certain period have an amount at the bank to their credit equal in amount to the capital they put into the scheme. If, further, the company places the amount allowed for depreciation to reserve, it will gradually accumulate funds which after a certain period will equal in amount the original capital, and so we come to the same point as with the municipalities.

1. In this case the capital has been repaid, in that the shareholders have received an increased interest. The fact that they do not treat the difference in interest as repayment of capital is altogether immaterial.

2. In this case also the capital is repaid by keeping the undertaking efficient out of revenue; and

3. It is a third time repaid by the amount allowed for depreciation and credited to reserve, such funds accumulating until the amount of the original capital is reached.

The period of time over which these operations are spread may be taken as equivalent to that determined by the Local Government Board for the repayment of capital borrowed by municipalities. One point of comparison may be dealt with here and now. A municipality is a public body, and it is impossible to contemplate that after it has repaid its capital, depreciated to obtain a fund in hand equal to its capital, and possessing an efficient going concern, that the then users should be charged to the same extent as the earlier users. But public opinion would have no weight and no effect upon a company, and so it is quite conceivable that a company's charges would remain unchanged and the shareholders would benefit to a considerable extent. It may be argued that the users have their remedy, but this is altogether fallacious, because all these undertakings which bring about contention between private and municipal enterprise are and must be of the nature of monopolies. Think of half-a-dozen competitors in water supply, or gas, or electric supply, in tramways or in telephones, or in power supply, and by thinking of extremes in competition, the inevitable conclusion must be reached that in these things competition is absolutely impracticable. It is all very well for one company to cry out to be allowed to compete with the municipality and let the weakest go to the wall, but if one, why not two, or twenty? There is no limit but the natural one of supply and demand in many directions, but there is a limit, and that not involving a system of trial and error to find out if the supply is greater than the demand, in those directions just mentioned.

Probably the view expressed by the author will be combated by the statement that maintenance is but partial and not complete as suggested, and that efficiency is not maintained in either case out of revenue in perpetuity. Even should this turn out to be so, it will merely modify what the author has said, unless the cost of maintenance and depreciation are so proportioned that together they only equal the repayments; but that this view is not correct can easily be shown by special examples. Aberdeen in four years paid to sinking fund 3380*l*., to depreciations and renewals 6774*l*. If this rate of repayment would terminate the loan in ten years, then the rate of depreciations and renewals would about doubly liquidate it. Bolton, again, has paid 16,397*l*. to sinking fund account, and 32,596*l*. to depreciations and renewals account. Glasgow goes one better, pays 112,038*l*. to sinking fund account, 407,805*l*. to depreciations and renewals account. The proportions are reversed at Liverpool where the sinking fund is credited with 143,311*l*., while depreciations and renewals only get 67,985*l*. However the question may be viewed, the examples quoted show a want of uniformity, and all the author endeavours to do is to raise the discussion in order to obtain some indication of the views generally held. It is abundantly clear that whether the accounts of a municipality or a company be considered, the profits they are endeavouring to make in this particular direction are excessive; the goods they have to sell are therefore too dear and the buyers are unduly penalised.

The average interest obtained for gilt-edged securities may be taken as the natural rate of interest, and if a higher rate than this be obtained it should be taken as repayment of capital required because of greater speculative character. Local authorities borrow at from 3 to 3¼ per cent. interest and reckon that 6 per cent. per annum will pay interest and the liquidating repayments of loan. If that is so it is surely just to look upon a company's 4, 5, or 6 per cent. dividend per annum in the like manner as part interest and part repayment of capital. The question here arising concerns the certainty of payment of interest and capital from local authorities, and the uncertainty as regards companies, the former having the rates to fall back upon, the latter being absolutely dependent upon the success of the undertaking.

A number of other questions could also be discussed involving those relating to the promotion of companies, watering of capital, excessive costs, no real responsibility, but the author has preferred to assume everything straight and above board, every man honest and feeling his responsibility, and to ask, all these things being above suspicion, whether, in connection with capital repayment, maintenance and depreciation, we are not creating a new phase in business accounts, and although the economic theory is that a thing is worth what it will fetch, the artificial restriction of competition may and does unduly penalise the present user?

The following communication from Mr. J. YOUNG, General Manager, Glasgow Tramways, was then read by the Secretary.

Referring to the author's remarks on the repayment of capital, I must say that it is a very difficult matter to determine during what period a particular loan should be repaid. He says that the fundamental principle on which the term of repayment should be fixed is, that no capital expenditure should exist at the period of extinction. That is to say, whenever the works fall to pieces or become useless, the last loan will be paid off. How is it possible for anyone to say when an undertaking which is maintained in perfect order out of revenue will become extinct? Again the author says, " the ideal method of repayment would, of course, be to pay an annual sum exactly equivalent to the deterioration in value of the work." If he means by "deterioration in value" the annual amount by which a piece of machinery or plant has depreciated, I cannot agree with him. I have always held that the sinking fund or repayment of debt has no connection with the depreciation which is, day by day, going on in every undertaking. This, I hold, must be provided for in addition out of revenue. In our undertaking, the Glasgow Corporation Tramways, the sinking fund is calculated to repay the debt in thirty-one years. This means a 2 per cent. sinking fund accumulated at 3 per cent. interest.

In the case of non-productive works to which he refers, viz. a sewage scheme, I think we may take it that there is really no depreciation, and in a case of this kind the longest term of years should be given for the repayment of the loan. An undertaking such as this might almost be put in the same category as a park. I would say that the repayment of the loan might be spread over a period of sixty, seventy or even eighty years. In a sewage scheme, however, we may have machinery for pumping, etc. Provision should certainly be made out of the annual revenue for the replacement of this machinery every thirty years or so. Some provision should also, of course, be made for depreciation on buildings.

As regards productive works, the author specially refers to a tramway undertaking under that heading. Possibly I had better state the principle which we have adopted in Glasgow in regard to our own undertaking. We, first of all, meet our ordinary working expenses, which include cost of electric current, traffic expenses, general management, expenses and maintenance. This last item of maintenance is simply the ordinary repairs which are found necessary to keep the plant in good working condition. In addition to the foregoing, we charge revenue with 450*l*. per mile of single track to meet the cost of renewals of permanent way; this sum is calculated on an average life of ten years. We also charge revenue with a sum to meet the annual depreciation of buildings, plant and equipment. This amount is calculated to keep up the undertaking to its original perfect condition. From the balance remaining after these charges have been met, we pay our interest on capital and sinking fund.

Coming now to the author's remarks regarding depreciation itself, I do not see how he arrives at the conclusion, that where depreciation is provided for in addition to the sinking fund, the result will be at the end of twenty years—should that be the life of the undertaking—(1) all loans repaid; (2) undertaking in perfect working order; and (3) money on hand equal to the original capital. I hold that at the end of the period named the final result will be: (1) all loans paid off; and (2) undertaking in perfect order, or cash available to put it in perfect order.

This difference may arise from the fact that, in the case stated by the author, the amount charged for maintenance is supposed to keep the undertaking in perfect order, whereas, in our case it takes both the annual expenditure on repairs, and the further sum charged for depreciation to ensure that the undertaking will always be kept in what we may term perfect order. I notice the author touches on this point, and he seems to indicate that the maintenance and depreciation charges together should equal the amount of the sinking fund. This cannot be, as, in the case of our permanent way, for instance, we may possibly require to renew the line three times over before the capital has been repaid through the operation of the sinking fund. I trust these remarks may be of service, although they have been hurriedly put together.

DISCUSSION.

The PRESIDENT, in moving a vote of thanks to the author, said that speaking broadly they had papers of two classes, namely: those which, like the one read at the last meeting, presented a comprehensive and even exhaustive treatise on some special branch of technical research, and those which, like that read this evening, were of a more brief and perhaps conversational nature, dealing with subjects which were more suitable for discussion. The author had given them an excellent example of the latter class of paper. The only difficulty arising in connection with this paper was that the author had lightly touched upon a large number of points arising out of his subject, and speakers might therefore be tempted to wander from the main points of the paper and thus produce a discussion which, through undue length and diversity, would lack something of interest and value. He would therefore ask those present to avoid wandering away into the many branches of the subject to which allusion had been made, but which were not specifically raised for the purpose of discussion. He would particularly ask them to avoid the question which was so very prominently before the public at the present time, viz. the general desirability or otherwise of municipalising speculative or trading undertakings. Some of the articles which had appeared in the *Times*, under the heading 'Municipal Socialism,' had dealt in detail with the point which the anthor had specifically dealt with in his paper, and therefore reference to any such article would be, of course, interesting and pertinent to the discussion, but he must ask speakers to avoid the general question of the expediency of fostering the development of municipal enterprise. The main points of the paper appeared to him to be, "What is the proper basis for dealing with the item 'depreciation' in municipal undertakings, and how far are company and municipal undertakings comparable with regard to this point; that was to say, in the allocation of revenue?" He had great pleasure in proposing a hearty vote of thanks to Mr. Biggs for bringing this interesting subject before them.

The vote of thanks was carried by acclamation.

Mr. H. SHERLEY-PRICE said that, so far from any apology being required for bringing this important subject before the Society of Engineers, the Society owed Mr. Biggs a debt of gratitude for affording them an opportunity of discussing this "burning question," even, as it must necessarily be, with the short time at their disposal, in an inadequate manner.

Passing by the interesting interlude as to the "idiosyncrasies of engineers," he would endeavour to come to the pith of the paper, namely, "the depreciation of plant and works." He would like to take the paper and make his comments seriatim; but that would occupy too much time and prevent other and more competent speakers following. At the outset the author desired a basis or common ground from which they could discuss the question of depreciation. He (Mr. Sherley-Price) thought that the author had, perhaps unconsciously, found it for them. Speaking of "a beam" and "a factor of safety," he used a phrase which most aptly and tersely formed, in his (the speaker's) opinion, the only solid ground upon which they might build approximate views of depreciation. He referred to the phrase "useful life"—a most comprehensive term, surely.

He should very much like to discuss what, with all deference, he conceived to be the fallacious reasoning by which the author arrived at conclusions respecting the accountancy part of this subject, conclusions which, if acted upon by corporations generally, would very soon have such evil effects upon municipal credit as to prove a veritable national disaster. To others more competent in accountancy he would, however, leave that part of the paper, and he would revert to the basis of "useful life," which must, perforce, form the groundwork both for the engineer and afterwards for the accountant. For brevity's sake, he would omit non-productive works, such as the supply of water, drainage, road-making, etc. They did not come into competition with private firms or companies to any serious extent.

Lest he should be tempted to dilate upon "municipal trading," with which the able paper before them was almost interwoven, he would treat both "municipal" and "company works" as being alike. Their objects were the same, namely, to supply a public need in the best possible manner, at the lowest cost, and—although that might be disputed—make a profit. The great difference between them was that the subscribers to the companies found the necessary capital voluntarily, relying upon the management to recoup them for their outlay; whereas the ratepayers had no voice, practically, in the matter, and must, perforce, supply the sinews of war.

To manufacture any article in quantity at a low cost necessitated using plant and machinery of the best of its kind. That was but a truism, and needed no enforcing at the present meeting, but it had a supreme importance in the question of depreciation. Depreciation not only comprised the lowering of value incidental to wear and tear, but it also included reduced value by reason of obsolescence. That latter quality, in the past, engineers had unfortunately greatly neglected, to their loss of profits and trade also; but to neglect that feature in modern municipal and company enterprise would be sheer madness.

A familiar example would illustrate that. Since the introduction of electrical tramways, there had been some thousands of tram-cars, formerly worked by steam and by horses, thrown upon the market, in perfect working order, but useless for any other purpose. Those had had to be sold for less than one-tenth the prices at which they stood in the books, after having had what was considered reasonable depreciation written off. That might be considered an extreme case, but, particularly in new industries, improvements were constantly being made, and depreciation must, in such cases, be sufficient to provide against that. Improvements in electric light and power station equipments, during the past six or eight years, had resulted in an economy of from 10 to 15 per cent. in costs of generating and distributing. Those costs varied in amount with the size of the undertaking; but assuming only a moderate concern of, say, 50,000l. turnover per annum, it would be seen that a saving of 5000l. to 7500l. annually would result from using machinery of the latest type. At the present moment there were, to his (the speaker's) knowledge, numerous plants, not ten years old, representing many thousands of horse-power, being taken out and replaced with new, for the sole reason of lack of economy. At the time these plants were installed they were quite up to date, but obviously, no matter what they cost, it did not pay to use them now.

Depreciation, again, not only had to provide for obsolescence, but for wear and tear. The rate of wear and tear, again, largely depend upon maintenance. It went without saying that the more perfectly plant, etc., were maintained, the less would be the wear and tear. One man had a high standard of efficiency always before him, and would see that everything was maintained up to the highest possible condition. Another would be quite content to run his plant "for all it was worth," getting out of it all that he possibly could, and paying scant attention to upkeep or maintenance. Hence the impossibility of fixing upon any rate of depreciation which was capable of uniform application. As to the rate of percentage to be written off— each "tub must stand upon its own bottom."

Addressing himself to the tramway undertaking referred to in the paper, he would observe that the author had divided that into sections as follows: (1) interest; (2) repayment of capital; (3) depreciation; (4) maintenance; and (5) working.

No. 1 represented in a private undertaking, profits; and that being a financial matter, he would, as before stated, leave it for accountants to deal with.

No. 2. Repayment of capital. It seemed to the author to be a hardship that that had to be repaid by corporations in thirty years, because he thought at the end of that period the original undertaking would come into possession of those who might not have contributed towards its cost. He had grave doubts about anything of the sort. If they examined any undertaking of the kind, they would find, every year, considerable sums expended in additions not contemplated in the loan, and, too often, nothing, or a very inadequate sum, set aside for replacements. There was very little in the shape of tramway plant or rolling stock that would last even fifteen years, or half the term allowed for repayment. And yet he found no reference to a fund for replacements.

Items 3 and 4. Depreciation and maintenance were, in his opinion, the same, and were, in fact, inseparable.

Without further labouring the point, in his opinion the rate for depreciation and maintenance should be sufficient to cover only the "useful life" of the objects depreciated. In other words, when the objects could no longer be used economically, no matter what their working condition, they should be discarded and replaced by up-to-date objects, and that provision for wear and tear, maintenance and obsolescence, must be made out of revenue, and not out of capital.

He was astonished to find the author did not allude to reserve fund except in a vague manner as being mixed up with depreciation. But that was quite a distinct matter. He had argued that depreciation should cover wear and tear, maintenance and obsolescence, but he had never yet heard any one advance the theory that it should also cover replacements. The replacement of worn-out or obsolete plant should be provided for by a reserve fund, and there was no other way, he ventured to suggest, in which they could be met on a sound economic basis.

Still keeping to their tramway, they would suppose that in seven or eight years, the exact period being immaterial, much of the line must be relaid with new rails. How were they to do this? The depreciation fund had done its part by reducing their value in the books to a minimum, but they now wanted cash to purchase new rails. If they added the cost of these new rails to capital account, they would be committing suicide from any sound economical point of view, because they would show in their capital account that they had now got the value of two sets of rails, and if that went on for thirty years—the length of the Act—they would show that they had three or four times the value of the rails, although they had then *de facto* only the same quantity in use as they originally started with. The same result would obtain for rolling stock, etc.

If Government had not made such a provision as to a reserve fund, he feared the majority of their corporations would speedily get into a parlous financial state, for they were none too eager to carry out the conditions laid down. In concluding his somewhat lengthy remarks, he would heartily thank Mr. Biggs for bringing so important a subject before them.

Professor HENRY ROBINSON said that in the early part of the paper the author classified engineers under three heads: the cobweb men, the mathematical men, and those who aimed at cheapness and disregarded posterity. In considering the question of depreciation nothing, he thought, could be more important than to take into account who it was that designed the work, whether it was the cobweb man or the mathematical man or the cheap man who did not care for posterity. For instance, whoever designed a wrought-iron structure should know that the factor of safety depended upon whether they took into account the effect of the reversal of stresses. A wrought-iron bar was able to bear a certain definite tensile stress, but if there was a reversal of the stress the whole of the conditions were varied. If a man designed a structure without that knowledge, he designed a structure which would not last as long as it would if the conditions were thoroughly understood. The question of vibration and the fatigue of metals must be taken into account. Before Mr. Young started his calculation of the number of years' life that the Local Government Board should allow for ironwork, it would be useful for the authorities to know whether the work was designed by an engineer who intended it to last, not for the immediate present only, but also, as far as possible, for the use of posterity. That involved the question of maintenance directly. If roofs and other wrought-iron structures were not properly maintained, and were subject to a great amount of deterioration, the life of them was obviously shorter than it would be if they were properly maintained. That point was obvious to everyone, but it was a point which had to be kept in mind in deciding the period to be allowed for the repayment of the loan.

Another point which had not been referred to either by Mr. Young or by Mr. Sherley-Price, was a very important matter. It was what he would term the human fallibility element. An engineer might design works skilfully and carefully, and they might be handed over either to a municipal authority or to the staff of a company and put in charge of some one who, at some time, became unfit for the purpose in consequence of mental aberration due to refreshments. The human fallibility part of the matter was one which must be kept in view, and they must allow for the possibility of some idiot running the plant for a time in a way which was not intended. Take, for instance, a secondary battery at an electric lighting station and its being over-discharged during a fog. The life of a plant, which might be set down at so many years, provided that human fallibility did not come in, might be materially shortened in consequence of being improperly worked. He had experience of a case where multitubular boilers cost far more for the renewal of the tubes than he had estimated, owing to an incompetent man neglecting to use a softening apparatus which he had provided, the water being hard. When he made an inspection he found men chipping out calcareous matter from the tubes with chisels and hammers, and in consequence the life of the tubes had been very much shortened by such a method of treatment. The human fallibility element must not be disregarded.

He was of the same opinion as Mr. Sherley-Price with regard to taking into account obsolescence, or the possibility of a plant becoming obsolete in the course of a few years. A tramway might become obsolete in ten or twenty years. It was impossible to say at the present time that a plant which now appeared to be perfectly right and up-to-date, would not, in the course of time, become obsolete, and for that reason he thought that it was very unwise not to lay by a sufficient reserve. Mr. Sherley-Price spoke of maintaining the plant without depreciation, and at the end of the period of time having enough capital in hand for another plant; but that was not what the author meant, though the paper read rather like it. It could not be the intention of the author to suggest that. The author spoke of an investment at 3½ per cent. on a "gilt-edged security," as he termed it. He (Professor Robinson) knew that if he were to invest a sum of money and get 3½ per cent., he should consider that to be his dividend to spend; but if he got 7 per cent., he should put the difference by in a safe place.

Alderman PEARSON (Bristol) said that the corporation of Bristol were very deeply interested in the question of depreciation, because they were very large municipal traders, and not traders of recent date either. They had about 2,300,000*l.* invested in their docks and plant, and with the object of bringing the undertaking up to date, they had, within the last few months, entered into a contract to spend another two millions. By the time the works were completed he expected that the amount invested would represent four millions and a half. They had, in addition to that, an electrical undertaking in which they had spent up to date 420,000*l.*, and notwithstanding the heated discussions upon municipal trading, the council had thought fit during the last three months to authorise a further expenditure to the extent of 129,000*l.* It would be seen that the question of finance in connection with the municipal trading in the city of Bristol, and particularly the question of depreciation, was one of the deepest interest to persons who, like himself, were concerned in the management of the municipal undertakings. He had therefore followed with very great care all the discussions which took place on that subject. He had read with interest many criticisms of municipal management, including some of the most venomous critics, because he believed that they might get ideas even from those who were not too polite in their criticisms, and those might enable them to avoid some of the pitfalls which beset their paths and which they were naturally very anxious to avoid.

Speaking broadly, he was prepared to support the views expressed in the paper. He thought Mr. Sherley-Price was under a misapprehension as to what the Local Government Board would do in their dealing with municipalities, and he had rather assumed that, because the period was twenty-five or thirty years for the repayment of corporation loans, the works were calculated to abide only for twenty-five or thirty years. But that was not the case. For that reason he had asked the Local Government Board inspector not to be so fond of equating the loans, and he had asked him to give a loan on a short period on the short-lived works, and for a long period on the long-lived works. Then they would know where they were. In discussing the subject they were apt to forget that the longer period was made up by cutting down the land period and the building period, and by levelling up the machinery. Therefore they were very apt to speak, as to-night, of the small life of tram rails. He should like the Local Government Board to give him a loan of thirty years for such works, but he knew they would not. It had been assumed that the Local Government Board did not understand their business; but his experience was that they had a very fair knowledge of it. The electrical accounts that they had to deal with were, of course, somewhat complicated, and although they were not subject to the audit of the Local Government Board, they were subject to Board of Trade criticism if they did not keep their accounts in the form that was required. As the chairman of the electric lighting committee, he considered that he had to hold, if he possibly could, the scale evenly between first, the ratepayers of to-day, second the ratepayers of the future, and thirdly, the gentleman who was not much thought of, and that was the consumer of to-day. They must have some regard to the claims which they made upon the consumer. In common honesty to the consumer, they ought to consider the least sum which they could take out of his pocket which would enable the municipality to carry on the concern without risk to the ratepayers of the future. When they had done that, they had done substantial justice to the consumer. He thought that they must call upon the consumer to protect the ratepayers of the future from any claim owing to having let off the consumer and ratepayers of the present day a little too lightly.

The Municipal Ratepayers' Association considered the question of depreciation, and they arrived at the conclusion that, so long as the Local Government Board proceeded upon the present line, and did not divide a loan on an undertaking into various parts according to the life of each separate part of the work and fixed the period to their own satisfaction, though not always to the satisfaction of the municipalities, practically speaking no depreciation was necessary; but the Association followed with a resolution that it was advisable that no sums of money should be appropriated for the benefit of the citizens by the reduction of rates until the reserve fund was absolutely closed.

Mention had been made of a reserve fund, but according to the Bristol orders the municipality could not provide a reserve fund of more than 10 per cent. on the amount of the capital. As far as the Bristol order was concerned, that was a common form of about ten years since, and he had not heard that it had been relaxed in subsequent orders. Therefore a municipality could not, under any circumstances, accumulate more than 10 per cent., and if the accounts showed that they were making more than 7 per cent. they were bound to decrease the price. Therefore the means of providing a depreciation fund and a reserve fund were cut down by the provisional orders.

Having regard to the management of the concern with which he was connected, he considered that they were, of course, to provide the light as cheaply as they possibly could, but they must first make provision for their interest, and they must next make provision for their sinking fund. And then he had always felt that, inasmuch as they could not get their work to depreciate precisely in order, each year being exactly the same amount, they ought to be able to put aside a reasonable sum from revenue to keep the work absolutely perfect for all time. He did not consider that the reserve should be touched for small renewals, and he hoped that his committee would agree that, in addition to providing for reserve fund, they should provide a fund (let it be called what it might) which would enable them to deal with any untoward accident which might occur out of that fund, without trenching upon the reserve or upon the pockets of the ratepayers. For the present purpose he would call it a depreciation fund. The ratepayers ought to be protected from claims such as that when they had given the municipality the benefit of the rates for the purpose of enabling them to raise money for the service of the public and for the undertaking at the cheapest possible rate.

He thought that they had no right to ask the consumer of to-day to provide for the absolute repayment of every penny that had been borrowed to start the work within the specified period from the date of the loan, and also ask him to keep the works in useful working order at the end of the period of the loan, and in addition to provide an accumulation in the shape of a depreciation fund. That would involve asking too much both of the ratepayers and the consumers of to-day. Of course in the first instance, it was the consumer of to-day that had to find the money. First, it would be wrong as being unjust to the consumer, and next, it would be wrong because it would tend to stop the demand. It had been found at Bristol that a very slight reduction of charge produced a very large increase in the demand, and that when the price was high the demand was not so great. He thought that in loading an undertaking, they ought to have regard in some degree to the persons who found the money, and, as he had said, the Local

Government Board would not allow them as a committee to do precisely as they liked, or to extract exactly what they thought fit from the consumer.

He thought that the meeting would agree with him that the Local Government Board were watching fairly and carefully, and he thought that most committees followed that view of the matter. He thought that the Local Government Board were dealing fairly and liberally with them, having regard to the fact that no one quite knew what was the life of a machine or what was the life of an undertaking. The ratepayers of to-day were willing to bear a considerable burden; but to attempt to add to that burden a huge depreciation to provide what he might call a sinking fund twice over was more than they ought to be asked to bear. He agreed in the main with the suggestions of the author of the paper.

Mr. SYDNEY MORSE said that the author in his paper said, "Those who insist that municipalities must depreciate as well as pay back capital ought to have very strong reasons to warrant such an extreme position"; and further, he went on to say, "The only possible motive for the suggestion is an insidious attempt to handicap a municipal undertaking because it is a municipal undertaking." Companies were in very much the same position as municipalities. During last week, before he knew that Mr. Biggs' paper was going to be read, he was asked to advise a company on the question. The secretary of the company came to him and said, "We have borrowed a quarter of a million of money for electric plant." He asked, "On what terms do you borrow the money?" The answer was, "On the terms that we should provide for its repayment by an annual sum sufficient to repay the total sum borrowed in thirty years. Shall we have to create a depreciation fund as well?" That was the position of municipalities. What was the depreciation fund for? Unless he was wrong, the depreciation fund was to provide the money to which resort could be had in case of need. The essence of a sinking fund was that the sinking fund should repay the capital, so that at the end of the time it might amount to the capital which would have to be repaid. In fact the sinking fund was a trust fund. It was marked and put aside as a sinking fund, and therefore no one could touch that fund. If depreciation was desirable they could not use the sinking fund. In talking to engineers he could not believe that any one would suggest that depreciation was not desirable. It was quite true that the Municipal Association Conference had come to a conclusion that depreciation was not desirable in addition to a sinking fund; but if a sinking fund could not be touched for depreciation they must have something which could be touched for depreciation.

Even, in reference to another part of the paper, he asked, was it reasonable to suggest that 3½ per cent. was a proper return for a very risky business? Electrical engineering was not to-day in its final form. They would not have ten years hence the system which they had to-day. Even since electrical engineering came into common use the systems had so changed that the plant put up in the early eighties had to be replaced very rapidly now. If that was so, the business was not one for which they could get money at 3½ per cent.

But he suggested that an undertaking must be considered entirely apart from the method or terms upon which the capital had been raised. The capital was one thing and the undertaking was another. If depreciation was necessary in order that the undertaking might have a fund by which the plant could be kept up, and in order to prevent an undue demand at a particular moment by an accident or anything of that sort, he thought that they would all see that depreciation was necessary and was not in any way to be confounded with a sinking fund for the repayment of capital.

For the reasons which he had stated he thought that they must conclude that repayment of capital was not depreciation, and that depreciation was necessary because plant did depreciate, and in some instances very rapidly.

Mr. ROBERT DONALD said that it seemed to him that municipalities had a great advantage over companies in the fact that they had a perpetual existence. Let them take the case which had been brought forward that evening. After seven years the Glasgow tramways department decided to replace the horse traction equipment by electric equipment. He hardly thought that any company would do such a thing, and certainly not in the same short period. Supposing that a company had only a lease for thirty years, it certainly would have been very foolish in the interests of the shareholders to do so. Therefore a municipality had an advantage in that respect by having perpetual existence. He did not think that any company was likely to get a perpetual lease. The result was that companies were unable to take advantage of new inventions and to keep their undertakings up to date to the same extent that corporations were. While he (Mr. Donald) agreed generally with Mr. Biggs' treatment of depreciation, he considered that corporations in dealing with tramways should be generous in provision for renewals.

Mr. W. WORBY BEAUMONT said that Mr. Biggs had put the main points in his paper categorically, and he might almost expect answers "yes" or "no" to them. In one respect that could not be done. He had not defined what he meant by "perfect and efficient maintenance." Upon that very much depended. Did he mean that a given plant should be as efficient at the end of the given period as it was at the beginning? Or did he mean that it should be as efficient as anything that is obtainable at the end of the period? If he meant, for instance, that a pumping engine used in a waterworks was maintained so that it would pump for a given consumption of steam as much water as it did at first, then the question would be answered in one way. If he meant that the pump had been found to be less efficient than something more modern, and had been replaced by the more modern machine, then they must answer the question in another way.

The author first of all referred to unproductive and to productive works. Of course they might all differ as to those terms. For instance, he would hold that the expenditure of money by a municipality on the construction of roads was a productive expenditure, though he knew that it was not considered so. It was indirectly as productive as most other things. It was a case in which the municipality would expend money which would be an expenditure in perpetuity, and there would be no such thing as depreciation or redemption in that case. But when they came to water-works, they came to the necessity for looking ahead with regard to the possible growth or possible decrease of the population of the place. Under some circumstances, even in this country, populations had decreased, and under some circumstances it would be manifestly unfair not to make some provision for the lessening of the amount of capital on which interest was to be paid.

On the other hand, coming directly to the author's questions, they had in the paper a summary. The first item was, "Is efficient maintenance, as is absolutely necessary in productive undertakings such as water-works, electrical works, tramways, etc., a capital or a revenue charge? If the latter, is not that one method of depreciation?" The answer to that question, as he had said, depended upon whether the author meant by "maintenance" merely the efficient maintenance of the thing as it was, or maintenance efficient from the point of view of the date of the end of the capital period, whatever that might be. In the latter case the answer would be "yes."

The second question was, "Is the repayment of capital depreciation? If not, what is it?" Taking that question alone, without the other, one was bound to admit that it was depreciation, and if the maintenance had been complete to the end of the time of paying off the capital the new generation which became possessor would say to itself, "We have got whatever there is handed over to us. This we have inherited. It has cost us nothing, and if we want to go on with it we must do as the old boys did in the past; we must now find the capital for working it, or for complete replacement if thought necessary." There was no reason whatever why the water-works, or whatever it was, should be given to them free of cost in complete working condition. The whole of it had been paid off and left in working order, "working order" meaning efficient maintenance up to the time at which the capital was paid off. Why should the first generation pay for something to be presented to the other generation—the complete thing with no payment whatever? He really could not see why, and therefore in answer to the second question he was bound to say that repayment of capital certainly was depreciation in the sense in which the author meant it. If the capital had been borrowed for sewerage, water-works or roads the term of repayment should be a very long one, or it should be in some cases perpetual; but for speculative work, such as electric supply or street tramways, it should be a short one, for a succeeding generation may not want the concern.

Coming to the question of companies' capital, the author said, "In this case also the capital is repaid by keeping the undertaking efficient out of revenue." Again the question occurred which he asked at first, namely, is the undertaking or plant to be maintained at the highest state of efficiency by replacement and not by mere maintenance? With regard to a company the answer would be "yes" as a rule. It was the duty of those who were managing to maintain in the most efficient condition, so as to make the greatest profit out of that which was sold, whatever was the product. It would have to be efficient in the sense that that which was most efficient at any one period must be that which the company adopted and paid for out of revenue.

Mr. Biggs said, with regard to competition, "Think of half a dozen competitors in water supply, or gas, or electric supply, or tramways and telephones, and so on." With regard to some of these matters it was of no use to think of such a number of competitors. It would be out of the question altogether that half a dozen water-supply companies should to-day compete for the supply of any one place. But, in talking of these things grouped together, tramways, telephones and so on, he was reminded of such questions as the replacement of rails. He took it that Mr. Biggs would include rails amongst the things which wear out in some places in from nine to eleven months, or in others in a few years. He would probably consider that such things as rails would be properly included in the maintenance, just as much as the oil or the wear and tear of bearings or anything else in machinery. So he did not see that the question of rails introduced any difficulty, but it emphasised again the necessity of something definite with regard to what is meant by efficient maintenance. Mr. Morse, from what he (Mr. Beaumont) could understand, seemed to want, even for a public company, first of all, the capital returned at a certain period, presumably for those who had provided it; and he wanted then also a sum provided which should be apparently equal to the capital; and it seemed to him that that was to be provided by one generation only, to be presented, for some reason which he could not understand, to some succeeding generation that might have been only in the position of customers. If the business or undertaking was likely even remotely to become obsolete, like street tramways, then provision for repayment of capital should of course be made.

The latter part of Mr. Biggs' paper, he could not help thinking, might be taken as an argument in favour of the creation of a community all in the employ of municipal and similar authorities, and dependent upon them, and therefore a community which would gradually get into the somnolent condition which they knew certain communities had got into in some countries abroad which might be mentioned, where the great aim of everybody, except the very few of marked individuality, was to become a state or municipal servant in some office or other; or a community whose subservience would first lead to expenditure beyond its means, then municipal revolution and repudiation, or ultimate dwindling of population in favour of less heavily rated places and freedom from municipal monopoly and restriction.

Mr. J. E. KINGSBURY said that Mr. Beaumont seemed to suggest that an enterprise might be carried on by a municipality up to a certain point, and if it had repaid its capital, further capital might be borrowed for the purpose of carrying it on. The subject was one of such interest that he did not hesitate to ask whether that was the object of the Local Government Board in fixing a date for the repayment of loans. He gathered that the Local Government Board made the loan on the understanding that it must be repaid at the end of a given period, and he took it that the idea was that not only should the loan be repaid, but that the work to be provided by it should remain in existence for the continued use of the community. They had to consider that, at present, engineers did not know the life of either electric generating machinery nor electric mains, nor of numerous other things which were used. He thought that that made it unquestionable that a depreciation fund must exist for the purpose of the replacement which, under ordinary circumstances, they might expect to become necessary. As they did not know absolutely what depreciation would be required, they must put aside something in the nature of insurance for such replacement of apparatus as might be necessary at the time of the repayment of the loan. The replacement question depended upon whether the continued use of the apparatus after the repayment of the loan was expected. If it was not expected that the enterprise would continue in use, or if it were assumed that a new loan might be obtained, then it was obvious that depreciation was not needed. All that they would have to do would be to pay off capital. But his impression was that continued use of the enterprise was contemplated, and an allowance for depreciation was therefore essential.

Mr. GEORGE E. ARNOLD said that he was generally in agreement with the author, except that he did not think the same rate of interest should be taken in comparing municipal and commercial undertakings. In the former the interest was guaranteed and commenced from the date the capital was obtained; in the latter there was no guarantee, and shareholders had to wait for their undertaking to be constructed, and then for it to be developed into a paying concern—very often a long period—so a portion of a company dividend should be looked upon as a dividend in arrear, before any part could be considered available for repaying capital. To cover that delay and the risk, the company shareholder was entitled to a higher rate of interest than the holder of an easily realised gilt-edged security.

Generally he thought that the Local Government Board were too considerate of the next generation of ratepayers in fixing too short periods for the repayment of capital. They were quite right to see that the next generation had not to pay for what had been consumed or used up by the present one, but in the case of freehold land in towns, why insist on repayment and so force up the present rates for the benefit of future ratepayers? The land does not wear out nor disappear, and generally appreciates in value.

He did not think that municipalities should set aside any sum for depreciation; so long as the capital was repaid within the useful life of the undertaking they were financially sound, and anything beyond that was an additional tax on the user of the tramway or the consumer of water. There was also no obligation on the part of the present generation to hand over to the future a going concern as a free gift, and a scheme as to how a part of the cost of benefits to posterity can be carried forward for it to pay would be welcomed by the present burdened ratepayers.

Mr. STEPHEN TERRY said that he noticed that Mr. Biggs called attention to some figures which might, perhaps, if they were not looked at from a particular point of view, be regarded as alarming. That was the enormously increased indebtedness of local authorities. He said, and no doubt quite truly, that at the present moment the indebtedness of local authorities extended to somewhere about 300,000,000*l*., and it was increasing at the rate of 10,000,000*l*. per annum for the last ten or twelve years. That was probably true, but it must be also borne in mind that the rateable value had increased, not, perhaps, in all the towns for which the loans had been granted, but as an average in the towns throughout the kingdom generally, probably, to that extent; and therefore the very heavy increase of indebtedness was also to some extent balanced and justified by the increased value on which the rates were made; so he did not think that from that point of view they had any cause for alarm.

He would touch only upon the point of depreciation, for the other points had been very ably discussed. In the matter of depreciation and renewal, Mr. Sherley-Price had hit the nail on the head. He said that what they had to watch for was obsolescence. That was no doubt a very important point in the sanction of loans for machinery. It applied, probably, more to machinery and structures of iron and steel than it would apply to other things. Where there were steel rails or steel bridges, or any other thing in which there was either mechanical movement or electrical action, there might be obsolescence at a comparatively early date. We had now more efficient machinery by which we got better returns for the power put into it, and that was certainly the case with regard to the steam engine. All those points showed that it was very necessary to watch for the particular item obsolescence, and that it was necessary to remember, in providing for depreciation, that there must be enough money provided, not simply for keeping a plant up to its going condition at the time, but even for its really being efficient and fit to hand over at the end of the time, for which purpose it must be not merely in working order, but modernised and kept up to date. That seemed to him to be a point which was not always quite carefully looked at. He remembered a matter vitally connected with the Peninsular and Oriental Steamship Company many years back, when they were in a bad financial position. The fact that the fleet was brought up to date was due entirely to Sir Thomas Sutherland who rose, he believed, from a humble position in the company to the proud position of chairman. He said, " You will have to provide new boilers to carry double the present pressures for nearly all your ships, and be satisfied with very small dividends for a few years until this has been done." There was a case in which, up to that date, obsolescence had not been considered. That was an all important point, which, although not culled from a municipal source, showed the necessity of providing for the absolute renewal of things on a basis which kept pace with the times. In that case there was a re-boilering of ships before the old boilers were nearly worn out, and it was necessary to provide proper means for make-up feed and for evaporators and other costly items in the modification of the propelling machinery. Those were points that had to be considered. Any attempt to cut down depreciation might have had a very disastrous financial bearing if such matters had not been properly looked into and provided for.

Mr. BIGGS, in replying upon the discussion, said that there was one thing for which Mr. Sherley-Price ought to be thankful, and that was in the discussion he had brought out a new word. They had been accustomed to talk about " antiquation " with regard to machinery, but now he supposed they were to choose between that and the term " obsolescence." He, however, did not care which was used.

He must plead guilty to one or two lapses. Whether he intended to allow those lapses to go or not might be a question. He would explain what he meant by maintenance, and he thought that when he had explained, the speakers would be pretty well all agreed. His idea of maintenance was Mr. Young's idea of maintenance, and he fancied it was the idea of a great many other people. They did not put down in their balance sheets the actual amount of money spent upon driving in a new nail, or giving a coat of new paint, or putting up new wires for a series of 10, 20, 30, or 40 miles of tramways as the maintenance charge. In Glasgow he believed the figures for the maintenance were 480*l*. a mile. That meant that they put aside annually, if he read their figures rightly, the sum of 480*l*. a mile, and that sum was to be used as the rails or apparatus required renewal. If the rails wore out three times in the period during which the loan was running, of course 480*l*. a mile would give them the opportunity, if they thought fit, of increasing the weight of rail (if too light rails had been initially been laid down) to 80 or 100 lb. rails. If their dynamos wore out, or if their wires broke down, they could renew with later types if so desired. They had the money to go to. That was what he intended to be meant by " maintenance being efficient." Surely that was a kind of reserve; not that the amount of money was spent annually, the sum of money per mile being put aside annually for the replacing of whatever was obsolete or worn out. He thought that that definition brought them pretty much on the same ground. He thought that it would bring Mr. Morse upon the same footing. Mr. Morse did not seem to recognise that maintenance was a necessity, and that maintenance must be efficient. Efficiency was also a necessity. If then, the maintenance was absolutely efficient, and the undertaking, whatever it might be, was up to date, what did they want further depreciation for? Initial capital was repaid after a certain period, and they had an undertaking efficient in every way. Surely that was enough to take out of revenue. He could not see why they should put another sum by. For what would it be? He did not know.

He was very glad to hear Alderman Pearson's remarks. As to the life of the various materials, he (Mr. Biggs) thought with Alderman Pearson, that it would be better if the Local Government Board would have a system of granting loans, not upon the average life, but upon the various lives, and if they would distinguish, more often than they did, between those parts of an undertaking which deteriorated very rapidly, or wore out very rapidly, and those which were of very much longer life. If a municipality had to buy land, and if a loan was obtained for the purpose of the land, say, for a recreation ground, they would have the land, and there was no necessity that it should be paid off in fifty years. It was just as much an advantage to the generations that came after the fifty years as it was to those who used it during the first fifty years; and it seemed a hardship, when they considered the demands that were made upon the pockets of the ratepayers, that they should be compelled to pay off the cost of the land.

Mr. SHERLEY-PRICE asked whose property the land would be if it was not paid for even in sixty years? He (Mr. Biggs) would say that they might have a lease of nine hundred and ninety-nine years, or nine thousand nine hundred and ninety-nine years, which vested in the authority.

Mr. SHERLEY-PRICE: Then the Government have to purchase all the land of the country?

Mr. BIGGS said that there was no necessity that that should be. The authority might have a loan for the purchase of freehold property. There was no necessity that that should be paid off within fifty years. They would have to pay the interest on it, undoubtedly, but they would use the land. Every generation would pay the interest, and every generation would use the land. It would be vested in the authority in perpetuity. Of course if the ratepayers all agreed to pay off, or if the Government insisted that they should pay off, that would be known when the money was borrowed. Whether it was right was another matter.

He thought that after his explanation of what he meant by maintenance, he had nothing more to say. He thanked the meeting for giving their attention to the paper and for discussing it, and he was glad that there might be some basis for future discussion of the question.

The following communication was received from Mr. BRIERLEY D. HEALEY, subsequently to the reading of the paper.

Generally, I agree with the author, but I write the result of my experience and express my opinions in slightly different terms. Capital for municipal purposes is obtained for the present ratepayers, with possible benefit in some cases to future ratepayers, and the period allowed for repayment should be equal to an average life as found by experienced actuaries.

To maintain the works in useful and safe working condition, it is necessary to put aside a certain sum annually, and this may be a reserve or suspense fund which can be drawn upon in any emergency. To put aside any sum annually for depreciation, with a view of providing a fund at the end of the loan period for renewing any plant or works, is not honest to present ratepayers.

Whenever new works or plant are required, whether to bring things up to date, or to replace any which have been used up, a new loan is the only equitable way, if we regard the possible benefit to future ratepayers. The granting of loans for initial plant and works, and of new loans for extensions or renewals of such works as are worn out, means no injustice to existing ratepayers or future ratepayers, and the accounts would be quite as easily managed as under the present system.

As regards capital for company purposes, assuming, as the author does, that all is square and true, and the management equal to anything of the kind under municipal control, the shareholders practically use their own money, and the payment of dividends is the only way of dividing the profits.

As to repayment of capital under company control, having once decided upon a certain fair rate of interest, then all dividends which exceed that rate include partial repayment of capital. It is, however, the duty of every directorate to set aside a sum annually as a renewal or reserve fund, out of which in due time the works may be brought up to date, and their usefulness thereby increased.

Although a few of the largest companies have such a fund, at the present time many of them find themselves sorely handicapped, because of the necessity of renewing their plants, and the lack of the sinews of war. For want of a reserve fund, many hitherto prosperous undertakings have been compelled to stop business. Mention might be made of certain works, where the dividends have ranged from 25 per cent. to nil, which are now obliged to be remodelled.

Obituary.

HENRY LUMLEY, whose death took place on January 2, 1902, was born in October 1838. He was educated at the City of London School, after leaving which he travelled for three years in America. On his return in 1853, he joined his brother Edward (deceased) and founded the firm of E. and H. Lumley. In 1862 he invented the rudder bearing his name, which was on the principle of angulation, and was adopted in several ships of the British Navy, notably the *Orontes*, *Columbine* and *Bullfinch*. It was also adopted in the Dutch and Italian navies. For services in connection with the latter, Mr. Lumley was made by Victor Emanuel I. a Chevalier of the Crown of Italy, being the first Englishman who received that decoration. In 1872 Mr. Lumley went to Palestine, for the purpose of tracing, defining and preparing plans for the restoration of the water supply to Jerusalem. He made a survey and prepared a scheme for restoring the aqueduct or conduit from its source at King Solomon's Pools to the city. A temporary repair was made, and the supply was successfully given for some short time, but the Bedouins and other nomads broke down the works and the Turkish Government took no trouble in the matter. He was an Associate of the Institute of Naval Architects, and a Member of several other Technical Institutions, and was elected a Member of the Society of Engineers in 1894.

FRANK GRIFFITHS PLUMPTON, whose death occurred on January 4, 1902, was born at Camberwell on February 1, 1876. He was educated at the Wilson Grammar School, and was afterwards articled to Mr. W. H. Allen, of Lambeth, mechanical engineer. He was subsequently engaged with Messrs. Maudslay, Sons and Field, of Lambeth, and was afterwards in the engineering department of the South-Eastern Railway Company. Mr. Plumpton was elected an Associate of the Society in 1900.

CHARLES STEWART BUZACOTT HARTLEY, who died at Brisbane on September 15, 1902, was born on April 30, 1880, at Rockhampton, Queensland. After a liberal education he was apprenticed, in 1896, to Messrs. Burns and Twigg, engineers, of Rockhampton. In 1898 he was appointed assistant to Mr. R. Schmidt, Chief Engineer to the Rockhampton Harbour Board, and at the time of his death he held the position of assistant engineer to the Board. He was elected an Associate of the Society of Engineers on June 2, 1902, and was a young engineer of great promise. At the time of his death he held the rank of Lieutenant in the Fifth Regiment, and was buried with military honours.

CHARLES GANDON, whose death took place on October 8, 1902, was born on June 13, 1837. On leaving school he was articled to the late Henry Palfrey Stephenson, one of the founders and the first President of the Society. Mr. Stephenson was then engaged chiefly in dock and railway engineering. About the fifties Mr. Stephenson took up gas work, and with Mr. Gandon as his assistant constructed gas-works at Naumburg, Ludwigsburg, Zeitz and Tilsit in Germany. Shortly afterwards Mr. Gandon was appointed an assistant engineer on the Great Northern Railway, but soon returned to the gas profession, becoming engineer and manager to the Ottoman Gas Company, and on his leaving Smyrna he was elected a member of the directorate.

During his residence in Smyrna, the town at that time being almost without water, he expressed his opinion that owing to the formation of the surrounding country an abundant underground supply existed. This at his suggestion was tested, with the result that an excellent supply of water was obtained, and to-day there are thousands of artesian wells in Smyrna. From Smyrna Mr. Gandon went to Bombay as engineer and manager to the Bombay Gas Company, and a few years after his return from there he was elected to a seat on the Board.

On leaving Bombay he was for a short time engineer to the Pará Gas Company. On his return to England in 1876, he was appointed engineer to the Crystal Palace District Gas Company, being again associated with Mr. H. P. Stephenson, who was a director of the company. The company at that time was only doing business in a small way, but on his leaving it 21 years later it was one of the leading gas concerns in the country. At the time of the gas labour troubles in London his tact and firmness, together with the esteem in which he was regarded by the men, averted a threatened strike at the Crystal Palace Gas Works, and immediately afterwards they presented him a piece of silver plate as a memento of the peace which followed the storm. On his retirement in 1897, beyond retaining his membership of the Boards of the Bombay, Ottoman, and Woking gas and water companies, Mr. Gandon gave up professional work.

Mr. Gandon was admitted as an Associate Member of the Institution of Civil Engineers in 1867, and was raised to membership in 1881. In 1876 he became a member of the old British Association of Gas Managers, and was President in 1888 under its new name of the Gas Institute. He was also a member, and in 1882 president, of the Southern Association.

For the Society of Engineers Mr. Gandon always expressed very great regard. He was elected a member in 1859, and was President of the Society in 1885. In 1881 he contributed a paper on "Gas Engines." He was also a member of the Institution of Gas Engineers, in which he took a great interest, being one of the originators of the Institution.

FORTY-EIGHTH ANNUAL GENERAL MEETING

HELD AT THE INSTITUTION OF MECHANICAL ENGINEERS, STOREY'S GATE, WESTMINSTER.

Monday, December 8th, 1902.

PERCY GRIFFITH, PRESIDENT, IN THE CHAIR.

THE Minutes of the Forty-seventh Annual General Meeting, held December 9th, 1901, were read, confirmed and signed.

The president nominated Mr. A. H. Smith and Mr. A. Burgess to act as scrutineers of the balloting lists for council and officers for the year 1903.

An address was then delivered by the president, in which, after referring to the advantage of issuing an annual report of the council to the members, he pointed out that owing to the date of the annual general meeting being fixed by the rules for December, it was impossible to present the annual report to the present meeting, and explained that he proposed to confine himself to such features of the year's work as might induce comment or discussion.

Dealing first with the papers read during the year, the president pointed out the wide range covered by them, claiming that in this respect the society was fulfilling the object with which it was established, viz. the advancement of the science and practice of engineering. After referring to the discussion following the papers, the president called attention to the excellence of both, and announced the names of the authors to whom premiums had been awarded (see the Report of the Council, page 310). The attendances at the ordinary meetings had shown a good average, and one that proved an increasing appreciation of the society's work.

The visits to works carried out during the summer vacation were then briefly reviewed, the president acknowledging the courtesy and hospitality of those who had afforded the members the opportunity of inspecting their works.

In regard to the increase in membership, the president announced that the total membership now exceeded its previous highest record (viz. 522) by one, this representing a net increase of 30 during the year.

x

Reference was then made to the reception and social reunion arranged to follow the meeting, which was practically the first function of the kind provided by the society. The question as to whether the reception should be associated with the annual general meeting in future years, as on this occasion, was then raised and the opinions of those present were invited.

In concluding his address, the president expressed his appreciation of the support he had received from the members generally during his year of office.

Responding to the president's invitation to those present to discuss points of interest in the society's work—

Mr. H. SHERLEY-PRICE said that as regarded the time of year at which the reception should be held, there were reasons both for and against its being held in December. He thought, however, that it was very desirable that the members should have an opportunity of meeting each other socially, and so knowing a little more of each other. The council were in this matter aiming at the welfare of the members. He thought it was a matter of considerable gratification that their ordinary meetings had been so well attended, considering the number of members, and the fact that so many of the members lived at considerable distances from London. He also thought that the tone of the discussions had been very good, and that the various papers had been well handled and very fully dealt with.

Mr. J. W. WILSON said that he thought they should consider whether it would be advisable in future to have the annual meeting on a different evening from the reception, and if so, when.

The PRESIDENT said that the date of the annual meeting was fixed absolutely by the rules.

Mr. J. W. WILSON suggested that the date of the reception might be altered. If it was held in the vacation, or at some time early in the autumn, it would leave December clear for the annual general meeting and for the dinner. Would it also not be possible to fix the dinner a little later?

The PRESIDENT said that he thought that their difficulty was in having the annual general meeting fixed by the rules. Personally he felt convinced that another function during the vacation would interfere with their visits to works, and further, it would be open to the most serious objection of competing to some extent with the social functions held by the larger institutions. The main point was whether they should hold the reception on the same night as the annual general meeting. It did not follow that because they altered the date of the reception they should necessarily go from winter to summer.

Mr. J. W. WILSON said that one alternative would be to

have the annual general meeting in the presence of the ladies and the other visitors attending the reception. The meeting might be confined to a short résumé of the work of the society by the president, and the usual votes of thanks.

Mr. FREDERICK HOVENDEN asked whether it was possible for the council to secure papers of a more mechanical character. He noticed that nearly all the papers were more or less connected with civil engineering, and it seemed to him that it was very seldom they had a paper on workshop practice. He was himself a mechanical and practical engineer, and therefore he should be very pleased to hear a practical paper, but one of a civil engineering character did not appeal to him.

Mr. WILSON said that he might point out that a few years ago the council were asked why they had so many mechanical papers and not more civil engineering papers. The society's transactions would show that they had a great many mechanical papers.

Mr. HOVENDEN said that he had been a member for about ten years, and had not come across many mechanical papers. As an example of what he meant, he suggested that if anyone could tell them about the working of piston-valves for locomotives it would be useful, as there seemed to be very little general knowledge on this subject.

The PRESIDENT said that the point raised was a very interesting one, and the council would certainly give it attention. They were very glad to receive the suggestion.

Mr. HOVENDEN asked whether it would be possible to arrange for a visit to Crewe.

The PRESIDENT replied that four or five years ago the society paid a visit to the locomotive works there, but, no doubt on account of the cost, the attendance was very small.

The PRESIDENT said that the next business was a matter which he was sure would receive the most cordial attention of those present, as well as of every member of the society. Not only had the building in which they were met been placed at their disposal, but the secretary of the Institution of Mechanical Engineers (Mr. Worthington) had taken a great personal interest in the details of the arrangements for this evening's meeting. He was sure that those present would be glad to have an opportunity of expressing to the president and council of the institution their appreciation of the courtesy which had been extended to the society. He proposed that the best thanks of the Society of Engineers be given to the President and Council of the Institution of Mechanical Engineers for their kindness in affording the society the use of their building for the purpose of the annual general meeting and the reception.

Mr. SHERLEY-PRICE seconded the resolution, which was carried by acclamation.

Mr. J. W. WILSON said that he had very great pleasure in moving a vote of thanks to the president (Mr. Percy Griffith) for his services to the society during the past year. They must all acknowledge that the president had very worthily discharged the duties which his year of office had entailed upon him, and had carried out the work of the society in a very successful manner. The president had brought not only ability, but enthusiasm, to bear upon his conduct of the society's affairs, and without enthusiasm ability was not of much use. They would all agree that very hearty thanks were due to Mr. Griffith for his services as president of the society.

Mr. HOLTTUM said that it gave him great pleasure to second the vote of thanks. He would heartily endorse what Mr. Wilson had said. He had been struck with the enthusiasm and pertinacity with which the president had carried out the onerous duties of his office.

The motion was put to the meeting by Mr. Wilson, and carried by acclamation.

The PRESIDENT, in reply, said that so far as the interests of the society had been advanced during his year of office he was extremely gratified, and no one would feel greater pleasure than himself in the results.

Mr. HOVENDEN moved that a hearty vote of thanks be given to the council and officers for their services during the past year.

Mr. WESTROPE seconded the motion. As one of the younger members of the society he very much appreciated the efforts made by the council to develop the social aspect of the society's work.

The motion was carried unanimously, and acknowledged by Mr. Sherley-Price.

The PRESIDENT then moved a vote of thanks to the secretary, Mr. Perry F. Nursey.

Mr. G. A. PRYCE CUXSON seconded the vote, which was unanimously carried, and duly acknowledged by Mr. Nursey.

The result of the ballot for the council and officers for 1903 was then announced, which concluded the business of the meeting.

ANNUAL REPORT OF THE COUNCIL, 1902.

The Council in reviewing the progress and work of the Society during the past year have much' pleasure in recording a steady advance in the membership, and the fact that on December 31, 1902, the number enrolled was greater than at any period in the society's history. The total as given in the following table (viz. 523) is one more than that on December 31, 1893, which was a considerable advance on any previous record. The present total is the more gratifying as the losses from death, resignation, and other causes during the year amounted to 29, the number of members and associates elected during the year having reached the satisfactory total of 59.

MEMBERSHIP.

The following table shows the numbers on the register of the Society at the end of 1902, and the two preceding years.

Class.	Dec. 31, 1900.	Dec. 31, 1901.	Dec. 31, 1902.
Honorary Members	18	20	20
Ordinary Members	246	252	273
Ordinary Associates	128	122	120
Foreign Members	64	69	78
Foreign Associates	30	30	32
Totals	486	493	523

FINANCE.

The accompanying statement of accounts (see pages 313 and 314) shows that in the important matters of finance the progress of the society has been of a gratifying character. Notwithstanding the extra expense entailed by the reception held on December 8, and by the visits and annual dinner, also the large amount of arrears of subscription written off as irrecoverable, the excess of income over expenditure gives the satisfactory sum of 50*l.* 3*s.* 3*d.* towards the accumulated fund. This fund (representing the cash assets of the Society) now amounts to nearly 700*l.* Of this, about 200*l.* is invested in railway stock, and 450*l.* stands on deposit at the society's bankers, the interest from both investments forming an appreciable addition to the annual income.

PAPERS READ.

The following is a list of papers read during the year:—
1. The President's Inaugural Address. By Mr. Percy Griffith.
2. British *v.* American Patent Law Practice and Engineering Invention. By Mr. Benjamin H. Thwaite.

3. Australian Timber Bridges and the Woods used in their Construction. By Mr. Herbert E. Bellamy.
4. Recent Blast Furnace Practice. By Mr. Brierley D. Healey.
5. Notes on some Twentieth Century Locomotives. By Mr. Charles Rous-Marten.
6. The Hennebique System of Ferro-Concrete Construction. By Mr. A. R. Galbraith.
7. The Effect of Segregation on the Strength of Steel Rails. By Mr. Thomas Andrews, F.R.S.
8. Depreciation of Plant and Works under Municipal and Company Management. By Mr. C. H. W. Biggs.

PREMIUMS.

The council have awarded premiums to the following authors for the papers named in the foregoing list, viz.:—

1. To Mr. Thomas Andrews, F.R.S., the President's Gold Medal.
2. To Mr. A. R. Galbraith, the Bessemer Premium of Books.
3. To Mr. Benjamin H. Thwaite, a Society's Premium of Books.
4. To Mr. Brierley D. Healey, a Society's Premium of Books.

ATTENDANCES AT ORDINARY MEETINGS.

It is satisfactory to find that the attendances at the ordinary meetings have improved during the past year, the average having been over 57 per meeting. The good attendance of visitors also is an indication of the value and interest of the papers read.

VISITS TO WORKS.

During the vacation the usual visits to works were made, and judging by the attendances, were much appreciated. The following is a list of the works visited:—

June 11, to the New Roof over the Centre Transept, the Pumping Stations in the Grounds, and the School of Practical Engineering in the South Tower of the Crystal Palace.

July 16, to the New Graving Dock and Ferro-Concrete Building Construction (in progress) at Southampton, and the Corporation Waterworks and Softening Plant at Otterbourne.

September 24, to Messrs. Joseph Baker and Son's works (manufacture of Industrial Machinery and Appliances), and the works of the Wicks Rotary Type Casting Company, both situate at Willesden.

By invitation of the last named company a number of members visited the Type Foundry in Blackfriars Road on September 30, when the type-casting machines were seen in operation.

In every instance the members were hospitably received and entertained by the respective hosts. The council desire to impress upon the junior members and associates the practical value of these visits to works, and to point out the desirability of attending them whenever circumstances permit.

Annual General Meeting and Reception.

On December 8 was held the forty-eighth annual general meeting of the society, and the report of the proceedings which appears on page 305 shows that points of interest and value to the society were discussed.

Following the meeting, a Reception was held by the President and Mrs. Percy Griffith, the Council of the Institution of Mechanical Engineers having kindly placed their premises at Storey's Gate, Westminster, at the disposal of the society for the purpose. The reception was well attended, and a musical programme was carried out during the evening. There was also a lecture, illustrated by lantern views, by Mr. Frederick Lambert, F.R.G.S., on "The Crystal Caves of New South Wales."

Annual Dinner.

The forty-eighth annual dinner was held at the Hotel Cecil on Wednesday, December 10, and proved very successful. Among the visitors present were Mr. James Mansergh, F.R.S., Past-President of the Institution of Civil Engineers, Mr. William Whitaker, F.R.S., Past-President of the Geological Society, Mr. James Swinburne, President of the Institution of Electrical Engineers, Mr. Arthur G. Charleton, President of the Institution of Mining and Metallurgy, Captain Wilson Moore, R.N., Col. Edward Raban, R.E., and Alderman Geo. W. Taylor, J.P., Mayor of Chelmsford.

Exchange Transactions.

The society continues the exchange of Transactions with the following Institutions, the volumes being available for reference by members and associates at the society's offices:—

The Institution of Civil Engineers.
The Institution of Mechanical Engineers.
The Institution of Electrical Engineers.
The Institution of Naval Architects.
The Iron and Steel Institute.
The Surveyors' Institution.
The Civil and Mechanical Engineers' Society.
The Junior Institution of Engineers.
The Institution of Mining and Metallurgy.
The Royal Engineers' Institute.
The Incorporated Gas Institute.
The Royal Institute of British Architects.
The Society of Arts.
The Liverpool Engineering Society.
The Cleveland Institution of Engineers.
The North East Coast Institution of Engineers and Shipbuilders.
The North of England Institute of Mining and Mechanical Engineers.
The South Wales Institute of Engineers.
The Institution of Engineers and Shipbuilders in Scotland.
The Institution of Civil Engineers of Ireland.
The French Institution of Civil Engineers.
The Canadian Civil Engineers' Society.
The Victorian Institute of Engineers.
The Engineering Association of New South Wales.
The American Society of Civil Engineers.
The Association of Engineering Societies.
The Smithsonian Institution.
The Franklin Institute.

JOURNALS AND MAGAZINES.

The library contains many books on engineering subjects, and in addition, the following journals and magazines are supplied gratuitously by their respective publishers, all being available for reference by the members.

American Machinist.
American Machinery.
Arms and Explosives.
Automotor.
British Architect.
Building News.
Builder.
Commerce.
Colliery Guardian.
Contract Journal.
Cassier's Magazine.
Electrical Review.
Electrician.
Engineer.
Engineering.
Engineering Magazine.
Engineers' Gazette.
Electro-Chemist and Metallurgist.
El Ingeniero Español.
Invention.
Indian Engineering.

Indian and Eastern Engineer.
Journal of Gas Lighting.
Local Government Journal.
L'Automobile.
Machinery Market.
Marine Engineer.
Mechanical World.
Mechanical Engineer.
Mechanical Progress.
Page's Magazine.
Public Health Engineer.
Railway Official Gazette.
Surveyor.
Sanitary Record.
Shipping World.
Society of Arts Journal.
The Quarry.
The Street.
The Tramway and Railway World.
Water.

In conclusion, the council would impress upon the members and associates the desirability of using their best endeavours to extend the usefulness of the society by contributing papers on engineering subjects of present day interest; by taking part in the discussions; by regular attendances at the meetings and visits, and by introducing such of their friends as may be eligible either as members or associates.

January 1903.

INCOME AND EXPENDITURE ACCOUNT for the Year ended 31st December, 1902.

Dr.

Expenditure.

	£	s.	d.
To Printing — less Sales	170	2	9
" Rent of Offices and Hire of Rooms	122	16	0
" Salary of Secretary	230	0	0
" Sundry and General Printing	30	17	3
" Reporting Meetings	17	8	9
" Refreshments for Mrs at Ordinary Meetings	7	11	0
" Fuel, Lighting, Office Cleaning and Repairs	15	12	5
" Postage, Telegrams, Carriage, and General Expenses	51	19	8
" Depreciation on Furniture	2	7	0
" Conversazion Invts, Reception, and Dinner deficiency	53	5	5
	702	0	3
" Balance carried to Accumulated Fund—Excess of Income over Expenditure for the year ended 31st December, 1902	50	3	3
	£752	3	6

Cr.

Income.

	£	s.	d.	£	s.	d.
By Subscriptions for the Year	715	1	0			
Less allowances and Bad Debts	73	9	6	641	11	6
" Life Subscriptions—proportion for 1902				8	11	9
" Admission Fees				89	5	0
" Interest on Deposits				8	1	8
" Interest on Investments				4	13	7
				£752	3	6

LIFE MEMBERSHIP FUND.

	£	s.	d.
To Revenue Account—proportion for the year 1902	8	11	9
" Balance to be carried to next Account	77	5	9
	£85	17	6

	£	s.	d.
By Balance brought from last Account, being the unexhausted Balance of the Fund on 31st December, 1901	85	17	6
	£85	17	6

PREMIUM FUND.

	£	s.	d.
To Premiums	16	2	0
" Balance to be carried to next Account	23	5	5
	£39	7	5

	£	s.	d.
By Balance brought forward from 31st December, 1901	18	8	3
" President's Donation	10	10	0
" Donation, H. E. Bellamy, Esq.	5	5	0
" Interest received from Bessemer Trust Fund Investment	5	4	2
	£39	7	5

BALANCE SHEET, 31st DECEMBER, 1902.

LIABILITIES.

	£	s.	d.
SUNDRY CREDITORS	54	5	3
LIFE MEMBERSHIP FUND	77	5	9
PREMIUM FUND	23	5	5
SUBSCRIPTIONS PAID IN ADVANCE	40	19	0
ACCUMULATED FUND:—			
Balance at 31st December, 1901 ... £643 19 10			
Add Excess of Income over Expenditure for the year ended 31st December, 1902 ... 50 3 3	694	3	1
	£889	18	6

ASSETS.

	£	s.	d.	£	s.	d.
SUBSCRIPTIONS IN ARREAR	£134	11	0			
ADMISSION FEES IN ARREAR	12	1	6			
	146	12	6			
Less reserved for Bad Debts	50	0	0			
				96	12	6
CASH AT BANK	30	13	1			
IN HAND	2	2	3			
ON DEPOSIT	450	0	0			
				482	15	4
INVESTMENT—£166 L. & N. W. By. Co. 3 per cent. Debenture Stock (cost)				199	5	8
OFFICE FURNITURE	35	0	0			
LIBRARY	40	0	0			
STOCK OF TRANSACTIONS	36	5	0			
				111	5	0
				£889	18	6

NOTE.—£185 North Eastern Railway Stock is held in Trust to secure payment of the late Sir Henry Bessemer's Premium. Audited and found correct,

(Signed) SAMUEL WOOD, F.C.A., *Hon. Auditor*,
30 Gracechurch Street,
London, E.C.

9th January, 1903.

SOCIETY OF ENGINEERS.

INDEX TO TRANSACTIONS

1857 to 1902.

AUTHORS.

Author.	Subject of Paper.	Transactions Year.	Page.
ABEL, C. D.	The Patent Laws	1865	25
ADAMS, H.	President's Address	1890	1
ADAMS, J. H.	President's Address	1875	1
ADAMS, T.	The Friction of the Slide-valve, and its Appendages.	1866	6
ADAMS, W.	President's Address	1870	1
ALFORD, C. J.	The Mineralogy of the Island of Sardinia.	1879	93
,, ,,	Appendix to Above: Laws relating to Mining.	1879	106
,, ,,	Engineering Notes on Cyprus	1880	131
ALFORD, R. F.	The Fell Engines on the Rimutakai Incline, New Zealand.	1882	121
ALLAN, G. W	Marine Insurance	1858	14
ALLANSON-WINN, R. G.	Foreshore Protection, with special reference to the Case System of Groyning	1899	133
ALLEN, A. T.	Concrete Subways for Underground Pipes.	1901	99
AMOS, E. C.	Machine Tools	1899	51
AMOS, JAMES	The Machinery employed in laying the Atlantic Cable.	1858	13
,, ,,	The New Hydraulic Machinery for lifting Vessels in the Thames Graving Dock.	1859	17
ANDERSON, CHRIS.	The Feasibility and Construction of Deep-sea Lighthouses.	1883	45
ANDRÉ, G. G.	The Ventilation of Coal Mines	1874	28
,, ,,	The Application of Electricity to the Ignition of Blasting Charges.	1878	123
ANDREWS, J. J. F.	Ship Caissons for Dock Basins and Dry Docks.	1890	189
ANDREWS, T.	The Strength of Wrought-iron Railway Axles.	1879	143
,, ,,	The Effect of Strain on Railway Axles, and the Minimum Flexion Resistance-Point in Axles.	1895	181
,, ,,	Effect of Segregation on the Strength of Steel Rails.	1902	209
ANSELL, W.	The Electric Telegraph	1857	11
ATKINSON, J.	Apparatus for Utilising the Waste Heat of Exhaust Steam.	1878	167
AULT, E.	The Shone Hydro-Pneumatic Sewerage System.	1887	68
BAILLIE, J. R.	President's Address	1889	1
BAKER, T. W	The Utilisation of Town Refuse for generating Steam.	1894	183

Author.	Subject of Paper.	Transactions. Year.	Page.
BALDWIN, T.	Single and Double Riveted Joints	1866	150
,, ,,	Experimental Researches into the Nature and Action of Safety Valves for Steam Boilers, &c.	1867	23
BAMBER, H. K.	Water, and its Effects on Steam Boilers	1867	65
BANCROFT, R. M.	Renewal of Roof over Departure Platform at King's Cross Term., G.N.R.	1887	125
BARBER, J. P.	Notes on the Proposed By-laws of the London County Council with respect to House Drainage.	1897	17
BARKER, C. M.	The Prevention of Leakage in Gas and Water Mains.	1869	79
BARKER, E. D.	Hydraulic Continuous and Automatic Brakes.	1879	75
BARKER, CAPT. W. B.	International System of Marine Course Signalling.	1884	47
BARNETT, S.	The Value of Exhibitions as Aids to Engineering Progress.	1883	75
BARTHOLOMEW, E. G.	Electric Telegraphy, irrespective of Telegraphic Apparatus.	1869	55
,, ,,	Electric Telegraph Instruments	1872	53
,, ,,	Electrical Batteries	1872	77
BEAUMONT, W. W.	President's Address	1898	1
,, ,,	Modern Steel as a Structural Material	1880	109
,, ,,	High Pressure Steam and Steam-engine Efficiency.	1888	221
BEAUMONT, W. W., and SELLON, S.	Notes on Parliamentary Procedure as affecting Light Railways & Tramways.	1895	33
BEHR, F. B.	The Lartigue Single-rail Railway	1886	163
BELLAMY, H. E.	Australian Timber Bridges	1902	63
BELLASIS, E. S.	The Roorkee Hydraulic Experiments	1886	41
BERNAYS, J.	President's Address	1880	1
,, ,,	The New Pits and Hauling Machinery for the San Domingos Mines in Portugal.	1879	29
BIGGS, C. H. W.	Depreciation of Plant and Works under Municipal and Company Management.	1902	271
BIGGS, C. H. W., and BEAUMONT, W. W.	Notes on Electric Light Engineering	1882	23
BINYON, A. H	Notes on Electric Traction	1900	103
BJORLING, P. B.	Direct-acting Pumping Engines	1877	79
BLACKBOURN, J.	The Action of Marine Worms, and Remedies applied in the Harbour of San Francisco.	1874	111
BOLTON, R.	The Application of Electricity to Hoisting Machinery.	1892	79
BOWER, G.	The Preservation and Ornamentation of Iron and Steel Surfaces.	1883	59
BOWERS, GEORGE	The Permanent Way of Railways	1857	11
BOYD, R. N	Collieries and Colliery Engineering	1893	169
,, ,,	A Deep Boring near Freistadt, Austria, by the Canadian System.	1894	119
BRIGG. T H	The Mechanics of Horse Haulage	1896	21

INDEX.

Author.	Subject of Paper.	Year.	Page.
BROTHERS, W. H.	Weighing Machinery and Automatic Apparatus in connection therewith.	1890	55
BROWNING, C. E.	The Strength of Suspension Bridges	1858	11
BROWNING, C. E	The Extension and Permanent Set of Wrought Iron.	1858	15
BRYANT, F. W.	Piling and Coffer Dams	1859	38
,, ,,	President's Address	1869	1
BUCKHAM, T.	The Drainage and Water Supply to the Town of Fareham.	1869	41
BURROWS, E.	Blake's Bridge, Reading	1893	71
BUTLER, D. B.	Portland Cement, some Points in its Testing, Uses and Abuses.	1895	61
,, ,,	The Effect of Admixtures of Kentish Ragstone,&c., upon Portland Cement.	1896	179
CAREY, R.	Hydraulic Lifts	1893	101
CARGILL, T.	President's Address	1877	1
,, ,,	Railway Bridge at the Place de l'Europe, Paris.	1866	191
,, ,,	Floating Breakwaters	1871	146
CARPENTER, C. C.	Modern Coal Gas Manufacture	1891	61
CARRINGTON, W. T.	Generation and Expansion of Steam	1858	15
,, ,,	Pumping Engines	1859	19
,, ,,	President's Address	1865	1
CARTER, J. H.	Roller-milling Machinery	1883	125
CHEESEWRIGHT, F. H.	Breakwater Construction	1890	83
CHURCH, JABEZ	President's Address	1872	1
,, ,,	President's Address	1873	1
CHURCH, JABEZ, jun.	President's Address	1882	1
,, ,,	President's Address	1883	1
COLAM, W. N.	Cable Tramways	1885	69
,, ,,	President's Address	1891	1
COLBURN, ZERAH	President's Address	1866	1
,, ,,	The Relation between the Safe Load and Ultimate Strength of Iron.	1863	35
,, ,,	Certain Methods of Treating Cast Iron in the Foundry.	1865	77
CONRADI, H.	Stone-sawing Machinery	1876	135
,, ,,	Cleaning of Tramway, and other Rails.	1893	47
COPLAND, H. S.	Modern Roadway Construction	1878	61
COWPER-COLES, S. O.	Protective Metallic Coatings for Iron and Steel.	1898	139
,, ,,	The Electrolytic Treatment of Complex Sulphide Ores	1899	203
COX, S. H.	Recent Improvements in Tin-dressing Machinery.	1874	11
,, ,,	Dry Crushing Machinery	1892	101
,, ,,	President's Address	1896	1
CRAMP, C. C.	Tramway Rolling-stock and Steam in connection therewith	1874	119
CRIMP, W. S.	The Wimbledon Main Drainage and Sewage Disposal Works.	1888	73
,, ,,	Sewer Ventilation	1890	139

z 2

Author.	Subject of Paper.	Transactions. Year.	Page.
CROLL, J.	Filter Presses for Sewage Sludge	1897	121
CROMPTON, E. S.	The Economics of Railway Maintenance.	1870	101
CULLEN, DR. E.	The Surveys of Proposed Lines for a Ship Canal between the Atlantic and Pacific Oceans.	1868	15
,, ,,	The Panama Railroad	1868	39
,, ,,	Isthmus of Darien and the Ship Canal	1868	52
CUNNINGHAM, J. H.	Pin-connected v. Riveted Bridges	1889	135
DANVERS, F. C.	Engineering in India	1868	137
DAVENPORT, J. G.	Wooden Jetties on the River Thames and their Improvement of the In-shore Navigation.	1859	24
DAVEY, H.	Pumping Engines for Town Water-supply.	1867	88
,, ,,	Milford Haven and its New Pier Work.	1872	89
,, ,,	Recent Improvements in Pumping Engines for Mines.	1873	182
,, ,,	The Underground Pumping Machinery at the Erin Colliery, Westphalia.	1876	119
DAY, ST. J. V.	Recent Arrangements of Continuous Brakes.	1875	87
DE SEGUNDO, E. C.	Power Distribution by Electricity, Water and Gas.	1894	143
DRURY, A. G.	The Shortlands and Nunhead Railway	1892	219
EDWARDS, E.	Mining Machinery	1859	35
,, ,,	Utilization of Waste Mineral Substances.	1861	82
ENGERT, A. C.	The Prevention of Smoke	1881	101
,, ,,	Defects of Steam Boilers and their Remedy.	1884	25
,, ,,	The Blow-pipe Flame Furnace	1884	185
FAIJA, H.	Portland Cement	1885	95
,, ,,	The Effect of Sea Water on Portland Cement.	1888	39
,, ,,	Forced Filtration of Water through Concrete.	1889	107
FARADAY, P. M.	The Rating of Engineering Undertakings.	1897	43
FELL, J. C.	President's Address	1899	1
,, ,,	Soft v. Hard Water for Manufacturing Purposes.	1884	103
FERRAR, W. G.	Practical Construction in the Colonies	1875	53
FOX, ST. G. LANE	The Lighting and Extinction of Street Gas Lamps by Electricity.	1878	105
FOX, W.	Reservoir Embankments	1898	23
FOX, W. H.	Continuous Railway Brakes	1872	112
FOX, W. H.	Continuous Railway Brakes	1873	33
FRASER, A.	Use of the Cornish Pumping Engine	1864	78

Author.	Subject of Paper.	Transactions. Year.	Page.
GALBRAITH, A. R.	The Hennebique System of Ferro-Concrete Construction.	1902	177
GANDON, C.	President's Address	1885	1
,, ,,	Gas Engines	1881	27
GIBBS J. D.	The Distribution of Electrical Energy by Secondary Generators.	1885	49
GLYNN, J., Junr.	Softening Water by Dr. Clarke's Process.	1859	29
,, ,,	The Chalk Formation	1860	46
GOODWIN, G. A.	President's Address	1894	1
GORE, H.	Modern Gas Works at Home and Abroad	1868	233
,, ,,	Horse Railways and Street Tramways	1873	113
GRAHAM, J.	The most Recent Improvements on the Injector.	1867	266
GRANTHAM, R. F.	Sea Defences	1897	143
,, ,,	The Closing of Breaches in Sea and River Embankments.	1900	21
GREIG, A.	Dundee Street Improvements and Drainage of Lochee.	1883	113
GRESHAM, J.	The most Recent Improvements on the Injector.	1867	266
GRIERSON, F. W.	The National Value of Cheap Patents	1880	145
GRIERSON, T. B.	The Treatment of Low-grade Iron Ores for the Smelting Furnace.	1901	69
GRIFFIN, S.	Modern Gas-Engine Practice	1889	169
GRIFFITH, P.	President's Address	1902	1
,, ,,	The Treatment and Utilisation of Exhaust Steam	1890	163
,, ,,	The Water Supply of Small Towns and Rural Districts.	1896	55
HAKEWILL, H.	Circular Tables	1865	147
HALL, C. E.	The Conversion of Peat into Fuel and Charcoal.	1876	151
,, ,,	Modern Machinery for Preparing Macadam for Roads.	1879	51
HARMAN, E. A.	Gas-Works Machinery	1898	97
HARRIS, G.	Water Supply to Country Mansions and Estates	1899	231
HARTLEY, F. W.	Methods employed in the Determination of the Commercial Value and the Purity of Coal Gas.	1869	100
,, ,,	An Improved Method of Charging and Drawing Gas Retorts.	1875	185
HEALEY, B. D.	The Economical Disposal of Town Refuse.	1900	65
,, ,,	Recent Blast Furnace Practice	1902	97
HENDERSON, R.	Paper-Making Machinery	1900	139
HENDRY, J.	Pump Valves	1859	25
HETHERINGTON, R. G.	The Main Drainage of Ilford	1901	173

Author.	Subject of Paper.	Year.	Page.
HOLLIS, S. A.	Preliminary Investigations for Water Supply.	1901	147
HOLTTUM, W. H.	The Use of Steel Needles in Driving a Tunnel at King's Cross.	1892	199
HOOPER, H. R.	The Practice of Foundry Work	1888	191
HORNER, J. G.	Certain Methods of Applying Screw Piles in a Bridge at Verona.	1867	52
HORSLEY, C.	President's Address	1881	1
HUTTON, R. J.	The Stability of Chimney Shafts	1887	150
JACOB, A.	The Designing and Construction of Storage Reservoirs.	1866	225
JENSEN, G. J. G...	Notes on Certain Details of Drainage Construction.	1901	19
JENSEN, P.	Incrustation in Marine Boilers	1866	117
,, ,,	Friction in Steam Cylinders	1870	17
JERRAM, G. B.	River Pollution caused by Sewage Disposal.	1886	217
JONES, A. W.	Modern Tramway Construction	1879	179
JUSTICE, P. S.	The Dephosphorisation of Iron in the Puddling Furnace.	1885	189
KERR, J.	Portable and Pioneer Railways	1891	89
KERSHAW, J. T.	Breakwaters and Harbours of Refuge	1859	31
,, ,,	The Pipe Sewers of Croydon and the Causes of their Failure.	1860	52
KING, G.	Irrigation with Town Sewage	1865	84
KINSEY, W. B.	The Arrangement, Construction, and Machinery of Breweries.	1881	173
KNAPP, H. F.	Harbour Bars: their Formation and Removal.	1878	139
KOCHS, W. E.	Strength and Rigidity	1865	69
LAILEY, C. N.	Acton Main Drainage Works	1888	125
LATHAM, B.	President's Address	1868	1
,, ,,	President's Address	1871	1
,, ,,	The Drainage of the Fens	1862	125
,, ,,	The Inundations of Marsh Lands	1862	173
,, ,,	The Supply of Water to Towns	1864	199
,, ,,	Utilisation of Sewage	1866	68
,, ,,	The Application of Steam to the Cultivation of the Soil.	1868	274
,, ,,	Ventilation of Sewers	1871	39
LAWFORD, G. M.	President's Address	1897	1
,, ,,	Fireproof Floors	1889	43
,, ,,	Drainage of Town Houses	1891	187
LAWFORD, W.	Light Railways	1888	181
LEE. J. B.	Photographic Surveying	1899	171
LE FEUVRE, W. H.	President's Address	1867	1
LE GRAND, A.	Tube Wells	1877	133

Author.	Subject of Paper.	Transactions.	
		Year.	Page.
LEWES, V. B.	Gas Substitutes	1893	141
LIGHT, C. J.	Railway Switches and Crossings	1857	13
,, ,,	Railway Turntables	1858	11
,, ,,	The Permanent Way of Railways	1859	21
,, ,,	The Need of further Experiments on the Strength of Materials.	1869	194
,, ,,	A new Method of setting-out the Slopes of Earthworks.	1873	216
,, ,,	Opening Bridges on the Furness Railway.	1885	119
LIGHT, C. L.	Street Railways	1860	56
LIGHTFOOT, T. B.	Refrigerating Machinery on Board Ship.	1887	105
,, ,,	A Trial of a Refrigerating Machine on the Linde System.	1891	39
LOUCH, J.	Surface Condensers	1861	96
MACGEORGE, W.	President's Address	1874	1
MAJOR, W.	The Priming of Steam-Boilers	1877	59
MARTIN, W.	The Strength of Flues in Lancashire and similar Boilers.	1882	157
MARTIN, W. A.	Induced v. Forced Draught for Marine Boilers.	1886	111
MASON, C.	President's Address	1901	1
,,	Street Subways for Large Towns	1895	91
MATHESON, E.	The Quality of Iron, as now used	1867	168
,, ,,	The Accumulator Cotton Press	1868	264
MAUDE, T.	The Government Brake Trials	1875	129
MAWBEY, E. G.	The Leicester Main Drainage	1893	25
McNAUGHT, W.	The Rolling of Ships	1876	187
MIDDLETON, R. E.	The Relative Value of Percolation Gauges.	1895	153
,, ,,	The Pollution of Water, and its Correction.	1897	173
MILES, W. P.	Locks and Fastenings	1860	26
MILLER, B. A.	The Cleansing and Ventilation of Pipe Sewers.	1892	167
MOLESWORTH, G. L.	The Utilisation of Water Power	1858	16
MOORE, L. G.	The Explosion at Erith, and the Repair of the River Bank.	1864	183
MORRIS, W.	The Machinery employed in raising Water from an Artesian Well.	1860	29
MOULTRIE, F. R.	Photography	1857	12
NURSEY, P. F.	President's Address	1886	1
,, ,,	Quartz Crushing Machinery	1860	17
,, ,,	The Chemical Extraction of Gold from its Ores.	1860	42
,, ,,	The Superheating of Steam, and the various Apparatus employed therein	1861	36
,, ,,	Steam Boiler Explosions	1863	1

Author.	Subject of Paper.	Year.	Page
Nursey, P. F.	Fuel	1864	1
,, ,,	Sugar-making Machinery	1865	155
,, ,,	Explosive Compounds for Engineering Purposes.	1869	10
,, ,,	English and Continental Intercommunication.	1869	154
,, ,,	Recent Improvements in Explosive Compounds.	1871	108
,, ,,	Economic Uses of Blast Furnace Slag	1873	196
,, ,,	Mechanical Puddling	1874	77
,, ,,	The Channel Railway	1876	17
,, ,,	Illumination by means of Compressed Gas.	1881	57
,, ,,	Modern Bronze Alloys for Engineering Purposes.	1884	127
,, ,,	Primary Batteries for Illuminating Purposes.	1887	185
,, ,,	Recent Developments in High Explosives.	1889	71
,, ,,	Fox's Patent System of Solid-pressed Steel Waggon and Carriage Frames.	1889	197
,, ,,	Pick's System of Manufacturing Salt in Vacuo.	1890	115
,, ,,	Some Practical Examples of Blasting	1893	201
,, ,,	The Preparation of Rhea Fibre for Textile Purposes.	1898	177
,, ,,	The Production of Metallic Bars and Tubes under Pressure.	1901	45
Oates, A.	The Utilisation of Tidal Energy	1882	81
O'Connor, H.	President's Address	1900	1
,, ,,	Machine Stokers for Gas Retorts	1891	117
,, ,,	Pile-Driving	1894	71
,, ,,	Automatic Gas Station Governors	1897	67
Olander, E.	The Enclosure of Lands from the Sea, and the Construction of Sea and other Banks.	1862	25
,, ,,	Bridge Floors: Their Design, Strength, and Cost.	1887	27
Olrick, H.	A new System of Treating Sewage Matter.	1883	23
Olrick, L.	Marine Governors	1862	104
,, ,,	Giffard's Injector	1865	280
Ordish, R. M.	Suspension Bridges	1857	13
,, ,,	The Forms and Strengths of Beams, Girders and Trusses.	1858	13
Page, G. G.	The Construction of Chelsea Bridge	1863	77
Parkes, M.	The Road Bridges of the Charing Cross Railway.	1864	143

Author.	Subject of Paper.	Year.	Page.
PARKES, M.	The Charing Cross Railway Bridge over the Thames.	1864	166
PARSEY, W.	Perspective	1859	26
,, ,,	Trussed Beams	1862	52
PAUL, J. H.	Corrosion in Steam Boilers	1891	147
PEARSE, J. W.	The Ventilation of Buildings	1876	97
,, ,,	The Mechanical Firing of Steam Boilers.	1877	31
,, ,,	Water Purification, Sanitary and Industrial.	1878	29
PEASE, E. L.	Gasholder Construction	1894	95
PEIRCE, W. G.	President's Address	1895	1
PENDRED, H. W.	Screw Propellers, their Shafts and Fittings.	1875	151
,, ,,	Distilling and Hoisting Machinery for Sea-going Vessels.	1880	55
,, ,,	Designs, Specifications, and Inspection of Ironwork.	1883	87
PENDRED, V.	President's Address	1876	1
,, ,,	Elastic Railway Wheels	1864	119
,, ,,	The Adhesion of Locomotive Engines, and certain Expedients for increasing or supplementing that Function.	1865	207
,, ,,	Water-tube Boilers	1867	109
,, ,,	Apparatus for Measuring the Velocity of Ships.	1869	215
,, ,,	Examples of Recent Practice in American Locomotive Engineering.	1872	19
PERRETT, E.	Filtration by Machinery	1888	106
PHILLIPS, J.	The Forms and Construction of Channels for the Conveyance of Sewage.	1874	145
PIEPER, C.	Ice-making Machinery	1882	139
POLLARD-URQUHART, M. A.	Examples of Railway Bridges for Branch Lines.	1896	117
RANSOM, H. B.	The Principles and Practice of Hydro-Extraction.	1894	227
READE, S. A.	The Redhill Sewage Works	1868	168
REDMAN, J. B.	Tidal Approaches and Deep Water Entrances.	1885	143
RIGG, A.	President's Address	1884	1
,, ,,	The Connection between the Shape of heavy Guns and their Durability.	1867	249
,, ,,	The Screw Propeller	1868	202
,, ,,	Sensitiveness and Isochronism in Governors.	1880	73
,, ,,	American Engineering Enterprise	1885	21
,, ,,	Obscure Influences of Reciprocation in High-speed Engines.	1886	83
,, ,,	Balancing of High-speed Steam Engines.	1891	23
,, ,,	Hydraulic Rotative Engines	1896	89

Author.	Subject of Paper.	Transactions. Year.	Page.
RILEY, E.	The Manufacture of Iron	1861	59
,, ,,	The Action of Peaty Water on a Boiler	1862	45
ROBERTS, W.	Steam Fire-engines, and the late Trials at the Crystal Palace.	1863	163
ROBERTSON, GEORGE	Mortars and Cements	1857	11
ROBINSON, H.	President's Address	1887	1
,, ,,	Sewage Disposal	1879	197
ROECHLING, H. A.	The Sewage Question during the last Century.	1901	193
ROOTS, J. D.	Petroleum Motor Vehicles	1899	95
ROUS-MARTEN, C.	Notes on English and French Compound Locomotives.	1900	173
,, ,,	Notes on Some Twentieth Century Locomotives.	1902	125
RUMBLE, T. W.	Armour Plates	1861	87
RYMER-JONES, T. M.	Railway Tunnelling in Japan	1882	101
SANDERSON, C.	The Bombay, Baroda, and Central India Railway.	1863	97
SCHÖNHEYDER, W.	Equalising the Wear in Horizontal Steam Cylinders.	1878	45
SEFI, M.	The Theory of Screw Propulsion	1870	82
SELLON, S.	Electrical Traction and its Financial Aspect.	1892	37
SELLON, S., and BEAUMONT, W. W.	Notes on Parliamentary Procedure as affecting Light Railways and Tramways.	1895	33
SHENTON, H C. H.	Recent Practice in Sewage Disposal	1900	217
SOMERVILLE, J.	Charging and Drawing Gas Retorts by Machinery.	1873	138
SPENCER, G.	State Railways and Railway Amalgamation.	1872	43
SPICE, R. P.	President's Address	1878	1
,, ,,	President's Address	1879	1
,, ,,	Some Modern Improvements in the Manufacture of Coal Gas.	1886	131
SPON, E.	The Use of Paints as an Engineering Material.	1875	69
STANDFIELD, J.	Floating Docks—the Depositing Dock, and the Double-power Dock.	1881	81
STEEL, J.	Air Compression	1876	65
STEPHENSON, H. P.	Screw Piles	1857	12
STEPHENSON, W. H.	The General Structure of the Earth	1858	12
,, ,,	Fire-clay Manufactures	1861	17
STOCKER, C. W.	Diving Apparatus	1860	32
STOPES, H.	The Engineering of Malting	1884	65
STOW, J. F.	Irrigation Works in South Africa	1901	159
STRACHAN, G. R.	The Construction and Repair of Roads	1889	17
SUCKLING, N. J.	Modern Systems of Generating Steam	1874	39
SUGG, W. T.	Apparatus employed for Illumination with Coal Gas.	1869	134
,, ,,	Ventilation and Warming	1895	235
SUTCLIFF, R.	Tube Wells for large Water Supplies	1877	149
TARBUTT, P. F.	Liquid Fuel	1886	193

Author.	Subject of Paper.	Year.	Page.
THRESH, J. C.	The Protection of Underground Water Supplies.	1898	47
THRUPP, E. C.	A New Formula for the Flow of Water in Pipes and Open Channels.	1887	224
THUDICHUM, G.	The Ultimate Purification of Sewage	1896	199
,, ,,	Bacterial Treatment of Sewage	1898	207
THWAITE, B. H.	British v. American Patent Law Practice.	1902	23
TREWMAN, H. C.	Flues and Ventilation	1876	91
TURNER, C.	The Timbering of Trenches and Tunnels.	1871	90
TWEDDELL, R. H.	Direct-Acting Hydraulic Machinery	1877	95
,, ,,	The Application of Water Pressure to Machine Tools and Appliances.	1890	35
UMNEY, H. W.	Safety Appliances for Elevators	1895	113
,, ,,	The Compression of Air by the Direct Action of Water.	1897	93
USILL, G. W.	Rural Sanitation	1877	113
VALON, W. A. M.	President's Address	1893	1
VARLEY, S. A.	Railway Train Intercommunication	1873	162
WAGSTAFF, E. W.	The Shan Hill Country and the Mandalay Railway	1899	21
WALES, W. G.	Discharging and Storing Grain	1896	149
WALKER, J.	Brewing Apparatus	1871	29
WALMISLEY, A. T.	President's Address	1888	1
,, ,,	Iron Roofs	1881	123
WANKLYN, J. A.	Cooper's Coal-liming Process	1884	167
WARD, G. M.	The Utilisation of Coal Slack in the Manufacture of Coke for Smelting.	1880	33
WARE, W. J., and WOODHEAD, G. S.	Disinfection of the Maidstone Water Service Mains.	1900	49
WARREN, A.	Steam Navigation on the Indus	1863	139
WESSELY, C. R. VON	Arched Roofs	1866	36
WILKINS, T.	The Machinery and Utensils of a Brewery.	1871	10
WILLCOCK, J.	Gas Meters and Pressure Gauges	1860	37
WILSON, A. F.	The Manufacture of Coal Gas	1864	35
WILSON, J. W.	The Construction of Modern Piers	1875	29
,, ,,	President's Address	1892	1
WISE, F.	Signalling for Land or Naval Purposes	1863	117
WISE, W. L.	The Patent Laws	1870	45
WOLLHEIM, A.	Foreign Sewage Precipitation Works	1892	123
WOODHEAD, G. S., and WARE, W. J.	Disinfection of the Maidstone Water Service Mains.	1900	49
WORSSAM, S. W.	Mechanical Saws	1867	196
YARROW, A. F.	Steam Carriages	1862	128
YOUNG, C. F. T.	The Use of Coal in Furnaces without Smoke.	1862	66

SUBJECTS.

Subject of Paper.	Author.	Year.	Page.
Accumulator Cotton Press	E. Matheson	1868	264
Action of Marine Worms and Remedies employed in Harbour of San Francisco.	J. Blackbourn	1874	111
Action of Peaty Water on a Boiler	E. Riley	1862	45
Acton Main Drainage Works	C. N. Lailey	1888	125
Address of President	W. T. Carrington	1865	1
,, ,,	Z. Colburn	1866	1
,,	W. H. Le Feuvre	1867	1
,,	B. Latham	1868	1
,,	F. W. Bryant	1869	1
,,	W. Adams	1870	1
,,	B. Latham	1871	1
,,	J. Church	1872	1
,,	,, ,,	1873	1
,,	W. Macgeorge	1874	1
,,	J. H. Adams	1875	1
,,	V. Pendred	1876	1
,,	T. Cargill	1877	1
,,	R. P. Spice	1878	1
,,	,, ,,	1879	1
,,	J. Bernays	1880	1
,,	C. Horsley	1881	1
,,	J. Church	1882	1
,,	,, ,,	1883	1
,,	A. Rigg	1884	1
,,	C. Gandon	1885	1
,,	P. F. Nursey	1886	1
,,	Professor H. Robinson	1887	1
,,	A. T. Walmisley	1888	1
,,	J. R. Baillie	1889	1
,,	H. Adams	1890	1
,,	W. N. Colam	1891	1
,,	J. W. Wilson	1892	1
,,	W. A. M. Valon	1893	1
,,	G. A. Goodwin	1894	1
,,	W. G. Peirce	1895	1
,,	S. H. Cox	1896	1
,,	G. M. Lawford	1897	1
,,	W. W. Beaumont	1898	1
,,	J. C. Fell	1899	1
,,	H. O'Connor	1900	1
,, ,,	C. Mason	1901	1
,, ,,	P. Griffith	1902	1
Adhesion of Locomotive Engines and certain Expedients for Increasing or Supplementing that Function.	V. Pendred	1865	207
Admixtures of Kentish Ragstone, &c., upon Portland Cement, The Effect of	D. B. Butler	1896	1 9

INDEX.

Subject of Paper.	Author.	Year.	Page.
Air Compression	J. Steel	1876	65
Air, Compression of, by the Direct Action of Water, The.	H. W. Umney	1897	93
American Engineering Enterprise	A. Rigg	1885	21
American Locomotive Engineering, Examples of recent Practice in	V. Pendred	1872	19
Apparatus for utilising the Waste Heat of Exhaust Steam.	J. Atkinson	1878	167
Apparatus for Measuring the Velocity of Ships.	V. Pendred	1869	15
Apparatus employed for Illumination with Coal Gas.	W. Sugg	1869	134
Application of Electricity to Hoisting Machinery.	R. Bolton	1892	79
Application of Steam to the Cultivation of the Soil.	B. Latham	1868	274
Application of Water Pressure to Machine Tools and Appliances.	R. H. Tweddell	1890	35
Arched Roofs	C. von Wessely	1866	36
Armour Plates	T. W. Rumble	1861	87
Arrangement, Construction, and Machinery of Breweries.	W. B. Kinsey	1881	173
Artesian Well, The Machinery employed in raising Water from an	W. Morris	1860	29
Atlantic Cable, The Machinery employed in laying the	J. Amos	1858	13
Australian Timber Bridges	H. E. Bellamy	1902	63
Automatic Gas Station Governors	H. O'Connor	1897	67
Axles, The Strength of Wrought-iron Railway	T. Andrews	1879	143
Bacterial Treatment of Sewage	G. Thudichum	1898	207
Bars and Tubes, The Production of Metallic, under Pressure	P. F. Nursey	1901	45
Batteries, Primary, for Illuminating Purposes.	P. F. Nursey	1887	185
Beams, Girders and Trusses, The Forms and Strengths of	R. M. Ordish	1858	13
Blake's Bridge, Reading	E. Burrows	1893	71
Blast Furnace Practice, Recent	B. D. Healey	1902	97
Blast-furnace Slag, The Economic Uses of	E. Burrows	1873	196
Blasting Charges, The Application of Electricity to the Ignition of	G. G. André	1878	123
Blasting, Some Practical Examples of	P. F. Nursey	1893	201
Blow-pipe Flame Furnace	A. C. Engert	1884	185
Boilers, Incrustation in Marine	P. Jensen	1866	117
Boilers, Strength of Flues in Lancashire and similar	W. Martin	1882	157
Boilers, The Defects of Steam, and their Remedy.	A. C. Engert	1884	25
Boilers, The Mechanical Firing of Steam	J. W. Pearse	1877	31
Boilers, The Priming of Steam	W. Major	1877	59
Boilers, Water and its Effects on Steam	H. K. Bamber	1867	65
Boilers, Water Tube	V. Pendred	1867	109

330 INDEX.

Subject of Paper.	Author.	Year.	Page.
Bombay, Baroda, and Central India Railway.	C. Sanderson	1863	97
Brake Trials, The Government	T. Maude	1875	129
Brakes, Continuous Railway	W. H. Fox	1872	112
" " "	" "	1873	33
Brakes, Hydraulic Continuous and Automatic	E. D. Barker	1879	75
Brakes, Recent Arrangements of Continuous	St. J. V. Day	1875	87
Breaches in Sea and River Embankments, The Closing of.	R. F. Grantham	1900	21
Breakwater Construction	F. H. Cheesewright	1890	83
Breakwaters and Harbours of Refuge	J. T. Kershaw	1859	31
Breakwaters, Floating	T. Cargill	1871	146
Breweries, the Arrangement, Construction, and Machinery of	W. B. Kinsey	1881	173
Brewery, Machinery and Utensils of a	T. Wilkins	1871	10
Brewing Apparatus	J. Walker	1871	29
Bridge, Blake's, Reading	E. Burrows	1893	71
Bridge, Railway, at the Place de l'Europe, Paris	T. Cargill	1866	191
Bridge Floors: Their Design, Strength and Cost.	E. Olander	1887	27
Bridge, Construction of Chelsea	G. G. Page	1863	77
Bridge over the Thames, Charing Cross Railway	M. Parkes	1864	166
Bridges of the Charing Cross Railway, The Road	M. Parkes	1864	143
Bridges, Opening, on Furness Railway	C. J. Light	1885	119
Bridges, The Strength of Suspension	C. E. Browning	1858	11
Bridges, Suspension	R. M. Ordish	1857	13
British v. American Patent Law Practice	B. H. Thwaite	1902	23
Bronze Alloys for Engineering Purposes, Modern	P. F. Nursey	1884	127
Buildings, The Ventilation of	J. W. Pearse	1876	97
Cable Tramways	W. N. Colam	1885	69
Caissons, Ship, for Dock Basins and Dry Docks.	J. J. F. Andrews	1890	189
Canadian System, A deep Boring near Freistadt, Austria, by the	R. N. Boyd	1894	119
Canal between the Atlantic and Pacific Oceans, The Surveys of Proposed Lines for a Ship	Dr. E. Cullen	1868	15
Canal, The Isthmus of Darien and the Ship	" "	1868	52
Carriage Frames, Fox's Patent System of Solid-pressed Steel Waggon and	P. F. Nursey	1889	197
Carriages, Steam	A. F. Yarrow	1862	128
Case System of Groyning, Foreshore Protection, with special reference to the	R. G. Allanson-Winn	1899	133
Cast-iron, Certain Methods of Treating, in the Foundry.	Z. Colburn	1865	77
Cement, Portland	H. Faija	1885	95

INDEX. 331

Subject of Paper.	Author.	Year.	Page.
Cements, Mortars and	G. Robertson	1857	11
Central India Railway, Bombay, Baroda, and	C. Sanderson	1863	97
Chalk Formation, The	J. Glynn, Junr.	1860	46
Channel Railway, The	P. F. Nursey	1876	17
Channels for the Conveyance of Sewage, The Forms and Construction of	J. Phillips	1874	145
Charcoal, The Conversion of Peat into Fuel and	C. E. Hall	1876	151
Charging and Drawing Gas Retorts by Machinery.	J. Somerville	1873	138
Charging and Drawing Gas Retorts, An Improved Method of	F. W. Hartley	1875	185
Charing Cross Railway Bridge over the Thames.	M. Parkes	1864	166
Charing Cross Railway, The Road Bridges of	,, ,,	1864	143
Cheap Patents, The National Value of	F. W. Grierson	1880	145
Chelsea Bridge, The Construction of	G. G. Page	1863	77
Chimney Shafts, Stability of	R. J. Hutton	1887	150
Circular Tables	H. Hakewill	1865	147
Cleansing and Ventilation of Pipe Sewers.	B. A. Miller	1892	167
Cleaning of Tramway and other Rails	H. Conradi	1893	47
Coal Gas, The methods employed in the determination of the Commercial Value and Purity of	F. W. Hartley	1869	100
Coal Gas, Apparatus employed for illumination with	W. Sugg	1869	134
Coal Gas, The Manufacture of	A. F. Wilson	1864	35
Coal Gas, Modern Manufacture of	C. C. Carpenter	1891	61
Coal Gas, Some Modern Improvements in the Manufacture of	R. P. Spice	1886	131
Coal-liming Process, Cooper's	J. A. Wanklyn	1884	167
Coal Mines, The Ventilation of	G. André	1874	23
Coal Slack, Utilisation of, in the Manufacture of Coke for Smelting.	G. M. Ward	1880	33
Coffer Dams, Piling and	F. W. Bryant	1859	38
Collieries and Colliery Engineering	R. N. Boyd	1893	169
Colliery, The Underground Pumping Machinery at the Erin, Westphalia.	H. Davey	1876	119
Colonies, Practical Construction in the	W. G. Ferrar	1875	53
Compressed Gas, Illumination by means of	P. F. Nursey	1881	57
Compression, Air	J. Steel	1876	65
Compression of Air by the Direct Action of Water, The.	H. W. Umney	1897	93
Concrete, Forced Filtration of Water Through	H. Faija	1889	107
Concrete Subways for Underground Pipes.	A. T. Allen	1901	99
Condensers, Surface	J. Louch	1861	96
Connection between the Shape of Heavy Guns and their Durability.	A. Rigg	1867	249
Construction, The, and Repair of Roads	G. R. Strachan	1889	17

Subject of Paper.	Author.	Year.	Page.
Construction, Modern Tramway	A. W. Jones	1879	79
Construction, Modern Roadway	H. S. Copland	1878	61
Construction of Chelsea Bridge	G. G. Page	1863	77
Construction of Modern Piers	J. W. Wilson	1875	29
Continental Intercommunication, English and	P. F. Nursey	1869	154
Continuous and Automatic Brakes, Hydraulic	E. D. Barker	1879	75
Continuous Brakes, Recent arrangements of	St. J. V. Day	1875	87
Continuous Railway Brakes	W. H. Fox	1872	112
,, ,, ,,	,, ,,	1873	33
Conversion of Peat into Fuel and Charcoal	C. E. Hall	1876	151
Cooper's Coal-liming Process	J. A. Wanklyn	1884	167
Cornish Pumping Engine, The Use of the	A. Fraser	1864	78
Corrosion in Steam Boilers	J. H. Paul	1891	147
Cotton Press, The Accumulator	E. Matheson	1868	264
Country Mansions and Estates, Water Supply to	G. Harris	1899	231
Crushing, Dry, Machinery	S. H. Cox	1892	101
Cylinders, Equalising the Wear in Horizontal Steam	W. Schonheyder	1878	45
Cylinders, Friction in Steam	P. Jensen	1870	17
Cyprus, Engineering Notes on	C. J. Alford	1880	131
Darien and the Ship Canal, The Isthmus of	Dr. E. Cullen	1868	52
Deep Boring near Freistadt, Austria, by the Canadian System.	R. N. Boyd	1894	119
Deep-Sea Lighthouses, Feasibility and Construction of	C. Anderson	1883	45
Deep-water Entrances, Tidal Approaches and	J. B. Redman	1885	143
Defects of Steam Boilers and their Remedy.	A. C. Engert	1884	25
Dephosphorisation of Iron in the Puddling Furnace.	P. S. Justice	1885	169
Depreciation of Plant and Works under Municipal and Company Management.	C. H. W. Biggs	1902	271
Design, Strength, and Cost of Bridge Floors.	E Olander	1887	27
Designing and Construction of Storage Reservoirs, The	A. Jacob	1866	225
Designs, Specifications, and Inspection of Ironwork.	H. W. Pendred	1883	87
Direct-acting Hydraulic Machinery	R. H. Tweddell	1877	95
Direct-acting Pumping Engines	P. B. Björling	1877	79
Discharging and Storing Grain	W. G. Wales	1896	149
Disinfection of the Maidstone Water Service Mains.	G. S. Woodhead and W. J. Ware	1900	49
Disposal of Sewage, The	H. Robinson	1879	197
Distilling and Hoisting Machinery for Sea-going Vessels.	H. W. Pendred	1880	55

Subject of Paper.	Author.	Transactions. Year.	Page.
Distribution of Electrical Energy by Secondary Generators.	J. D. Gibbs	1885	49
Diving Apparatus	C. W. Stocker	1860	32
Dock Basins and Dry Docks, Ship Caissons for	J. J. F. Andrews	1890	189
Docks, Floating.—The Depositing Dock and the Double-power Dock.	J. Standfield	1881	81
Drainage and Water Supply to the Town of Fareham.	T. Buckham	1869	41
Drainage Construction, Notes on certain details of	G. J. G. Jensen	1901	19
Drainage of Town Houses	G. M. Lawford	1891	187
Drainage Works, Main, Acton	C. N. Lailey	1888	125
Drainage of the Fens, The	B. Latham	1862	154
Dry Crushing Machinery	S. H. Cox	1892	101
Dundee Street Improvements and Drainage of Lochee.	A. Greig	1883	113
Earth, The General Structure of	W. H. Stephenson	1858	12
Earthworks, A New Method of Setting out the Slopes of	C. J. Light	1873	216
Economics of Railway Maintenance	E. S. Crompton	1870	101
Economic Uses of Blast Furnace Slag	P. F. Nursey	1873	196
Effect of Sea Water on Portland Cement	H. Faija	1888	39
Effect of Strain on Railway Axles, and the Minimum Flexion Resistance-Point in Axles, The	T. Andrews	1895	181
Elastic Railway Wheels	V. Pendred	1864	119
Electric Light Engineering, Notes on	{ C. H. W. Biggs and W. W. Beaumont }	1882	23
Electric Telegraph, The	W. Ansell	1857	11
Electric Telegraph Instruments	E. G. Bartholomew	1872	53
Electric Telegraphy, irrespective of Telegraphic Apparatus.	,, ,,	1869	55
Electric Traction, Notes on	A. H. Binyon	1900	103
Electrical Batteries	,, ,,	1872	77
Electrical Energy, Distribution of, by Secondary Generators.	J. D. Gibbs	1885	49
Electrical Traction, and its Financial Aspect.	S. Sellon	1892	37
Electricity, the Application of, to Hoisting Machinery.	R. Bolton	1892	79
Electricity, The application of, to the ignition of Blasting Charges.	G. G. André	1878	123
Electricity, The Lighting and Extinction of Street Gas Lamps by	St. G. Lane Fox	1878	105
Electricity, Water and Gas, Power Distribution by	E. C. de Segundo	1894	143
Electrolytic Treatment of Complex Sulphide Ores	S. O. Cowper-Coles	1899	203
Elevators, Safety Appliances for	H. W. Umney	1895	113
Embankments, Reservoir	W. Fox	1898	23
Embankments, The Closing of Breaches in Sea and River.	R. F. Grantham	1900	21

Subject of Paper.	Author.	Year.	Page.
Enclosure of Lands from the Sea, and the Construction of Sea and other banks.	E. Olander	1862	25
Engineering, Collieries and Colliery	R. N. Boyd	1893	169
Engineering Enterprise, American	A. Rigg	1885	21
Engineering Notes on Cyprus	C. J. Alford	1880	131
Engineering in India	F. C. Danvers	1868	137
Engineering of Malting	H. Stopes	1884	65
Engines, Direct-acting Pumping	P. B. Björling	1877	79
Engines for Mines, Recent Improvements in Pumping.	H. Davey	1873	182
Engines for Town Water Supply, Pumping	„ „	1867	88
Engines, Gas	C. Gandon	1881	27
Engines, Hydraulic Rotative	A. Rigg	1896	89
Engines, Obscure Influences of Reciprocation in High Speed	A. Rigg	1886	83
Engines on the Rimutakai Incline, New Zealand, The Fell	R. J. Alford	1882	121
Engines, Pumping	W. T. Carrington	1859	19
Engines, Steam Fire, and the late trial at the Crystal Palace.	W. Roberts	1863	163
Engines, The Adhesion of Locomotive, and certain expedients for increasing or supplementing that Function.	V. Pendred	1865	207
Engines, Use of the Cornish Pumping	A. Fraser	1864	78
English and Continental Intercommunication.	P. F. Nursey	1869	154
English and French Compound Locomotives, Notes on.	C. Rous-Marten	1900	173
Equalising the Wear in Horizontal Steam Cylinders.	W. Schönheyder	1878	45
Examples of Railway Bridges for Branch Lines	M. A. Pollard-Urquhart	1896	117
Examples of Recent Practice in American Locomotive Engineering.	V. Pendred	1872	19
Exhaust Steam, Apparatus for Utilising the Waste Heat of	J. Atkinson	1878	167
Exhaust Steam, Treatment and Utilisation of	P. Griffith	1890	163
Exhibitions as Aids to Engineering Progress, Value of The	S. Barnett	1883	75
Experimental Researches into the Nature and Action of Steam Boilers.	T. Baldwin	1867	23
Experiments on the Strength of Materials, The Need of further	C. J. Light	1869	194
Explosion at Erith, and the Repair of the River Bank.	L. G. Moore	1864	183
Explosions, Steam Boiler	P. F. Nursey	1863	1
Explosive Compounds for Engineering Purposes.	„ „	1869	10
Explosive Compounds, Recent Improvements in	„ „	1871	108
Explosives, High, Recent Developments	„ „	1889	75
Extension and Permanent Set of Wrought Iron, The	C. E. Browning	1858	15

Subject of Paper.	Author.	Year.	Page.
Extraction of Gold, The Chemical	P. F. Nursey	1860	42
Fareham, Drainage and Water Supply of the Town of	T. Buckham	1869	41
Feasibility and Construction of Deep-sea Lighthouses.	C. Anderson	1883	45
Fell Engines on the Rimutakai Incline, New Zealand.	R. J. Alford	1882	121
Fens, The Drainage of the	B. Latham	1862	154
Forro-Concrete Construction, The Hennebique System of.	A. R. Galbraith	1902	177
Filter Presses for Sewage Sludge	J. Croll	1897	121
Filtration by Machinery	E. Perrett	1888	106
Filtration, Forced, of Water through Concrete	H. Faija	1889	107
Fire-clay Manufactures	W. H. Stevenson	1861	17
Fire Engines, Steam	W. Roberts	1863	163
Fireproof Floors	G. M. Lawford	1889	43
Floating Breakwaters	T. Cargill	1871	146
Floating Docks: The Depositing Dock and the Double-power Dock.	J. Standfield	1881	81
Floors, Fireproof	G. M. Lawford	1889	43
Flow of Water, New Formula for, in Pipes and Open Channels.	E. C. Thrupp	1887	224
Flues and Ventilation	A. H. C. Trewman	1876	91
Flues in Lancashire and similar Boilers, The Strength of	W. Martin	1882	157
Foreign Sewage Precipitation Works	A. Wollheim	1892	123
Foreshore Protection, with special reference to the Case System of Groyning.	R. G. Allanson-Winn	1899	133
Forms and Construction of Channels for the Conveyance of Sewage.	J. Phillips	1874	145
Forms and Strengths of Beams, Girders and Trusses, The	R. M. Ordish	1858	13
Foundry Work, Practice of	H. R. Hooper	1888	191
Fox's Patent System of Solid-pressed Steel Waggon and Carriage Frames.	P. F. Nursey	1889	197
Friction in Steam Cylinders	P. Jensen	1870	17
Friction of the Slide Valve and its Appendages.	T. Adams	1866	6
Fuel	P. F. Nursey	1864	1
Fuel and Charcoal, The Conversion of Peat into.	C. E. Hall	1876	151
Fuel, Liquid	P. F. Tarbutt	1886	193
Furnace, The Blowpipe Flame	A. C. Engert	1884	185
Gas, Apparatus employed for Illumination with Coal	W. Sugg	1869	134
Gas Engines	C. Gandon	1881	27
Gas-Engine, Modern, Practice	S. Griffin	1889	169
Gasholder Construction	E. L. Pease	1894	95
Gas, Illumination by means of Compressed	P. F. Nursey	1881	57
Gas Meters and Pressure Gauges	J. Willcock	1860	37

Subject of Paper.	Author.	Year.	Page.
Gas, Modern Improvements in the Manufacture of Coal	R. P. Spice..	1886	131
Gas Retorts, An Improved Method of Charging and Drawing	F. W. Hartley	1875	185
Gas Retorts, Charging and Drawing by Machinery.	J. Somerville	1873	138
Gas Retorts, Machine Stokers for..	H. O'Connor	1891	117
Gas Station Governors, Automatic	H. O'Connor	1897	67
Gas Substitutes	V. B. Lewes	1893	141
Gas, The Manufacture of Coal	A. F. Wilson	1864	35
Gas, The Methods employed in the determination of the Commercial Value and Purity of Coal	F. W. Hartley	1869	100
Gas and Water Mains, The Prevention of Leakage in	C. M. Barker	1869	79
Gas Works at Home and Abroad, Modern	H. Gore	1868	233
Gas Works Machinery	E. A. Harman	1898	97
General Structure of the Earth, The	W. H. Stephenson	1858	12
Generation and Expansion of Steam	W. T. Carrington	1858	15
Giffard's Injector	L. Olrick	1865	280
Gold, The Chemical Extraction of	P. F. Nursey	1860	42
Government Brake Trials	T. Maude	1875	129
Governors, Marine	L. Olrick	1862	104
Governors, Sensitiveness and Isochronism in	A. Rigg	1880	73
Grain, Discharging and Storing	W. G. Wales	1896	149
Guns, The connection between the Shape of Heavy, and their Durability.	A. Rigg	1867	249
Harbour Bars: their Formation and Removal.	H. F. Knapp	1878	139
Harbours of Refuge, Breakwaters and	J. T. Kershaw	1859	31
Hauling Machinery, The new Pit and, for the San Domingos Mine in Portugal.	J. Bernays	1879	29
Hennebique System of Ferro-Concrete Construction, The	A. R. Galbraith	1902	177
High Pressure Steam and Steam Engine Efficiency.	W. W. Beaumont	1888	221
Hoisting Machinery, The Application of Electricity to	R. Bolton	1892	79
Horse Haulage, the Mechanics of	T. H. Brigg	1896	21
Horse Railways and Street Tramways.	H. Gore	1873	113
House Drainage, Notes on the Proposed By-laws of the London County Council with respect to.	J. P. Barber	1897	17
Houses, Town, Drainage of	G. M. Lawford	1891	187
Hydraulic Continuous and Automatic Brakes	E. D. Barker	1879	75
Hydraulic Experiments, the Roorkee	E. S. Bellasis	1886	41
Hydraulic Lifts	R. Carey..	1893	101
Hydraulic Machinery, Direct-acting	R. H. Tweddell..	1877	95

Subject of Paper.	Author.	Year.	Page.
Hydraulic Machinery for lifting Vessels in the Thames Graving Dock, The New	J. Amos	1859	17
Hydraulic Rotative Engines..	A. Rigg	1896	89
Hydro-extraction, The Principles and Practice of	H. B. Ransom	1894	227
Ice-making Machinery	C. Pieper	1882	139
Ignition of Blasting Charges, The application of Electricity to	G. G. André	1878	123
Ilford, The Main Drainage of	R. G. Hetherington	1901	173
Illumination by means of Compressed Gas.	P. F. Nursey	1881	57
Illumination with Coal Gas, Apparatus employed for	W. Sugg	1869	134
Improved Method of Charging and Drawing Gas Retorts.	F. W. Hartley	1875	185
Improvements in Explosive Compounds, Recent	P. F. Nursey	1871	108
Improvements in Pumping-engines for Mines, Recent	H. Davey	1873	182
Improvements in Tin Dressing Machinery, Recent	S. H. Cox	1874	11
Incrustation in Marine Boilers	P. Jensen	1866	117
India, Engineering in	F. C. Danvers	1868	137
Induced v. Forced Draught for Marine Boilers.	W. A. Martin	1886	111
Indus, Steam Navigation on the	A. Warren	1863	139
Injector, Giffard's	L. Olrick	1865	280
Injector, The most Recent Improvements on the	J. Gresham	1867	266
Instruments, Electric Telegraph	E. G Bartholomew	1872	53
Insurance, Marine	G. W. Allan	1858	14
Intercommunication, English and Continental	P. F. Nursey	1869	154
Intercommunication, Railway Train	S. A. Varley	1873	162
International System of Marine Course Signalling.	Capt. W. B. Barker	1884	47
Inundations of Marsh Land	B. Latham	1862	173
Iron Ores, The Treatment of Low-Grade, for the Smelting Furnace.	T. B. Grierson	1901	69
Iron, The Dephosphorisation of, in the Puddling Furnace.	P. S. Justice	1885	169
Iron, The Manufacture of	E. Riley	1861	59
Iron, The Relation between the Safe Load and ultimate Strength of	Z. Colburn	1863	35
Iron, The quality of, as now used	E. Matheson	1867	168
Iron Roofs	A. T. Walmisley	1881	123
Iron and Steel, Protective Metallic Coatings for.	S. O. Cowper-Coles	1898	139
Iron and Steel Surfaces, The Preservation and Ornamentation of	G. Bower	1883	59
Ironwork, Designs, Specifications, and Inspection of	H. W. Pendred	1883	87
Irrigation with Town Sewage	G. King	1865	84

Subject of Paper.	Author.	Year.	Page.
Irrigation Works in South Africa	J. F. Stow	1901	159
Isochronism in Governors, Sensitiveness and	A. Rigg	1880	73
Isthmus of Darien and the Ship Canal	Dr. E. Cullen	1868	52
Japan, Railway Tunnelling in	T. M. Rymer-Jones	1882	101
Joints, Single- and Double-riveted	T. Baldwin	1866	150
Land and Naval Purposes, Signalling for	F. Wise	1863	117
Lartigue Single-rail Railway, The	F. B. Behr	1886	163
Laws, Decrees and Regulations relating to Mining Operations in the old kingdom of Sardinia.	C. J. Alford	1879	106
Leakage in Gas and Water Mains, The prevention of	C. M. Barker	1869	79
Leicester Main Drainage, The	E. G. Mawbey	1893	25
Lifts, Hydraulic	R. Carey	1893	101
Light Railways	W. Lawford	1888	181
Light Railways and Tramways, Parliamentary Procedure as affecting	W. W. Beaumont and S. Sellon.	1895	33
Lighthouses, The feasibility and construction of Deep-sea	C. Anderson	1883	45
Lighting and Extinction of Street Gas-Lamps by Electricity.	St. G. Lane Fox	1878	105
Linde System, Trial of a Refrigerating Machine on the	T. B. Lightfoot	1891	39
Liquid Fuel	P. F. Tarbutt	1886	193
Lochee, Dundee Street Improvements and Drainage of	A. Greig	1883	113
Locks and Fastenings	W. P. Miles	1860	26
Locomotive Engineering, Examples of recent Practice in American	V. Pendred	1872	19
Locomotive Engines, The adhesion of	,, ,,	1865	207
Locomotives, English and French Compound, Notes on.	C. Rous-Marten	1900	173
Locomotives, Notes on some Twentieth Century.	C. Rous-Marten	1902	125
Macadam for Roads, Modern Machinery for preparing	C. E. Hall	1879	51
Machine, Refrigerating, on the Linde System, Trial of	T. B. Lightfoot	1891	39
Machinery, Dry, Crushing	S. H. Cox	1892	101
Machinery, Refrigerating, on Board Ship	T. B. Lightfoot	1887	105
Machinery, Filtration by	E. Perrett	1888	106
Machinery and Utensils of a Brewery	T. Wilkins	1871	10
Machinery, Gas Works	E. A. Harman	1898	97
Machinery, Quartz Crushing	P. F. Nursey	1860	17
Machine Stokers for Gas Retorts	H. O'Connor	1891	117
Machine Tools	E. C. Amos	1899	51
Machine Tools and Appliances, application of Water Pressure to	R. H. Tweddell	1890	35
Machinery employed in laying the Atlantic Cable, The	J. Amos	1858	13

INDEX.

Subject of Paper.	Author.	Year.	Page.
Machinery employed in raising Water from an Artesian Well, The	W. Morris	1860	29
Machinery, Mining	E. Edwards	1859	35
Machinery, Paper-Making	R. Henderson	1900	139
Machinery, Weighing, and Automatic Apparatus in connection therewith.	W. H. Brothers	1890	55
Maidstone Water Service Mains, Disinfection of the	G. S. Woodhead and W. J. Ware	1900	49
Main Drainage and Sewage Disposal Works, Wimbledon	W. S. Crimp	1888	73
Main Drainage of Ilford	R. G. Hetherington	1901	173
Main Drainage, The Leicester	E. G. Mawbey	1893	25
Main Drainage Works, Acton	C. N. Lailey	1888	125
Maintenance, Economics of Railway	E. S. Crompton	1870	101
Malting, The Engineering of	H. Stopes	1884	65
Mandalay Railway, The Shan Hill Country and the	E. W. Wagstaff	1899	21
Manufacture, Modern Coal Gas	C. C. Carpenter	1891	61
Manufacture of Coal Gas	A. F. Wilson	1864	35
Manufacture of Iron	E. Riley	1861	59
Marine Boilers, Incrustation in	P. Jensen	1866	117
Marine Boilers, Induced v. Forced Draught for	W. A. Martin	1886	111
Marine Course Signalling, International System of	Capt. W. B. Barker	1884	47
Marine Governors	L. Olrick	1862	104
Marine Insurance	G. W. Allan	1858	14
Marine Worms, Action of and Remedies for, at San Francisco.	J. Blackbourn	1874	111
Marsh Lands, The Inundations of	B. Latham	1862	173
Mechanical Firing of Steam Boilers	J. W. Pearse	1877	31
Mechanical Puddling	P. F. Nursey	1874	77
Mechanical Saws	S. W. Worssam	1867	196
Mechanics of Horse Haulage	T. H. Brigg	1896	21
Metallic Bars and Tubes, The Production of, under Pressure.	P. F. Nursey	1901	45
Methods employed in the Determination of the Commercial Value and Purity of Coal Gas.	F. W. Hartley	1869	100
Milford Haven and its New Pier Works.	H. Davey	1872	89
Milling Machinery, Roller	J. H. Carter	1883	125
Mineralogy of the Island of Sardinia	C. J. Alford	1879	93
Mining Machinery	E. Edwards	1859	35
Mining Operations in the old kingdom of Sardinia, Laws, &c., relating to	,, ,,	1879	106
Modern Bronze Alloys for Engineering purposes.	P. F. Nursey	1884	127
Modern Coal Gas Manufacture	C. C. Carpenter	1891	61
Modern Gas-Engine Practice	S. Griffin	1889	169
Modern Gasworks at Home and Abroad	H. Gore	1868	233
Modern Improvements in the Manufacture of Coal Gas.	R. P. Spice	1886	131
Modern Roadway Construction	H. S. Copland	1878	61
Modern Steel as a Structural Material	W. W. Beaumont	1880	109

Subject of Paper.	Author.	Year.	Page.
Modern Systems of Generating Steam	N. J. Suckling	1874	39
Modern Tramway Construction	A. W. Jones	1879	179
Mortars and Cements	G. Robertson	1857	11
Motor Vehicles, Petroleum	J. D. Roots	1899	95
National Value of Cheap Patents	F. W. Grierson	1880	145
Navigation on the Indus, Steam	A. Warren	1863	139
Need of further Experiments on the Strength of Materials.	C. J. Light	1869	194
New Formula for the Flow of Water in Pipes and Open Channels.	E. C. Thrupp	1887	224
Notes on Cyprus, Engineering	C. J. Alford	1880	131
Notes on Electric Light Engineering	C. H. W. Biggs and W. W. Beaumont	1882	23
Notes on some Twentieth Century Locomotives.	C. Rous-Marten	1902	125
Notes on the Proposed By-laws of the London County Council with respect to House Drainage.	J. P. Barber	1897	17
Nunhead, Shortlands and, Railway	A. G. Drury	1892	219
Obituary		1887	265
,,		1888	253
,,		1889	219
,,		1890	211
,,		1891	215
,,		1892	243
,,		1893	237
,,		1894	253
,,		1895	269
,,		1896	241
,,		1897	201
,,		1898	233
,,		1899	259
,,		1900	267
,,		1901	231
,,		1902	301
Obscure Influences of Reciprocation in High-speed Engines.	A. Rigg	1886	83
Opening Bridges on the Furness Railway.	C. J. Light	1885	119
Ores, The Electrolytic Treatment of Complex Sulphide	S. O. Cowper-Coles	1899	203
Paints, The Use of, as an Engineering Material.	E. Spon	1875	69
Panama Railroad, The	Dr. E. Cullen	1868	39
Paper-Making Machinery	R. Henderson	1900	139
Paris Exhibition, Visit to the		1867	151
Parliamentary Procedure as affecting Light Railways and Tramways, Notes on	W. W. Beaumont and S. Sellon.	1895	33
Patent Law Practice, British v. American	B. H. Thwaite	1902	23
Patent Laws, The	C. D. Abel	1865	25
,, ,, ,,	W. L. Wise	1870	45
Patents, The National Value of Cheap	F. W. Grierson	1880	145

Subject of Paper.	Author.	Year.	Page.
Peat, The Conversion of, into Fuel and Charcoal.	C. E. Hall	1876	151
Peaty Water, The Action of, on a Boiler	E. Riley	1862	45
Percolation Gauges, Relative Value of	R. E. Middleton	1895	153
Permanent Way of Railways, The	C. J. Light	1859	21
Permanent Way of Railways, The	G. Bowers	1857	11
Perspective	W. Parsey	1859	26
Petroleum Motor Vehicles	J. D. Roots	1899	95
Photographic Surveying	J. B. Lee	1899	171
Photography	F. R. Moultrie	1857	12
Pick's System of Manufacturing Salt in Vacuo.	P. F. Nursey	1890	115
Piers, The Construction of Modern	J. W. Wilson, junr.	1875	29
Pier Works, Milford Haven and its New	H. Davey	1872	89
Pile-Driving	H. O'Connor	1894	71
Piles, Screw	H. P. Stephenson	1857	12
Piling and Coffer Dams	F. W. Bryant	1859	38
Pin-connected v. Riveted Bridges	J. H. Cunningham	1889	131
Pioneer and Portable Railways	J. Kerr	1891	89
Pipe Sewers of Croydon and the Causes of their Failure, The	J. T. Kershaw	1860	52
Pits and Hauling Machinery for the San Domingos Mines in Portugal.	J. Bernays	1879	29
Pollution of Water, and its Correction	R. E. Middleton	1897	173
Portable and Pioneer Railways	J. Kerr	1891	89
Portland Cement	H. Faija	1885	95
Portland Cement, Effect of Sea Water on	H. Faija	1888	39
Portland Cement, some Points in its Testing, Uses and Abuses.	D. B. Butler	1895	61
Portland Cement, The Effect of Admixtures of Kentish Ragstone, &c., upon	D. B. Butler	1896	179
Power Distribution by Electricity, Water and Gas.	E. C. de Segundo	1894	143
Practical Construction in the Colonies	W. G. Ferrar	1875	53
Practical Examples of Blasting, Some	P. F. Nursey	1893	201
Practice of Foundry Work	H. R. Hooper	1888	191
Preparation of Rhea Fibre for Textile Purposes, The	P. F. Nursey	1898	177
Preservation and Ornamentation of Iron and Steel Surfaces.	G. Bower	1883	59
Pressure Gauges, Gas Meters and	J. Willcock	1860	37
Prevention of Smoke	A. C. Engert	1881	101
Prevention of Leakage in Gas and Water Mains.	C. M. Barker	1869	79
Primary Batteries for Illuminating Purposes.	P. F. Nursey	1887	185
Priming of Steam Boilers	W. Major	1877	59
Principles and Practice of Hydro-extraction, The	H. B. Ransom	1894	227
Propeller, The Screw	A. Rigg	1868	202
Propellers, Screw; their Shafts & Fittings	H. W. Pendred	1875	151
Propulsion, The Theory of Screw	M. Sefi	1870	82
Protection of Underground Water Supplies, The	J. C. Thresh	1898	47

Subject of Paper.	Author.	Year.	Page.
Protective Metallic Coatings for Iron and Steel.	S. O. Cowper-Coles ..	1898	139
Puddling Furnace, The Dephosphorisation of Iron in	P. S. Justice	1885	169
Puddling, Mechanical	P. F. Nursey	1874	77
Pumping Engine, The Use of the Cornish	A. Fraser	1864	78
Pumping Engines	W. T. Carrington ..	1859	19
Pumping Engines, Direct-acting ..	P. B. Björling	1877	79
Pumping Engines for Mines, Recent Improvements in	H. Davey	1873	182
Pumping Engines for Town Water Supply.	,, ,, ...	1867	88
Pumping Machinery, The Underground, at the Erin Colliery, Westphalia.	,, ,, ...	1876	119
Pump Valves	J. Hendry	1859	25
Quality of Iron, as now used	E. Matheson	1867	168
Quartz Crushing Machinery	P. F. Nursey	1860	17
Railway Amalgamation, State Railways and	G. Spencer	1872	43
Railway Axles, The Strength of Wrought-iron	T. Andrews	1879	143
Railway Axles, The Effect of Strain on, and the Minimum Flexion-Resistance Point in Axles.	,, ,, ...	1895	181
Railway Brakes, Continuous	W. H. Fox	1872	112
,, ,, ,, ...	,, ,, ...	1873	33
Railway Bridge at the Place de l'Europe	T. Cargill	1866	191
Railway Bridge over the Thames, The Charing Cross	M. Parkes	1864	166
Railway Bridges for Branch Lines, Examples of	M. A. Pollard-Urquhart	1896	117
Railway Maintenance, The Economics of	E. S. Crompton	1870	101
Railway Switches and Crossings	C. J. Light	1857	13
Railways, Street	C. L. Light	1860	56
Railways, The Permanent Way of ..	G. Bowers	1857	11
,, ,, ,, ..	C. J. Light	1859	21
Railway, The Shortlands and Nunhead	A. G. Drury	1892	219
Railway, The Channel	P. F. Nursey	1876	17
Railway Train Intercommunication ..	S. A. Varley	1873	162
Railway Tunnelling in Japan	T. M. Rymer-Jones ..	1882	101
Railway Turntables	,, ,,	1858	11
Railway Wheels, Elastic	V. Pendred	1864	119
Railways, Horse, and Street Tramways	H. Gore	1873	113
Railways, Light	W. Lawford	1888	181
Railways, Portable and Pioneer	J. Kerr	1891	89
Rating of Engineering Undertakings	P. M. Faraday	1897	43
Recent Blast Furnace Practice	B. D. Healey	1902	97
Recent Developments in High Explosives.	P. F. Nursey	1889	71
Recent Practice in Sewage Disposal ..	H. C. H. Shenton ..	1900	217
Reciprocation, Obscure Influences of, in High-speed Engines.	A. Rigg	1886	83

Subject of Paper.	Author.	Year	Page
Redhill Sewage Works, The	S. A. Reade	1868	168
Refrigerating Machine on the Linde System, Trial of	T. B. Lightfoot	1891	39
Refrigerating Machinery on Board Ship	T. B. Lightfoot	1887	105
Relation between the Safe Load and ultimate Strength of Iron.	Z. Colburn	1863	35
Relative Value of Percolation Gauges	R. E. Middleton	1895	153
Renewal of Roof over Departure Platform at King's Cross Terminus, G.N.R.	R. M. Bancroft	1887	125
Repair of the River Bank, The Explosion at Erith and the	L. G. Moore	1864	183
Reservoir Embankments	W. Fox	1898	23
Reservoirs, Storage, The Designing and Construction of	A. Jacob	1866	225
Rhea Fibre for Textile Purposes, The Preparation of	P. F. Nursey	1898	177
Rigidity, Strength and	W. E. Kochs	1865	69
River Pollution caused by Sewage Disposal.	G. B. Jerram	1886	217
River Thames, Wooden Jetties on the	J. G. Davenport	1860	24
Riveted, Pin-connected Bridges v.	J. H. Cunningham	1889	135
Road Bridges of Charing Cross Railway	M. Parkes	1864	143
Roads, The Construction and Repair of	G. R. Strachan	1889	17
Roadway Construction, Modern	H. S. Copland	1878	61
Roller-milling Machinery	J. H. Carter	1883	125
Rolling of Ships	W. McNaught	1876	187
Roof, Renewal of, over Departure Platform at King's Cross Terminus, G.N.R.	R. M. Bancroft	1887	125
Roofs, Arched	C. R. von Wessely	1866	36
Roots, Iron	A. T. Walmisley	1881	123
Roorkee Hydraulic Experiments	E. S. Bellasis	1886	41
Rural Sanitation	G. W. Usill	1877	113
Safe Load and Ultimate Strength of Iron, Relation between	Z. Colburn	1863	35
Safety Appliances for Elevators	H. W. Umney	1895	113
Safety Valves for Steam Boilers, &c.	T. Baldwin	1867	23
Salt, Pick's System of Manufacturing, in Vacuo.	P. F. Nursey	1890	115
Sanitary and Industrial, Water Purification	J. W. Pearse	1878	29
Sardinia, The Mineralogy of the Island of	C. J. Alford	1879	93
Saws, Mechanical	S. W. Worssam	1867	196
Screw Piles	H. P. Stephenson	1857	12
Screw Piles in the Construction of a Wrought-iron Bridge at Verona.	J. G. Horner	1867	52
Screw Propeller, The	A. Rigg	1868	202
Screw Propellers, their Shafts and Fittings.	H. W. Pendred	1875	151
Screw Propulsion, The Theory of	M. Sefi	1870	82
Sea and other Banks, Construction of	E. Olander	1862	25
Sea Defences	R. F. Grantham	1897	143
Sea Water, Effect on Portland Cement	H. Faija	1888	39
Secondary Generators, Distribution of Electrical Energy by	J. D. Gibbs	1885	49

Subject of Paper.	Author.	Year.	Page.
Segregation, Effect of, on the Strength of Steel Rails.	T. Andrews	1902	209
Sensitiveness and Isochronism in Governors.	A. Rigg	1880	73
Sewage, Bacterial Treatment of	G. Thudichum	1898	207
Sewage Disposal	H. Robinson	1879	197
Sewage Disposal, Recent Practice in	H. C. H. Shenton	1900	217
Sewage Disposal, River Pollution caused by	G. B. Jerram	1886	217
Sewage Disposal Works, and Main Drainage, Wimbledon.	W. S. Crimp	1888	73
Sewage, Irrigation with Town	G. King	1865	84
Sewage Matter, A new System of treating	H. Olrick	1883	23
Sewage Precipitation Works, Foreign	A. Wollheim	1892	123
Sewage Question during the last Century, The	H. A. Roechling	1901	193
Sewage Sludge, Filter Presses for	J. Croll	1897	121
Sewage, The Forms and Construction of Channels for the Conveyance of	J. Phillips	1874	145
Sewage, Ultimate Purification of	G. Thudichum	1896	199
Sewage, Utilisation of	B. Latham	1866	68
Sewage Works, Redhill	S. A. Reade	1868	168
Sewerage System, The Shone Hydro-Pneumatic.	E. Ault	1887	68
Sewer Ventilation	W. S. Crimp	1890	139
Sewers, Pipe, The Cleansing and Ventilation of	B. A. Miller	1892	167
Sewers, Ventilation of	B. Latham	1871	39
Shan Hill Country and the Mandalay Railway, The	E. W. Wagstaff	1899	21
Ship Caissons for Dock Basins and Dry Docks.	J. J. F. Andrews	1890	189
Ship Canal between the Atlantic and Pacific Oceans, The Surveys for	Dr. E. Cullen	1868	15
Ship Canal, The Isthmus of Darien and the	,, ,,	1868	52
Ships, Apparatus for Measuring the Velocity of	V. Pendred	1869	215
Ships, The Rolling of	W. McNaught	1876	187
Shone Hydro-Pneumatic Sewerage System.	E. Ault	1887	68
Shortlands and Nunhead Railway	A. G. Drury	1892	219
Signalling for Land and Naval purposes	F. Wise	1863	117
Single and Double-riveted Joints	T. Baldwin	1866	150
Slag, Economic Use of Blast-furnace	P. F. Nursey	1873	196
Slide-valve, The Friction of, and its appendages.	T. Adams	1866	6
Slopes of Earthwork, A new Method of Setting-out	C. J. Light	1873	216
Small Towns and Rural Districts, Water Supply of	P. Griffith	1896	55
Smoke, Prevention of	A. C. Engert	1881	101
Soft v. Hard Water for Manufacturing purposes.	J. C. Fell	1884	103

Subject of Paper.	Author.	Transactions.	
		Year.	Page.
Soil, Application of Steam to the Cultivation of the	B. Latham	1868	274
Stability of Chimney Shafts	R. J. Hutton	1887	150
State Railways and Railway Amalgamation.	G. Spencer	1872	43
Steam, Apparatus for Utilising the Waste Heat of Exhaust	J. Atkinson	1878	167
Steam Boilers, Corrosion in	J. H. Paul	1891	147
Steam Boilers, The Defects of, and their Remedy.	A. C. Engert	1884	25
Steam Boiler Explosions	P. F. Nursey	1863	1
Steam Boilers, Experimental Researches into the Nature and Action of	T. Baldwin	1867	23
Steam Boilers, The Priming of	W. Major	1877	59
Steam Boilers, Mechanical Firing of	J. W. Pearse	1877	31
Steam Boilers, Water and its effect on	H. K. Bamber	1867	65
Steam Carriages	A. F. Yarrow	1862	128
Steam Cylinders, Equalising the Wear in Horizontal.	W. Schönheyder	1878	45
Steam Cylinders, Friction in	P. Jensen	1870	17
Steam Fire Engines, and the late trial at the Crystal Palace.	W. Roberts	1863	163
Steam, High Pressure, and Steam Engine Efficiency	W. W. Beaumont	1888	221
Steam, Modern Systems of Generating	N. J. Suckling	1874	39
Steam Navigation on the Indus	A. Warren	1863	139
Steam, Generation and Expansion of	W. T. Carrington	1858	15
Steam, The Utilisation of Town Refuse for generating	T. W. Baker	1894	183
Steel, Modern, as a structural Material	W. W. Beaumont	1880	109
Steel Needles, The Use of, in driving a Tunnel at King's Cross.	W. H. Holttum	1892	199
Steel Rails, Effect of Segregation on the Strength of.	T. Andrews	1902	209
Steel Waggon and Carriage Frames, Fox's System of Solid-pressed	P. F. Nursey	1889	197
Stokers, Machine, for Gas Retorts	H. O'Connor	1891	117
Stone-sawing Machinery	H. Conradi	1876	135
Storage Reservoirs, Designing and Construction of	A. Jacob	1866	225
Street Railways	C. L. Light	1860	56
Street Subways for Large Towns	C. Mason	1895	91
Street Tramways, Horse Railways and	H. Gore	1873	113
Strength and Rigidity	W. E. Kochs	1865	69
Strength of Flues in Lancashire and similar Boilers.	W. Martin	1882	157
Strength of Materials, Need of further Experiments on the	C. J. Light	1869	194
Strength of Suspension Bridges, The	C. E. Browning	1858	11
Strength of Wrought-iron Railway Axles	T. Andrews	1879	143
Substitutes, Gas	V. B. Lewes	1893	141
Subways, Concrete, for Underground Pipes.	A. T. Allen	1901	99
Subways for Large Towns, Street	C. Mason	1895	91

Subject of Paper.	Author.	Transactions.	
		Year.	Page.
Sugar-making Machinery	P. F. Nursey	1865	155
Super-heating of Steam, and the various Appliances employed therein.	,, ,,	1861	36
Supply of Water to Towns	B. Latham	1864	199
Surface Condensers	J. Louch	1861	96
Suspension Bridges	R. M. Ordish	1857	13
Switches and Crossings, Railway	C. J. Light	1857	13
Tables, Circular	H. Hakewill	1865	147
Telegraph Instruments, Electric	E. G. Bartholomew	1872	53
Telegraph, The Electric	W. Ansell	1857	11
Telegraphy, Electric, irrespective of Telegraphic apparatus.	,, ,,	1869	55
Testimonial to Mr. Alfred Williams		1868	161
Thames Graving Dock, The New Hydraulic Machinery for lifting Vessels in the	J. Amos	1859	17
Theory of Screw Propulsion	M. Sefi	1870	82
Tidal Approaches and Deep-water Entrances.	J. B. Redman	1885	143
Tidal Energy, Utilisation of	A. Oates	1882	81
Timber Bridges, Australian	H. E. Bellamy	1902	63
Timbering of Trenches and Tunnels	C. Turner	1871	90
Tin-dressing Machinery, Recent Improvements in	S. H. Cox	1874	11
Tools, Machine	E. C. Amos	1899	51
Town Houses, Drainage of	G. M. Lawford	1891	187
Town Refuse, The Economical Disposal of	B. D. Healey	1900	65
Town Sewage, Irrigation with	G. King	1865	84
Towns, Supply of Water to	B. Latham	1864	199
Town Water Supply, Pumping Engines for	H. Davey	1867	88
Traction, Electrical, and its Financial Aspect.	S. Sellon	1892	37
Tramway Construction, Modern	A. W. Jones	1879	179
Tramway Rolling Stock, and Steam in connection therewith.	C. C. Cramp	1874	119
Tramway and other Rails, The Cleaning of	H. Conradi	1893	47
Tramways, Cable	W. N. Colam	1885	69
Treating Sewage Matter, A new system of	H. Olrick	1883	23
Treatment and Utilisation of Exhaust Steam.	P. Griffith	1890	163
Trial of a Refrigerating Machine on the Linde System	T. B. Lightfoot	1891	39
Trussed Beams	W. Parsey	1862	52
Tube Wells	A. Le Grand	1877	133
Tube Wells for large Water Supply	R. Sutcliff	1877	149
Tunnelling, Railway, in Japan	T. M. Rymer-Jones	1882	101
Tunnels, Timbering of Trenches and	C. Turner	1871	90
Tunnel, The Use of Steel Needles in driving, at King's Cross.	W. H. Holttum	1892	199
Turntables, Railway	C. J. Light	1858	11
Ultimate Purification of Sewage	G. Thudichum	1896	199

INDEX.

Subject of Paper.	Author.	Year.	Page.
Underground Pumping Machinery at the Erin Colliery, Westphalia.	H. Davey	1876	119
Use of Coal in Furnaces without Smoke	C. F. T. Young	1862	66
Use of Steel Needles in driving a Tunnel at King's Cross.	W. H. Holttum	1892	199
Utensils of a Brewery, Machinery and	T. Wilkins	1871	10
Utilisation of Coal Slack in the Manufacture of Coke for Smelting.	G. M. Ward	1880	33
Utilisation of Sewage	B. Latham	1866	68
Utilisation of Tidal Energy	A. Oates	1882	81
Utilisation of Town Refuse for generating Steam.	T. W. Baker	1894	183
Utilisation of Waste Mineral Substances.	E. Edwards	1861	82
Utilisation of Water Power, The	G. L. Molesworth	1858	16
Vacation Visits		1878	119
,, ,,		1879	141
,, ,,		1880	89
,, ,,		1881	121
,, ,,		1882	119
,, ,,		1883	85
,, ,,		1884	119
,, ,,		1885	113
,, ,,		1886	153
,, ,,		1887	145
,, ,,		1888	148
,, ,,		1889	131
,, ,,		1890	133
,, ,,		1891	111
,, ,,		1892	157
,, ,,		1893	133
,, ,,		1894	177
,, ,,		1895	145
,, ,,		1896	137
,, ,,		1897	115
,, ,,		1898	127
,, ,,		1899	161
,, ,,		1900	133
,, ,,		1901	137
,, ,,		1902	155
Value of Exhibitions as Aids to Engineering Progress.	S. Barnett	1883	75
Value of Cheap Patents, The National	E. W. Grierson	1880	145
Valves, Pump	J. Hendry	1859	25
Velocity of Ships, Apparatus for Measuring the	V. Pendred	1869	215
Ventilation and Warming	W. T. Sugg	1895	235
Ventilation, Cleansing and, of Pipe Sewers.	B. A. Miller	1892	167
Ventilation, Flues and	A. H. C. Trewman	1876	91
Ventilation of Buildings	J. W. Pearse	1876	97
Ventilation of Coal Mines	G. G. André	1874	23
Ventilation of Sewers	B. Latham	1871	39
Ventilation, Sewer	W. S. Crimp	1890	139

Subject of Paper.	Author.	Year.	Page.
Verona, Screw Piles in the construction of wrought-iron Bridge at	J. G. Horner	1867	52
Visit to the Paris Exhibition		1867	151
Waggon and Carriage Frames, Fox's Patent System of Solid-pressed Steel.	P. F. Nursey	1889	197
Waste Heat of Exhaust Steam, Apparatus for Utilising the	J. Atkinson	1878	167
Waste Mineral Substances, The Utilisation of	E. Edwards	1861	82
Water and its effects on Steam Boilers	H. K. Bamber	1867	65
Water, The Pollution of, and its Correction.	R. E. Middleton	1897	173
Water Power, The Utilisation of	G. L. Molesworth	1858	16
Water Pressure, Application of, to Machine Tools and Appliances.	R. H. Tweddell	1890	35
Water Purification, Sanitary and Industrial.	J. W. Pearse	1878	29
Water, Soft v. Hard, for Manufacturing purposes.	J. C. Fell	1884	103
Water Softening, by Dr. Clarke's Process	J. Glynn, Junr.	1859	29
Water Supplies, The Protection of Underground.	J. C. Thresh	1898	47
Water, Supply of, to Towns	B. Latham	1864	199
Water Supply, Preliminary Investigations for	S. A. Hollis	1901	147
Water Supply, Pumping-engines for Town	H. Davey	1867	88
Water Supply to the Town of Fareham, Drainage and	T. Buckham	1869	41
Water Supply of Small Towns and Rural Districts	P. Griffith	1896	55
Water Supply to Country Mansions and Estates	G. Harris	1899	231
Water-tube Boilers	V. Pendred	1867	109
Wear in Horizontal Steam Cylinders, Equalising the	W. Schönheyder	1878	45
Weighing Machinery and Automatic Apparatus in connection therewith.	W. H. Brothers	1890	55
Wells, Tube	A. Le Grand	1877	133
Wells, Tube, for large Water Supply	R. Sutcliff	1877	149
Wheels, Elastic Railway	V. Pendred	1864	119
Wimbledon Main Drainage and Sewage Disposal Works.	W. S. Crimp	1888	73
Wooden Jetties on the River Thames	J. G. Davenport	1860	24
Worms, The Action of Marine, and Remedies employed in the Harbour of San Francisco.	J. Blackbourn	1874	111
Wrought Iron, The Extension and Permanent Set of	C. E. Browning	1858	15

TA
1
S67
1902

Society of Engineers,
London
    Journal

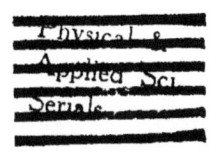

Engineering

PLEASE DO NOT REMOVE
CARDS OR SLIPS FROM THIS POCKET

UNIVERSITY OF TORONTO LIBRARY